# Analysis of Correlated Data with SAS and R
## Fourth Edition

# Analysis of Correlated Data with SAS and R

## Fourth Edition

Mohamed M. Shoukri

CRC Press
Taylor & Francis Group
Boca Raton London New York

CRC Press is an imprint of the
Taylor & Francis Group, an **informa** business

A CHAPMAN & HALL BOOK

CRC Press
Taylor & Francis Group
6000 Broken Sound Parkway NW, Suite 300
Boca Raton, FL 33487-2742

First issued in paperback 2020

ISBN 13: 978-0-367-73495-4 (pbk)
ISBN 13: 978-1-138-19745-9 (hbk)

| Library of Congress Cataloging-in-Publication Data |
| --- |
| Names: Shoukri, M. M. (Mohamed M.), author. \| Shoukri, M. M. (Mohamed M.). Analysis of correlated data with SAS and R. Title: Statistical analysis of health data using SAS and R / Mohamed M. Shoukri. Description: Fourth edition. \| Boca Raton : CRC Press, 2018. \| Previous edition: Analysis of correlated data with SAS and R / Mohamed M. Shoukri (Boca Raton : Chapman & Hall/CRC, 2007). Identifiers: LCCN 2017050107 \| ISBN 9781138197459 (hardback) Subjects: LCSH: Epidemiology–Statistical methods. \| Mathematical |

**Visit the Taylor & Francis Web site at**
**http://www.taylorandfrancis.com**

**and the CRC Press Web site at**
**http://www.crcpress.com**

*Success is not final. Failure is not fatal. It is the courage to continue that counts.*

**Sir Winston Churchill**

*To my loving wife Suhair.*

# Contents

# *Preface*

A decade has passed since the publication of the third edition. There has been an increasing demand by biological and clinical research journals to use state-of-the-art statistical methods in the data analysis components of submitted research papers. It is not uncommon to read papers published in high-impact journals such as *The Lancet*, *New England Journal of Medicine*, and *British Medical Journal* that deal with sophisticated topics such as meta-analysis, multiple imputations of missing data, and competing risks in survival analysis. The fourth edition results in part from continuous collaboration between the author and biomedical and clinical researchers inside and outside the King Faisal Specialist Hospital. I want to emphasize that there are fundamental differences, both qualitative and quantitative, between this and the previous editions.

The fourth edition has five additional chapters that cover a wide range of applied statistical methods. Chapter 1 covers the basics of study design and the concept of effect size. Chapter 2 deals with comparing group means when the assumptions of normality and independence of measured outcomes are not satisfied. The last three chapters titled "Introduction to Propensity Score Analysis" (Chapter 10), "Introductory Meta-Analysis" (Chapter 11), and "Missing Data" (Chapter 12) are also new. The other chapters from the third edition have been extensively revised to keep up with new developments in the respective areas. For example, topics such as competing risk, time-dependent covariates, joint modeling of longitudinal and survival data, and forecasting with exponential smoothing are introduced. Equally important is the use of the extensive library of packages in R (4.3.3), throughout the book we use packages produced by the R Development Core Team (2006, 2010, 2014), and the new developments in the SAS (version 9.4) output capabilities. Every topic is illustrated by examples. Data, with both R and SAS codes are also given.

Special thanks go to Mr. Abdelmonein El-Dali, Mrs. Tusneem Al-Hassan, and Mr. Parvez Siddiqui for their valued help. I am also grateful to Mrs. Kris Hervera and Mrs. Cielo Mendiola who both typed the entire volume.

Last but not the least, special thanks go to Mr. David Grubbs, Ms. Shelly Thomas, Mr. Jay Margolis, Ms. Adel Rosario, and the entire production team at CRC Press. Without the help of these people, the timely completion of this work would have been impossible.

*Additional material is available from the CRC Web site:* https://www.crcpress.com/9781138197459.

# 1

## Study Designs and Measures of Effect Size

## 1.1 Study Designs

### 1.1.1 Introduction

A research design is the framework or guide used for the planning, implementation, and analysis of data related to a specific study. Scientific questions are often formulated in the form of hypotheses. Different types of questions or hypotheses demand different types of studies. Therefore, it is quite important to have a thorough understanding of the different types of research designs available. In this chapter we shall briefly discuss the most commonly used study designs in biomedical and epidemiological research. Our focus will be on quantitative research designs. These types of design are mainly concerned with quantifying relationships between or among several variables. These variables are classified into two groups: *independent* or predictor variable(s), and *dependent* or outcome variable(s). Quantitative research designs are classified as either *nonexperimental* or *experimental*. Nonexperimental designs are used to describe, differentiate, or examine associations, as opposed to direct relationships between or among variables, groups, or situations. There is no random assignment, control groups, or manipulation of variables, as these designs use observations only. Whether the study is experimental or observational, the researcher should address several relevant questions:

1. What are the outcome variables of interest?
2. How many independent variables are being included in the study?
3. How many groups of subjects are being tested?
4. What is the sampling strategy to include subjects in the study?
5. Can subjects be randomly assigned to groups?
6. Is interest focused on establishing similarities or differences among the groups?

### 1.1.2 Nonexperimental or Observational Studies

These designs are used in situations where manipulations of an independent variable, control, or randomization of subjects are not involved. These designs are used to describe and measure independent and dependent variables and the relationship among them. They are sometimes called descriptive research designs. It is almost impossible to prove the causal relationship between exposure and outcome using these nonexperimental designs. The goal is to describe phenomena and explore and explain relationships between variables.

### 1.1.3 Types of Nonexperimental Designs

#### 1.1.3.1 Descriptive/Exploratory Survey Studies

These exploratory studies are used to examine differences or similarities between variables. This is a comparatively weak quantitative design, often used when little is known about a topic or to initially explore a research question. Similar to the nonexperimental design there is no random assignment or control. The design can be used with both quantitative and qualitative methods. Remember that because of the differences between qualitative and quantitative research in philosophy, and the order in which steps are taken, the two studies would be very different.

Survey designs gather information from a segment of the population. Random selection of subjects for the survey can increase the ability to generalize information. (*Random selection is not the same as random assignment; assignment occurs before the experimental condition is experienced.*) Commonly used instruments to gather information under this type design include interviews (telephone and in person) and questionnaires (mailed or administered in person).

#### 1.1.3.2 Correlational Studies (Ecological Studies)

A correlation is an examination of the strength of the relationship(s) between two or more variables. If subjects were assigned to treatment and control groups, that group assignment and manipulation of the independent variable would turn this into an experimental or quasi-experimental study (depending on the level of control of extraneous variables). We shall discuss in the propensity score chapter (Chapter 10) how this manipulation might strengthen the statistical inference based on this study design.

> If there is no manipulation of the independent variable and there is no assignment of groups, then it is a nonexperimental design. Data can be collected through the use of a questionnaire, interviews, or it can be measured with a variety of instruments.

- Advantages of nonexperimental correlational designs are they are straightforward, usually inexpensive, and quick. They may be used as pilot research or preliminary research for future studies.
- A disadvantage is only the association between variables can be determined, not causation.

These types of nonexperimental studies are also classified according to the timing of data collection. The first type is called cross-sectional. In a *cross-sectional study*, variables are identified one point in time and place, and the relationships between them are determined. The second type is called *longitudinal study*, where data are collected at different points over time. In observational studies one is interested in the relationship between the external and the dependent variables.

### 1.1.3.3 Cross-Sectional Studies

The main objective of using a cross-sectional type of design is to examine data across different groups at one point in time.

- Advantage—Less time consuming or expensive
- Disadvantage—Less control of extraneous variables

### 1.1.3.4 Longitudinal Studies

In longitudinal studies we collect data from one group of individuals over different points in time.

- Advantage—A subject serves as his/her own control, which greatly controls extraneous variables.
- Disadvantage—Costly and time consuming; may take decades to finish study.

### 1.1.3.5 Prospective or Cohort Studies

In a *prospective study*, or cohort study, potential factors and variables are determined to answer a specific question of scientific interest. In environmental biology we may be interested in investigating exposure to an environmental insult such as inhaling benzene and try to link this exposure to potential diseases that may occur in the future. Most, if not all, clinical trials are prospective by design. Typically, patients are randomized to receive either experimental or standard drugs. After a follow-up period of certain length, the responses are collected from the patients and the interest may be to compare the disease rates in both groups.

- **Advantages**
  - Can test cause-and-effect relationships.
  - Provides better control of the experiment and minimizes threats to validity.
- **Disadvantages**
  - For many studies, all extraneous variables cannot be identified or controlled.
  - Not all variables can be experimentally manipulated.
  - Some experimental procedures may not be practical or ethical in a clinical setting.

### 1.1.3.6  Case-Control Studies

Case-control studies involve a description of cases with and without a pre-existing exposure. The cases, subjects, or units of study can be an individual, a family, or a group. Case-control studies are more feasible than experiments in cases in which an outcome is rare or takes years to develop.

Simply put, a *case-control study* is a study that compares patients who have a disease or outcome of interest (cases) with subjects who do not have the disease or outcome (controls), and looks back retrospectively to compare the frequency of exposure to a risk factor in each group to determine the relationship between the risk factor and the disease.

Case-control studies are observational because no intervention is attempted and no attempt is made to alter the course of the disease. The goal is to retrospectively determine the exposure to the risk factor of interest from each of the two groups of individuals: cases and controls. These studies are designed to estimate odds of exposure, and not the odds of disease.

Case-control studies are also known as retrospective studies.

- **Advantages**
  - Good for studying rare conditions or diseases.
  - Less time needed to conduct the study because the condition or disease has already occurred.
  - These studies allow us to simultaneously look at multiple risk factors.
  - Useful as initial studies to establish an association.
  - Can answer questions that could not be answered through other study designs.

- **Disadvantages**
  - These studies are subject to what is known as "recall bias." They also have more problems with data quality because they rely on memory.
  - Not good for evaluating diagnostic tests because it is already clear that the cases have the condition and the controls do not.
  - In many situations, finding a suitable control group may be difficult.

---

### Design Pitfalls to Look Out For

Care should be taken to avoid confounding, which arises when an exposure and an outcome are both strongly associated with a third variable. Controls should be subjects who might have been cases in the study but are selected independent of the exposure. Although matching can be a good way to control for confounding, as much as we can "overmatching" should be avoided.

Is the control group appropriate for the population? Does the study use matching or pairing appropriately to avoid the effects of a confounding variable? Does it use appropriate inclusion and exclusion criteria?

---

#### *1.1.3.7 Nested Case-Control Study*

A nested case-control study is within a cohort study. At the beginning of a cohort study, the time is $t = 0$. Thereafter, members of the cohort are assessed for risk factors. Cases and controls are identified subsequently at time $t = 1$. The control group is selected from the risk set (cohort members who do not meet the case definition at $t = 1$).

- **Advantages**
  - Efficient, not all members of parent cohort require diagnostic testing.
  - Flexible, allows testing of hypotheses not anticipated when the cohort was drawn.
  - Reduces selection bias since cases and controls sampled from same population.
- **Disadvantages**
  - Reduction in the statistical power of validating the null hypothesis because of the reduction in the sample size.

The following example illustrates a nested case-control study design: In a cohort study of risk factors for cancer, Lin et al. (1995) recruited 9775 men. Blood samples were taken and frozen at recruitment into the cohort study. Over a follow-up of around seven years, 29 incident cases of gastric cancer were identified. By this time, a number of authors had published studies suggesting that *Helicobacter pylori* might be a risk factor for gastric cancer. Since *H. pylori* can be ascertained through an assay on blood samples, Lin and colleagues decided to exploit their stored samples to test this emerging theory through a nested case-control study. Between five and eight controls were sampled for each case, giving 220 controls altogether. Laboratory work was much reduced compared with analyzing the entire cohort: 249 assays rather than 9775 were required.

Another example of nested case-control study is concerned with a study investigating the potential health effects of occupational exposures to power frequency electric and magnetic fields. Sahl et al. (1993) collected a sample of work force that represented a variety of different occupations and work locations, and collected 776 days of magnetic field measurements exposure information and derived exposure scores by linking job history data to measured magnetic fields. In job title analysis, they compared "electrical workers" with other field and craft occupations, office, and technical support staff. Age-specific cancer rates were then compared.

### 1.1.3.8 Case-Crossover Study

A case-crossover study design is useful when the risk factor/exposure is transient. For example, cell phone use or sleep disturbances are transitory occurrences. Each case serves as its own control, meaning that the study is self-matched. For each person, there is a period of time during which the person was a case, and a period of time associated with not being a case. Risk exposure during the case time frame is compared to risk exposure during the control time frame.

- **Advantages**
  - Efficient self-matching
  - Efficient (reduction in variation) because we select only cases
  - Can use multiple controls for one case window
- **Disadvantages**
  - Information bias, inaccurate recall of exposure during control window (can be overcome by choosing control window to occur after case window)
  - Requires careful selection of time period during which the control window occurs
  - Requires careful selection of the length and timing of the windows (e.g., in an investigation of the risk of cell phone usage

on auto accidents, cell phone usage that ceases 30 minutes before accident unlikely to be relevant to accident)

In an Italian case-crossover study by Valent et a1. (2001) of sleep disturbance and injury among children, each child was asked about her or his sleep in the 24 hours before the injury occurred (the case window) and in the 24 hours before that (the control window). Among 181 boys, 40 had less than 10 hours sleep on both days concerned; 111 had less than 10 hours sleep on neither day; 21 had less than 10 hours sleep only on the day before the injury; and 9 had less than 10 hours sleep only on the penultimate day before the injury. Measures of association between sleep disturbance and injury, which will be discussed in Chapter 4, are odds ratios. The odds ratio for injury, comparing days without and with 10 hours or more sleep, is 2.33 (95% confidence interval; 1.02, 5.79).

### 1.1.4 Quasi-Experimental Designs

Quasi-experimental, like true-experimental designs, examine cause-and-effect relationships among independent and dependent variables. However, one of the characteristics of true-experimental design is missing, typically the random assignment of subjects to groups. Although quasi-experimental designs are useful in testing the effectiveness of an intervention and are considered closer to natural settings, these research designs are exposed to a greater number of threats of internal and external validity, which may decrease confidence and generalization of the study's findings. An example of quasi-experimental design will be presented and discussed in the "Exercises" section.

### 1.1.5 Single-Subject Design (SSD)

Single-subject research is experimental rather than descriptive, and its purpose is to establish a causal relationship between independent and dependent variables. Single-subject research employs within- and between-subjects comparisons to control for major threats to internal validity and requires systematic replication to enhance external validity (Martella, Nelson, and Marchand-Martella, 1999). We shall now outline some of the basic features of single-subject design (SSD).

Single-subject designs may involve only one participant, but typically include multiple participants (e.g., three to seven) in a single study. Each participant serves as his or her own control. Values of the dependent variables after intervention are compared to their values prior to intervention. Single-subject designs employ at least one dependent variable. These variables should be a valid and consistent assessment of the variable. It should be noted that the repeated measurement of individual responses is critical for comparing the performance of each participant with his or her own prior performance. Independent variables in SSD include intervention, and some of the subjects' demographics or behavioral mechanisms under investigation. Some of these independent variables can be manipulated.

As a final remark on this type of design, unlike the research questions often addressed by studies using traditional group designs, studies employing SSD can address the effects that intervention strategies and environmental variables have on performance at the subject level. One advantage of the SSD methodology is that it permits flexibility within the study to modify the independent variables when the expected results are not achieved. As a result, SSD methodology provides a useful alternative to randomized control trials (RCTs; and quasi-experimental group designs) for the goal of empirically demonstrating that an intervention is effective.

### 1.1.6 Quality of Designs

The various designs we discussed thus far differ in the quality of evidence they produce for a cause-and-effect relationship between variables. A well-designed cross-sectional or case-control study can provide good evidence for the absence of a relationship. But if such a study does reveal a relationship, it generally represents only suggestive evidence of a causal connection. A cross-sectional or case-control study is therefore a good starting point to decide whether it is worth proceeding to better designs. Prospective studies are more difficult and time-consuming to perform, but they produce more convincing conclusions about cause and effect. Experimental studies provide the best evidence about how some variables affect other variables, and double-blind randomized controlled trials are the gold standard in the field of experimental science.

### 1.1.7 Confounding

The confounding phenomenon is a potential problem in nonexperimental studies that try to establish cause and effect. Confounding occurs when part or all of a significant association between two variables arises through both being causally associated with a third variable. For example, in a population study you could easily show a negative association between smoking and heart disease. But older people are less active, and older people are more diseased, so you are bound to find an association between activity and disease without one necessarily causing the other. To get over this problem you have to control for potential confounding factors. For example, make sure all your subjects are the same age or create several age strata to try to remove its effect on the relationship between the other two variables.

### 1.1.8 Sampling

In most epidemiological investigations we have to work with a sample of subjects rather than the full population. Since the public is interested in the population under study we should be able to generalize our results from the sample to the population. To generalize from the sample to the population, the sample has to be representative of the population. The safest way to ensure that it is representative is to use a random selection procedure. Other strategies for sampling to control confounding and reduce heterogeneity may be followed. For example, we may use a stratified random sampling procedure to make sure that we have proportional representation of population subgroups (e.g., sexes, races, regions).

When the sample is not representative of the population, selection bias is a possibility. A typical source of bias in population studies is age, gender, level of education, or socioeconomic status. It is noted that the very young and the very old tend not to take part in the studies. Thus a high compliance (the proportion of people contacted who end up as subjects) is important in avoiding bias.

Failure to randomize subjects to control and treatment groups in experiments can also produce bias. If we select patients who are willing to participate in a clinical trial with a promising new drug or if we select the groups in any way that makes one group different from another, then any result we get might reflect the group difference rather than an effect of the intervention. For this reason, it is important to randomly assign subjects in a way that ensures the groups are balanced in terms of important variables that could modify the effect of the treatment (e.g., age, gender, physical performance). Human subjects may not be happy about being randomized, so we need to state clearly that it is a condition of taking part.

Often the most important variables to balance are the baseline or preintervention value of the dependent variable itself.

### 1.1.9 Types of Sampling Strategies

We all agree that the golden rule is to sample subjects that are representative of the population from which they have been recruited. Several sampling strategies are available, and depending on the structure of the population and the study objectives a specific strategy is adopted to achieve the study objectives. Here we briefly outline the most often used sampling techniques. The details can be found in the book by Cochran (1977).

A *simple random sample* is a subset of a statistical population in which each member of the subset has an equal probability of being chosen. An example of a simple random sample would be the names of 20 nurses being chosen out of a hat from a hospital of 200 nurses. In this case, the population is all 200 nurses, and the sample is random because each nurse has an equal chance of being selected.

*Stratified random sampling* is a method of sampling that involves the division of a population into smaller subgroups known as strata. In stratified random sampling, the strata are formed based on members' shared characteristics. A random sample from each stratum is taken in a number proportional to the stratum's size when compared to the population. These subsets of the strata are then pooled to form a random sample. Similar to a weighted average, this method of sampling produces characteristics in the sample that are proportional to the overall population. Stratified sampling works well for populations with a variety of characteristics.

*Cluster sampling* (also known as one-stage cluster sampling) is a technique in which aggregates or clusters of participants that represent the population are identified and included in the sample. This is popular in conducting household or family studies. The main aim of cluster sampling can be specified as cost reduction and increasing the levels of efficiency of sampling. This specific technique can also be applied in integration with multistage sampling.

*Multistage sampling* (also known as multistage cluster sampling) is a more complex form of cluster sampling that contains two or more stages in sample selection. In multistage sampling, large clusters of the population are divided into smaller clusters in several stages in order to make primary data collection more manageable. It has to be acknowledged that multistage sampling is not as efficient as true random sampling; however, it addresses certain disadvantages associated with true random sampling such as being overly expensive and time consuming.

A major difference between cluster and stratified sampling relates to the fact that in cluster sampling a cluster is perceived as a sampling unit, whereas in stratified only specific elements of strata are accepted as a sampling unit. Accordingly, in cluster sampling a complete list of clusters represent the sampling frame. Then, a few clusters are chosen randomly as the source of primary data.

An advantage of multistage sampling is that it is effective in primary data collection from geographically dispersed populations when face-to-face contact in required (e.g., semistructured in-depth interviews). This type of sampling can also be cost effective and time effective. A disadvantage of multistage sampling is to effectively account for the heterogeneity, group-level information should be available.

### 1.1.10 Summary

The selection of a research design is based on the research question or hypothesis and the phenomena being studied. A true-experimental design is considered the strongest or most rigorous with regard to establishing causal effects and *internal validity*. Internal validity is the control of factors within the study that might influence the outcomes besides the experimental intervention or treatment. A nonexperimental design is generally the weakest in this respect. However, this does not mean that nonexperimental designs are weak designs overall. They are weak only with respect to assessing cause-effect relationships and the establishment of internal validity. In fact, the simplest form of nonexperiment, the one-time survey design that consists of one single observation, is one of the most common forms of research and for some research questions, especially descriptive ones, is clearly a strong and most appropriate design.

## 1.2 Effect Size

### 1.2.1 What Is Effect Size?

In medical research studies that compare different treatment interventions, effect size (ES) is the magnitude of the difference between two or more groups. For example, if we compare two drugs that are believed to lower systolic blood pressure levels, the absolute effect size is the difference between the average reductions. As another example, if a treatment modality for cancer resulted in the improvement of patients' quality of life scores measured by a VAS (visual analogue scale) by an average total of 20 as compared to that of another intervention that improves the average quality of life scores by 40, the absolute effect size is 20. Absolute effect size does not take into account the variability in scores, in that not every subject achieved the average outcome.

Thus, effect size can refer to the raw difference between group means, or absolute effect size, as well as standardized measures of effect, which are calculated to transform the effect to an easily understood scale. Absolute effect size is useful when the variables under study have intrinsic meaning (e.g., length of stay after surgery). Calculated indices of effect size are useful when the measurements have no intrinsic meaning, such as numbers on a Likert scale, when studies have used different scales so no direct comparison is possible, or when effect size is examined in the context of variability in the population under study.

### 1.2.2 Why Report Effect Sizes?

The effect size is considered the main finding of a quantitative study. While a *p*-value can inform the reader when an effect exists, the *p*-value will not reveal the size of the effect. In reporting and interpreting studies, both the effect size and statistical significance (*p*-value) are essential results to be reported.

### Why Isn't the *p*-value Enough?

Statistical significance is the least interesting thing about the results. You should describe the results in terms of measures of magnitude—not just, does a treatment affect people, but how much does it affect them.

**Gene V. Glass, 1976 (in Kline, 2004)**

The primary product of a research inquiry is one or more measures of effect size, not p values.

**Jacob Cohen, 1965**

The *p*-value is the probability that the observed difference between two groups is due to chance. If the *p*-value is larger than the alpha level chosen (e.g., .05), then any observed difference is assumed to be explained by sampling variability. With a sufficiently large sample, a statistical test will almost always demonstrate a significant difference, unless there is no effect whatsoever, that is, when the effect size is exactly zero; yet very small differences, even if significant, are often meaningless. Thus, reporting only the significant *p*-value for an analysis is not adequate for readers to fully understand the results. For example, if a sample size is 10,000, a significant *p*-value is likely to be found even when the difference in outcomes between groups is negligible and may not justify an expensive or time-consuming intervention over another. The level of significance by itself does not predict effect size. Unlike significance tests, effect size is independent of sample size. Statistical significance, on the other hand, depends upon both sample size and effect size. For this reason, *p*-values are considered to be confounded because of their dependence on sample size. Sometimes a statistically significant result means only that a huge sample size was used.

A commonly cited example of this problem is the Physicians Health Study of aspirin to prevent myocardial infarction (MI) (Bartolucci et al., 2011). In more than 22,000 subjects over an average of five years, aspirin was associated with a reduction in MI (although not in overall cardiovascular mortality) that was highly statistically significant: p = .00001. The study was terminated early due to the conclusive evidence, and aspirin was recommended for general prevention. Another example was given by Lipsey and Wilson (1993).

### 1.2.3 Measures of Effect Size

Cohen (1965) remarked that in research concerned with comparisons among treatment means, investigators nonetheless typically confined themselves to reporting test statistics such as $t$ or $F$ and did not attempt to derive measures of effect size. This section reviews research on the development and practical value of different measures of effect size.

As a start let us consider the mean difference and the standardized mean difference between two independent groups, with the primary focus on the derivation and estimation of the latter as a measure of effect size and on its concomitant advantages and disadvantages. But as we will see in the third chapter of this book, other effect sizes are used as effect sizes depending on objectives of the study and the nature of the dependent variables.

Effect size measures either measure the sizes of associations or the sizes of differences. You already know the most common effect-size measure—the correlation/regression coefficients $r$ and $R$ are actually measures of effect size. Because $r$ covers the whole range of relationship strengths, from no relationship whatsoever (zero) to a perfect relationship (1 or –1), it is telling us exactly how large the relationship really is between the variables we have studied—and is independent of how many people were tested. Cohen (1988, 1990) provided rules of thumb for interpreting these effect sizes, suggesting that an $r$ of |.1| represents a "small" effect size, |.3| represents a "medium" effect size, while |.5| represents a "large" effect size.

Another common measure of effect size is $d$, sometimes known as Cohen's $d$ (as you might have guessed by now, Cohen was quite influential in the field of effect sizes). This can be used when comparing two means, as when you might do a $t$-test, and is simply the difference in the two groups' means divided by the average of their standard deviations. This means that if we see a $d$ of 1, we know that the two groups' means differ by one standard deviation; a $d$ of .5 tells us that the two groups' means differ by half a standard deviation; and so on. Cohen suggested that $d = 0.2$ be considered a "small" effect size, 0.5 represents a "medium" effect size, and 0.8 a "large" effect size.

There are various other measures of effect size, but the only other one we need to know for now is partial eta-squared ($\eta_2$), which can be used in ANOVA. Partial eta-squared is a measure of variance, like $r$-squared. It tells us what proportion of the variance in the dependent variable is attributable to the factor in question.

### 1.2.4 What Is Meant by "Small," "Medium," and "Large"?

In Cohen's terminology, a small effect size is one in which there is a real effect, that is, something is really happening in the world but which you can only see through careful study. A "large" effect size is an effect that is big enough, and/or consistent enough, that you may be able to see it with the naked eye. For example, just by looking at a room full of people, you would probably be able to tell that on average, the men were taller than the women. This is what is meant by an effect that can be seen with the naked eye (actually, the $d$ for the gender difference in height is about 1.4, which is really large, but it serves to illustrate the point). A large effect size is one that is very substantial.

There is a wide array of formulas used to measure effect size. For the occasional reader of meta-analysis studies, this diversity can be confusing. One of my objectives in putting together this set of lecture notes was to organize and summarize the various measures of effect size. In general, effect size can be measured in two ways:

1. As the standardized difference between two means

$$d = \frac{mean(group1) - mean(group2)}{pooled\ standard\ deviation}$$

2. As the correlation between the independent variable classification and the individual scores on the dependent variable. This correlation is called the *effect size correlation* (Rosenthal, 1994; Rosnow and Rosenthal, 1996).

### The Standardized Mean Difference

In the simplest situation, two samples of size $n_1$ and $n_2$ (where $n_1 + n_2 = N$) are drawn independently and at random from populations whose mean response rates are $\pi_1$ and $\pi_2$, respectively, and whose standard deviations are $\sigma_1$ and $\sigma_2$, respectively. Suppose that the two samples are found to have means of $p_1$ and $p_2$ and variances of $s_1 = p_1(1 - p_1)/n_1$ and $s_2 = p_2(1 - p_2)/n_2$, respectively, for $p_1$ and $p_2$. The simplest index of effect size is the difference between the two population mean rates $(\pi_1 - \pi_2)$. This measure has two useful features. First, it is expressed in terms of the original units of measurement, and thus it is intuitively meaningful to researchers. Second, although it is a parameter based on the underlying populations and hence is typically unknown, it has an unbiased estimate in the difference between the sample means $(p_1 - p_2)$. In order to make meaningful comparisons among studies employing different procedures, Cohen (1965) pointed out that this could be achieved if the difference between the two population mean rates is standardized against the population within group standard deviation. This yields an effect size index $\delta$, defined as

$$\delta = \frac{(p_1 - p_2)}{\sqrt{(s_1 + s_2)}}$$

In other words, $\delta$ is the mean difference that would be obtained if the dependent variables were scaled to have unit variance within both populations.

### 1.2.5 Summary

As Winer et al. (1971) pointed out, an experimental design that achieves a numerically high level of statistical power can lead to the rejection of the null hypothesis even though the treatment effects are quite trivial from a practical or theoretical point of view. The measures of effect size described in this section represent different attempts to evaluate the importance of the observed effects in a way that is independent of the level of statistical significance that they attain.

Measures of effect size were developed partly to compare and evaluate results obtained across different studies in the research literature, but criticisms have been expressed by various authors regarding the weaknesses and limitations of meta-analytic techniques. However, these criticisms do not in themselves call into question the usefulness of measures of effect size in reporting or interpreting the findings obtained in single studies. Cohen (1965, p. 106) and Hays (1963, p. 328) recommended that researchers routinely report measures of effect size as well as test statistics and significance levels as a matter of good practice, but this is not of course to imply that such measures should be used uncritically.

### 1.2.6 American Statistical Association (ASA) Statement about the *p*-value

In the previous sections we discussed two topics that are fundamental to the presentation of the results of medical research. The first is the *p*-value and the second is the effect size. Both quantities must be reported and the following arguments by leaders in the statistical communities support this idea. Here we present summaries of the debated issues. Wasserstein and Lazar (2016) wrote:

> In February, 2014, George Cobb, Professor Emeritus of Mathematics and Statistics at Mount Holyoke College, posed these questions to an ASA discussion forum:
>
> Q: Why do so many colleges and grad schools teach p = .05?
> A: Because that's still what the scientific community and journal editors use.
> Q: Why do so many people still use p = 0.05?
> A: Because that's what they were taught in college or grad school.

Cobb's concern was a long-worrisome circularity in the sociology of science based on the use of bright lines such as P < 0.05 : "We teach it because it's what we do; we do it because it's what we teach." This concern was brought to the attention of the ASA Board.

A *Science News* article (Siegfried, 2010) on March 27, 2010, stated "statistical techniques for testing hypotheses ... have more flaws than Facebook's privacy policies." A week later, statistician and "Simply Statistics" blogger Jeff Leek responded. "The problem is not that people use p-values poorly," Leek (2014) wrote, "it is that the vast majority of data analysis is not performed by people properly trained to perform data analysis."

The statistical community has been deeply concerned about issues of *reproducibility* and *replicability* of scientific conclusions. Misunderstanding or misuse of statistical inference is only one cause of the "reproducibility crisis" (Peng, 2015), but to the community of statisticians it is an important one.

Although the *p*-value can be a useful statistical measure, it is commonly misused and misinterpreted. This has led to some scientific journals discouraging the use of *p*-values, and some scientists and statisticians recommending their abandonment, with some arguments essentially unchanged since *p*-values were first introduced.

In this context, the American Statistical Association (ASA) believes that the scientific community could benefit from a formal statement clarifying several widely agreed upon principles underlying the proper use and interpretation of the *p*-value. The issues touched on here affect not only research, but research funding, journal practices, career advancement, scientific education, public policy, journalism, and law. This statement does not seek to resolve all the issues relating to sound statistical practice, nor to settle foundational controversies. Rather, the statement articulates in nontechnical terms a few select principles that could improve the conduct or interpretation of quantitative science, according to widespread consensus in the statistical community.

In summary, the best definition of a *p*-value is articulated in the following statement:

Researchers often wish to turn a *p*-value into a statement about the truth of a null hypothesis, or about the probability that random chance produced the observed data. The *p*-value is neither. It is a statement about data in relation to a specified hypothetical explanation, and is not a statement about the explanation itself.

## Exercises

1.1. Explain how the different scales of measurement affect the outcome of interest and the power of testing a hypothesis. For example, instead of comparing the mean systolic blood pressures in two groups, we categorize the blood pressure levels into normal and abnormal, and then compare the percentage of abnormalities in the two groups.

1.2. What are the advantages and disadvantages of experimental designs when conducting applied research?

1.3. True or false? A research design is considered experimental if there is random assignment, control, manipulation, and measurements.

1.4. Develop an experimental design that utilizes all three features of experimental designs (pre-/post-), control, and between-within subject variations, and discuss the issue of generalizability.

1.5. What are the advantages and disadvantages of quasi-experimental designs when conducting field research?

1.6. Tokelau Islands migrant studies (Shoukri and Ward, 1989): Islands in the South Pacific are frequently hit with devastating hurricanes. The populations of these islands lived a traditional lifestyle. In the early 1970s, the government of New Zealand decided to move the people of these islands to the main island where European immigrants have settled. Before the move, a team of researchers (doctors, epidemiologists) visited the islands and collected demographic and anthropometric measurements. A few years later, the same data were collected on the same people in order to quantify the effect of the new environment on them. What kind of design is this? What is the research question? Formulate a proper hypothesis that needs be tested. What are the difficulties the researchers are expected to face while conducting the research and reporting their findings?

1.7. When researchers have difficulty establishing a causal pathway between exposure to risk factor and disease, would retrospective studies be a possible approach? What are the statistical challenges that researchers face? Suggest appropriate statistical techniques to overcome these difficulties.

1.8. Describe and name a study design (naturally nonexperimental) that may be used to establish correlation between fat intake and prostate cancer mortality.

1.9. Differentiate between researcher questions and hypotheses.

1.10. What are the most serious difficulties a researcher is expected to face when conducting longitudinal or repeated measures study? What is the potential impact of these difficulties on the power of the study, and how can a researcher reduce the impact of the difficulties on the expected outcomes of the study.

1.11. Robust effect size: When the sample contains outliers, both the mean and the standard deviations are affected and they may not be good representatives of the measurements. In this case we use the median as a measure of location, and the median absolute deviation (MAD) as a measure of spread. Provide a numerical data example where outliers affect the mean and the standard deviation, but not the median.

1.12. Computing the median absolute deviation: The formula is MAD = median($|y$ – median(y)$|$) where $x$ represents the collection of numbers. In words, the formula says:

- Find the median of y
- Subtract this median from each value in y
- Take the absolute value of these differences
- Find the median of these absolute differences

A robust effect size can be constructed as

$$\beta = \frac{median(x) - median(y)}{MAD(x) + MAD(y)}$$

This would be contrasted with Cohen's effect size:

$$d = \frac{mean(x) - mean(y)}{\sqrt{var(x) + var(y)}}$$

Given the two groups of data $x$ = c(2,5,7,10,13,14,11) and $y$ = c(1,4,6,10,11,16,19), calculate the ES using the $\beta$ and $d$.

In group $y$, replace 10 with 100. Repeat the calculations. What do you observe?

# 2

# Comparing Group Means When the Standard Assumptions Are Violated

## 2.1 Introduction

The majority of statistical experiments are conducted so as to compare two or more groups. It is understood that the word *group* is a generic term used by statisticians to label and distinguish the individuals who share a set of experimental conditions. For example, diets, breeds, age intervals, and methods of evaluations are groups. In this chapter we will be concerned with comparing group means. When we have two groups, that is, when we are concerned with comparing two means $\mu_1$ and $\mu_2$, the familiar Student's $t$ statistic is the tool that is commonly used by most data analysts. If we are interested in comparing several group means, that is, if the null hypothesis is $H_0: \mu_1 = \mu_2 = \cdots = \mu_k$, a problem known as the analysis of variance (ANOVA), we use the F-ratio to seek the evidence in the data and see whether it is sufficient to justify the hypothesis.

In performing an ANOVA experiment, we are always reminded of three basic assumptions:

Assumption 1—That the observations in each group are a random sample from a normally distributed population with mean $\mu_i$ and variance $\sigma_i^2$ $(i = 1, 2, \cdots, k)$.

Assumption 2—The variances $\sigma_1^2, \cdots, \sigma_k^2$ are all equal. This is known as the variance homogeneity assumption.

Assumption 3—The observations within each group should be independent.

The following sections will address the effect of the violation of each of these three assumptions on the ANOVA procedure, how it manifests itself, and possible corrective steps that can be taken.

## 2.2 Nonnormality

It has been reported (Miller, 1986, p. 80) that a lack of normality has very little effect on the significance level of the F-test. The robustness of the F-test improves with increasing the number of groups being compared, together with an increase in group size $n_i$. The reason for this is a rather technical issue and will not be discussed here. However, the investigators should remain aware that although the significance level may be valid, the F-test may not be appreciably powerful. Transformations such as the square root or the logarithm to improve normality can improve the power of the F-test when applied to nonnormal samples. To provide the reader with the proper tool to detect the nonnormality of the data, either before or after performing the proper trans-formation, we need some notation to facilitate the presentation. Let $y_{ij}$ denote the $j$th observation in the $i$th group, where $j = 1,2,\cdots,n_i$ and $i = 1,2,\cdots,k$. Moreover, suppose that $y_{ij} = \mu_i + e_{ij}$, where it is assumed that $e_{ij}$ are identically, indepen-dently, and normally distributed random variables with $E(e_{ij}) = 0$ and variance $(e_{ij}) = \sigma^2$ (or $e_{ij} \sim iidN(0,\sigma^2)$). Hence, $y_{ij} \sim iidN(\mu_i,\sigma^2)$ and the assumption of nor-mality of the data $y_{ij}$ need to be verified before using the ANOVA F-statistic.

Miller (1986, p. 82) recommended that a test of normality should be per-formed for each group. One should avoid omnibus tests that utilize the combined groups such as combined goodness-of-fit $\chi^2$ test or a multiple sample Kolmogorov-Smirnov. A review of these tests is given in D'Agostino and Stephens (1986, chapter 9). They showed that the $\chi^2$ test and Kolmogorov test have poor power properties and should not be used when testing for normality.

Unfortunately, the aforementioned results are not widely known to nonstatisticians. Major statistical packages, such as SAS and R, perform the Shapiro-Wilk W test for group sizes up to 50. For larger group sizes, SAS provides us with poor power Kolmogorov test. This package also produces measures of skewness and kurtosis, though strictly speaking these are not the actual measures of skewness $\sqrt{b_1}$ and kurtosis $b_2$ defined as

$$\sqrt{b_1} = \frac{1}{n}\sum_{j=1}^{n}\left(y_j - \bar{y}\right)^3 \Big/ \left[\frac{1}{n}\sum_{j=1}^{n}\left(y_j - \bar{y}\right)^2\right]^{3/2}$$

$$b_2 = \frac{1}{n}\sum_{j=1}^{n}\left(y_j - \bar{y}\right)^4 \Big/ \left[\frac{1}{n}\sum_{j=1}^{n}\left(y_j - \bar{y}\right)^2\right]^2.$$

In a recent article by D'Agostino et al. (1990) the relationship between $b_1$ and $b_2$ and the measures of skewness and kurtosis produced by SAS is established. Also provided is a simple SAS macro that produces an excellent, informative analysis for investigating normality. They recommended that, as descriptive statistics, values of $b_1$ and $b_2$ close to 0 and 3, respectively, indicate normality. More precisely, the expected values of these are 0 and $3(n - 1)/(n + 1)$ under normality.

To test for skewness, the null hypothesis, $H_0$: *underlying population is normal* is tested as follows (D'Agostino and Pearson, 1973):

1. Compute $\sqrt{b_1}$ from the data.
2. Compute

$$u = \sqrt{b_1} \left\{ \frac{(n + 1)(n + 3)}{6(n - 2)} \right\}^{1/2}$$

3. Compute

$$\beta_2(\sqrt{b_1}) = \frac{3(n^2 + 27n - 70)(n + 1)(n + 3)}{(n - 2)(n + 5)(n + 7)(n + 9)}$$

$$w^2 = -1 + \{2[\beta_2(\sqrt{b_1}) - 1]\}^{1/2},$$

$$\delta = (\ln w)^{-1/2}, \text{ and } \alpha = \{2/(w^2 - 1)\}^{1/2}.$$

$$Z(\sqrt{b_1}) = \delta \ln(u/\alpha + \{(u/\alpha)^2 + 1\}^{1/2}).$$

$Z(\sqrt{b_1})$ has approximately standard normal distribution, under the null hypothesis.

To test for kurtosis, a two-sided test (for $\beta_2 \neq 3$) or one-sided test (for $\beta_2 > 3$ or $\beta_2 < 3$) can be constructed:

1. Compute $b_2$ from the data:
$$E(b_2) = \frac{3(n - 1)}{n + 1}$$

2. Compute
$$Var(b_2) = \frac{24n(n - 2)(n - 3)}{(n + 1)^2(n + 3)(n + 5)}$$
and
$$x = \frac{(b_2 - E(b_2))}{\sqrt{Var(b_2)}}$$

3. Compute
$$\sqrt{\beta_1(b_2)} = \frac{6(n^2 - 5n + 2)}{(n + 7)(n + 9)} \sqrt{\frac{6(n + 3)(n + 5)}{n(n - 2)(n - 3)}}$$

4. Compute
$$A = 6 + \frac{8}{\sqrt{\beta_1(b_2)}} \left[ \frac{2}{\sqrt{\beta_1(b_2)}} + \sqrt{1 + \frac{4}{\beta_1(b_2)}} \right]$$

5. Compute

$$Z(b_2) = [2/(9A)]^{-1/2} \left\{ (1 - \tfrac{2}{9A}) - \left[ \frac{1-2/A}{1+x\sqrt{2/(A-4)}} \right]^{1/3} \right\}.$$

$Z(b_2)$ has approximately standard normal distribution, under the null hypothesis. D'Agostino and Pearson (1973) constructed a test based on both $\sqrt{b_1}$ and $b_2$ to test for

$$K^2 = \left( Z\left(\sqrt{b_1}\right) \right)^2 + (Z(b_2))^2$$

skewness and kurtosis. This test is given by and has approximately a chi-square distribution with 2 degrees of freedom.

For routine evaluation of the statistic $K^2$, the SAS package produces measures of skewness ($g_1$) and kurtosis ($g_2$) that are different from $\sqrt{b_1}$ and $b_2$. After obtaining $g_1$ and $g_2$ from PROC UNIVARIATE in SAS, we evaluate $\sqrt{b_1}$ as

$$\sqrt{b_1} = \frac{(n-2)}{\sqrt{n(n-1)}} g_1$$

and

$$b_2 = \frac{(n-2)(n-3)}{(n+1)(n-1)} g_2 + \frac{3(n-1)}{n+1}$$

and one then should proceed to compute $Z(\sqrt{b_1})$ and $Z(\sqrt{b_2})$ and hence $K^2$.

Note that when $n$ is large, $\sqrt{b_1} \to g_2$ and $b_2 \to g_2$.

An equally important tool that should be used to assess normality is graphical representation of the data. A very effective technique is the empirical quantile-quantile plot (Wilk and Gnanadesikan, 1968). It is constructed by plotting the quantiles of the empirical distribution against the corresponding quantiles of the normal distribution. If the quantiles plot is a straight line, it indicates that the data is drawn from a normal population.

A good complete normality test would consist of the use of the statistic $K^2$ and the normal quantile plot.

Note that if the empirical quantile-quantile plot follows a straight line, then one can be certain, to some extent, that the data is drawn from a normal population. Moreover, the quantile-quantile plot provides a succinct summary of the data. For example, if the plot is curved this may suggest that a logarithmic transformation of the data points could bring them closer to normality. The plot also provides us with a useful tool for detecting outlying observations.

In addition to the quantile-quantile plot, a box plot (Tukey, 1977) should be provided for the data. In the box plot the 75th and the 25th percentiles of the data are portrayed by the top and bottom of a rectangle, and the median is portrayed by a horizontal line segment within the rectangle. Dashed lines extend from the ends of the box to the adjacent values, which are defined as follows. First, the interquantile range (IQR) is computed as

$$IQR = 75\text{th percentile} - 25\text{th percentile}$$

The upper adjacent value is defined to be the largest observation that is less than or equal to the 75th percentile plus (1.5) IQR. The lower adjacent value is defined to be the smallest observation that is greater or equal to the 25th percentile minus (1.5) IQR. If any data point falls outside the range of the two adjacent values, it is called an outside value (not necessarily an outlier) and is plotted as an individual point.

The box plot displays certain characteristics of the data. The median shows the center of the distribution, and the lengths of the dashed lines relative to the box show how far the tails of the distribution extend. The individual outside values give us the opportunity to examine them more closely and to subsequently identify them as possible outliers.

**Example 2.1**

One hundred twenty observations representing the average milk production per day/herd in kilograms were collected from 10 Ontario farms. All plots are produced by SAS. Histograms and the Q-Q plots are shown in Figures 2.1 and 2.2.

The summary statistics produced by SAS UNIVARIATE PROCEDURE are given in Table 2.1.

The results of the D'Agostino et al. (1990) procedure is summarized in Table 2.2.

The value of the omnibus test (6.528) when compared to a chi-squared value with 2 degrees of freedom (5.84) is found to be significant. Thus, based on this test, we can say that the data are not normally distributed.

In attempting to improve the normality of the data, it was found that the best results were obtained through squaring the milk yield values. This solved the skewness problem and resulted in a nonsignificant $K^2$ value (Table 2.3). Combined with the normal probability plot (Figure 2.2) produced by the D'Agostino et al. (1990) macro, one can see that the data has been normalized.

## 2.3 Heterogeneity of Variances

The effect of unequal variances on the F-test of the ANOVA problem had been investigated by Box (1954a, 1954b). He showed that in balanced experiments (i.e., when $n_1 = n_2 = \cdots = n_k$) the effect of heterogeneity is not too serious.

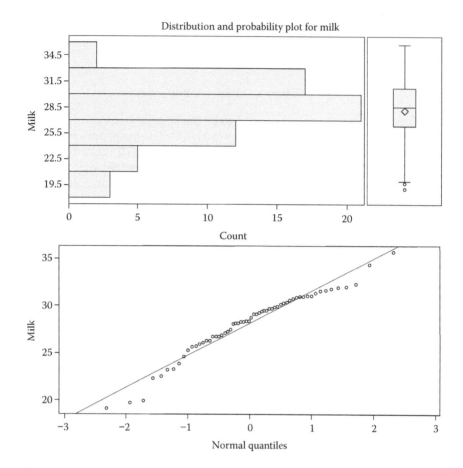

**FIGURE 2.1**
Histogram and normal quantile plots for the milk from large farms.

When we have unbalanced experiments, then the effect can be more serious, particularly if the large $\sigma_i^2$ are associated with the small $n_i$. The recommendation is to balance the experiment whenever possible, for then unequal variances have the least effect.

Detecting heterogeneity or testing the hypothesis $H_0 : \sigma_1^2 = \sigma_2^2 = \cdots = \sigma_k^2$ is a problem that has applications in the field of measurement errors, particularly reproducibility and reliability studies, as will be seen in the next chapter. In this section we introduce some of the widely used tests of heterogeneity.

### 2.3.1 Bartlett's Test

Bartlett (1937) modified the likelihood ratio statistic on $H_0 : \sigma_1^2 = \sigma_2^2 = \cdots = \sigma_k^2$ by introducing a test statistic widely known as Bartlett's test on homogeneity of variances. This statistic is

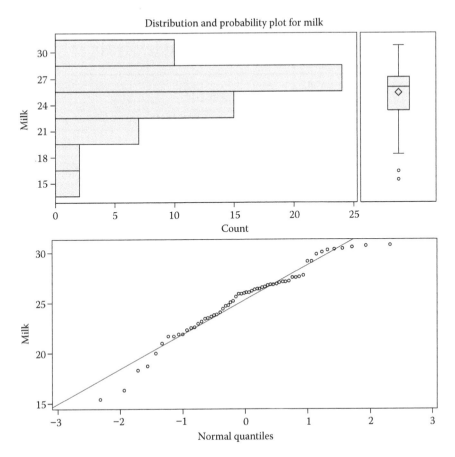

**FIGURE 2.2**
Histogram and normal quantile plots for the milk from small farms.

$$B = M/C$$

where

$$M = v \log s^2 - \sum_{i=1}^{k} v_i \log s_i^2,$$

$$v_i = n_i - 1, v = \sum_{i=1}^{k} v_i, s^2 = \frac{1}{v} \sum_{i=1}^{k} v_i s_i^2,$$

$$s_i^2 = \frac{1}{v_i} \sum_{j=1}^{n_i} \left( y_{ij} - \bar{y}_1 \right)^2, \bar{y}_1 = \frac{1}{n_i} \sum_{j=1}^{n_i} y_{ij}$$

**TABLE 2.1**

Summary Statistics from SAS UNIVARIATE PROCEDURE
on the Original Data

| Variable = Milk | | | |
|---|---|---|---|
| Moments | | | |
| N | 120 | Sum Wgts | 120 |
| Mean | 26.74183 | Sum | 3209.02 |
| Std dev | 3.705186 | Variance | 13.7284 |
| **Skewness** | **−0.54127** | **Kurtosis** | **0.345688** |
| USS | 87448.76 | CSS | 1633.68 |
| CV | 13.85539 | Std Mean | 0.338236 |
| T:Mean = 0 | 79.06274 | Prob> \|T\| | 0.001 |
| Sgn rank | 3630 | Prob> \|S\| | 0.001 |
| Num ^=0 | 120 | | |
| W:Normal | 0.970445 | Prob<W | |

**TABLE 2.2**

Results of a Normality Test on the Original Data Using an SAS Macro Written
by D'Agostino et al. (1990)

| Skewness | $g_1 = 0.54127$ | $b_1 = 0.53448$ | $Z(\sqrt{b_1}) = -2.39068$ | $p = 0.0168$ |
|---|---|---|---|---|
| Kurtosis | $g_2 = 0.34569$ | $b_2 = 3.28186$ | $Z(\sqrt{b_2}) = 0.90171$ | $p = 0.3672$ |
| Omnibus test | $K^2 = 6.52846$ (*against chisq 2df*) | | | $p = 0.0382$ |

**TABLE 2.3**

Results of a Normality Test on the Transformed Data (***Milk$^2$***)

| Skewness | $g_1 = -0.11814$ | $b_1 = 0.11666$ | $Z = -0.54876$ | $p = 0.5832$ |
|---|---|---|---|---|
| Kurtosis | $g_2 = -0.03602$ | $b_2 = 2.91587$ | $Z = 0.00295$ | $p = 0.9976$ |
| Omnibus test | $K^2 = 0.30115$ | | | $p = 0.860$ |

$$C = 1 + \frac{1}{3(k-1)} \left[ \left( \sum_{i=1}^{k} \frac{1}{v_i} \right) - \frac{1}{v} \right]$$

The hypothesis of equal variances is rejected if $B > \chi^2_{\alpha,k-1}$.

One of the disadvantages of Bartlett's test is that it is quite sensitive to departure from normality. In fact, if the distributions from which we sample are not normal, this test can have an actual test size several times larger than its nominal level of significance.

### 2.3.2 Levene's Test (1960)

The second test was proposed by Levene (1960). It entails doing a one-way ANOVA on the variables $L_{ij} = |y_{ij} - \bar{y}_i|$, $(i = 1, 2, \cdots k)$.

If the F-test is significant, homogeneity of the variances is in doubt. Brown and Forsythe (1974a) conducted extensive simulation studies to compare between tests on homogeneity of variances and reported the following:

1. If the data are nearly normal, as evidenced by the previously outlined test of normality, then we should use Bartlett's test, which is equally suitable for balanced and unbalanced data.

2. If the data are not normal, Levene's test should be used. Moreover, if during the step of detecting normality we find that the value of $\sqrt{b_1}$ is large, that is, if the data tend to be very skewed, then Levene's test can be improved by replacing $y_i$ with $\tilde{y}_i$, where $\tilde{y}_i$ is the median of the $i$th group. Therefore $\tilde{L}_{ij} = |y_{ij} - \bar{y}_i|$, and an ANOVA is done on $\tilde{L}_{ij}$. The conclusion would be either to accept or reject $H_0 : \sigma_1^2 = \cdots = \sigma_k^2$. If the hypothesis is accepted, then the main hypothesis of equality of several means can be tested using the familiar ANOVA.

If the heterogeneity of variance is established, then it is possible to take remedial action by employing transformation on the data. As a general guideline in finding an appropriate transformation, first plot the k pairs $(y_i, s_i^2) i = 1, 2 \cdots, k$ as a mean to detect the nature of the relationship between $y_i$ and $s_i^2$. For example, if the plot exhibits a relationship that is approximately linear we may try $log y_{ij}$. If the plot shows a curved relationship, then we suggest the square root transformation. Whatever transformation is selected, it is important to test the variance heterogeneity on the transformed data to ascertain if the transformation has in fact stabilized the variance.

If attempts to stabilize the variances through transformation are unsuccessful, several approximately valid procedures to compare between-group means in the presence of variance heterogeneity have been proposed.

## 2.4 Testing Equality of Group Means

### 2.4.1 Welch's Statistic (1951)

$$WL = \frac{1}{k-1} \sum_{i=1}^{k} \frac{w_i(\bar{y}_i - \bar{\bar{y}})^2}{c_1}$$

where

$$c_1 = 1 + \frac{2(k-1)}{(k^2-1)} \sum_{i=1}^{k}(n_i-1)^{-1}\left(1-\frac{w_i}{w}\right)^2$$

$$w_i = n_i/s_i^2, w = \sum_{i-1}^{k} w_i,$$

and

$$\bar{\bar{y}} = \sum_{i=1}^{k} w_i \bar{y}_i / w$$

When all population means are equal (even if the variances are unequal), WL is approximately distributed as an F-statistic with $k-1$ and $f_1$ degrees of freedom, where

$$f_1 = \left[\frac{3}{(k^2-1)} \sum_{i=1}^{k}(n_i-1)^{-1}\left(1-\frac{w_i}{w}\right)^2\right]^{-1}$$

### 2.4.2 Brown and Forsythe Statistic (1974b) for Testing Equality of Group Means

$$BF = \sum_{i=1}^{k} n_i(\bar{y}_i - \bar{y})^2 / c_2$$

$$c_2 = \sum_{i=1}^{k}\left(1-\frac{n_i}{N}\right)s_i^2$$

$$\bar{y} = \frac{1}{N}\sum_{i=1}^{k}\sum_{j=1}^{n_i} y_{ij}, \text{ and } N = n_1 + n_2 + \ldots + n_k$$

$$f_2 = \left[\sum_{i=1}^{k}(n_i-1)^{-1}a_i^2\right]^{-1}$$

and

$$a_i = \left(1-\frac{n_1}{N}\right)s_i^2 / c_2$$

### Remarks

1. The approximations *WL* and *BF* are valid when each group has at least 10 observations and are not unreasonable for *WL* when the group size is as small as 5 observations per group.

2. The choice between *WL* and *BF* depends upon whether extreme means are thought to have extreme variance, in which case *BF* should be used. If extreme means have smaller variance, then *WL* is preferred. Considering the structure of *BF*, Brown and Forsythe recommend the use of their statistic if some sample variances appear unusually low.

### 2.4.3 Cochran's (1937) Method of Weighing for Testing Equality of Group Means

To test the hypothesis $H_0 : \mu_1 = \mu_2 = \cdots = \mu_k = \mu$, where $\mu$ is a specified value, Cochran suggested

$$G = \sum_{i=1}^{k} w_i (\bar{y}_i - \mu)^2$$

the weighted sum of squares, as a test statistic on the hypothesis.

The quantity $G$ follows (approximately) a $\chi_k^2$ distribution. However, if we do not wish to specify the value of $\mu$, it can be replaced with $\bar{y}$, then $G$ becomes

$$\hat{G} = \sum_{i=1}^{k} w_i \left( \bar{y}_i - \bar{\bar{y}} \right)^2$$

$$= \sum_{i=1}^{k} w_i \bar{y}_i^2 - \left( \sum_{i=1}^{k} w_i \bar{y}_i \right)^2 / w$$

It can be shown that, under the null hypothesis of equality of means, $\hat{G}$ is approximately distributed as $\chi_{k-1}^2$; the loss of 1 degree of freedom is due to the replacement of $\mu$ with $\bar{y}$. Large values of $\hat{G}$ indicate evidence against the equality of the means.

#### Example 2.2

In this example we analyze the milk data (untransformed). The main objective of this example is to assess the heterogeneity of variances amongst the 10 farms. Further summary statistics are provided in Table 2.4.

**TABLE 2.4**

Summary Statistics for Each of the 10 Farms

| Farm | Mean | $n_i$ | Std. Dev. | $s_i^2$ |
|------|------|-------|-----------|---------|
| 1 | 30.446 | 12 | 1.7480 | 3.0555 |
| 2 | 29.041 | 12 | 2.9314 | 8.5931 |
| 3 | 24.108 | 12 | 3.3302 | 11.0902 |
| 4 | 27.961 | 12 | 3.2875 | 10.8077 |
| 5 | 29.235 | 12 | 1.7646 | 3.1138 |
| 6 | 27.533 | 12 | 2.1148 | 4.4724 |
| 7 | 28.427 | 12 | 1.9874 | 3.9498 |
| 8 | 24.472 | 12 | 1.8912 | 3.5766 |
| 9 | 24.892 | 12 | 1.7900 | 3.2041 |
| 10 | 21.303 | 12 | 3.8132 | 14.5405 |

The value of the Bartlett's statistics is $B = 18.67$, and from the table of chi-square at 9 degrees of freedom, the $p-value \approx 0.028$. Therefore, there is evidence of variance heterogeneity.

However, after employing the square transformation on the data (see summary statistics in Table 2.5) the Bartlett's statistic is $B = 14.62$ with $p - value \approx 0.102$.

Hence, by squaring the data we achieve two objectives: First, the data were transformed to normality and, second, the variances were stabilized.

**TABLE 2.5**

Summary Statistics on the Transformed Data ($Milk^2$) for Each of the 10 Farms

| Farm | Mean | $n_i$ | Std. Dev. | $s^2$ |
|------|------|-------|-----------|-------|
| 1 | 929.75 | 12 | 109.020 | 11885.36 |
| 2 | 851.25 | 12 | 160.810 | 25.859.86 |
| 3 | 591.34 | 12 | 158.850 | 25233.32 |
| 4 | 791.72 | 12 | 195.020 | 38032.80 |
| 5 | 857.54 | 12 | 102.630 | 10532.92 |
| 6 | 762.18 | 12 | 116.020 | 13460.64 |
| 7 | 811.74 | 12 | 113.140 | 12800.66 |
| 8 | 602.22 | 12 | 92.774 | 8607.02 |
| 9 | 622.53 | 12 | 88.385 | 7811.91 |
| 10 | 467.13 | 12 | 165.080 | 27251.41 |

**Example 2.3**

The following data are the results of a clinical trial involving four groups of dogs assigned at random to each of four therapeutic drugs believed to treat compulsive behavior. The scores given in Table 2.6 are measures of the severity of the disorder after 21 days of treatment. Before comparing the mean scores, we test the homogeneity of variances using Levene's test.

The estimated variances are $s_1^2 = 12.97, s_2^2 = 3.5, s_3^2 = 7.37, s_4^2 = 6.27$.
Table 2.7 gives the location free scores $L_{ij} = |y_{ij} - \bar{y}_i|$.
In Table 2.8 we provide the result of the ANOVA on $L_{ij}$. The F-statistic is 2.11 with a *p*-value of 0.133. This indicates that there is no significant difference between the variances of the four groups even though visually the estimated variances seem quite different from one another.

**TABLE 2.6**

Scores of Four Drugs Given to Dogs in a Clinical Trial

| Drug | Score | | | | | |
|------|------|------|------|------|------|------|
| 1 | 5 | 10 | 2 | 10 | 10 | 4 |
| 2 | 1 | 2 | 5 | 2 | 5 | |
| 3 | 8 | 10 | 4 | 5 | 10 | 10 |
| 4 | 9 | 8 | 7 | 9 | 3 | 10 |

**TABLE 2.7**

Location Free Scores $L_{ij} = |y_{ij} - y_i|$ for the Data of Table 2.6

| Drug | $L_{ij}$ | | | | | |
|------|------|------|------|------|------|------|
| 1 | 1.83 | 3.17 | 4.83 | 3.17 | 3.17 | 2.83 |
| 2 | 2 | 1 | 2 | 1 | 2 | |
| 3 | 0.17 | 2.17 | 3.83 | 2.13 | 2.17 | 2.17 |
| 4 | 1.33 | 0.33 | 0.67 | 1.33 | 4.67 | 2.33 |

**TABLE 2.8**

Results of the ANOVA of $L_{ij}$

| Source | df | Sum Square | Mean Square | F Value | Pr > F |
|--------|-----|-----------|-------------|---------|--------|
| Model | 3 | 8.47824 | 2.82608 | 2.11 | 0.1326 |
| Error | 19 | 25.4350 | 1.33868 | | |
| Corrected total | 22 | 33.9132 | | | |

Levene's test could not detect this heterogeneity because the sample sizes and the number of groups are small. This lack of power is a common feature of many nonparametric tests such as Levene's.

## 2.5 Nonindependence

A fundamental assumption in the analysis of group means is that the samples must be independent, as are the observations within each sample. For now, we shall analyze the effect of within-sample dependency with emphasis on the assumption that the samples or the groups are independent of each other.

Two types of dependency and their effect on the problem of comparing means will be discussed in this section. The first type is caused by a sequence effect. The observations in each group may be taken serially over time in which case observations that are separated by shorter time intervals may be correlated because the experimental conditions are not expected to change, or because the present observation may directly affect the succeeding one. The second type of correlation is caused by the blocking effect. The $n_i$ data points $y_{i1} \cdots y_{ini}$ may have been collected as clusters. For example, the $y$'s may be collected from litters of animals, or from "herds." The possibility that animals or subjects belonging to the same cluster may respond in a similar manner (herd or cluster effect) creates a correlation that should be accounted for. The cluster effect can be significantly large if there are positive correlations between animals within clusters or if there is a trend between clusters. This within herd correlation is known as the intracluster correlation and will be denoted by $\rho$.

Before we examine the effect of serial correlation on comparing group means, we shall review its effect on the problem of testing one mean, that is, its effect on the $t$-statistic.

The simplest time sequence effect is of a serial correlation of lag 1. Higher order lags are discussed in Albers (1978). Denoting $y_1 \cdots y_n$ as the sample values drawn from a population with mean $\mu$ and variance $\sigma^2$, we assume that

$$\text{Cov}\left(y_i, y_{i+j}\right) = \rho^j \sigma^2 \, j = 0, 1, \text{ else it is } 0$$

$$var\left(\bar{y}\right) = \frac{\sigma^2}{n}[1 + 2\rho]$$

Since the parameters $\sigma^2$ and $\rho$ are unknown, we may replace them by their moment estimators:

$$\hat{\sigma}^2 = s^2 = \frac{1}{n-1} \sum_{i=1}^{n} (y_1 - \bar{y})^2$$

and

$$\hat{\rho} = \frac{1}{n-1} \sum_{i=1}^{n-1} (y_i - \bar{y})(y_{i+1} - \bar{y}) / s^2$$

Note that the var $(\bar{y})$ in the presence of serial correlation is inflated by the factor $d = 1 + 2\rho$. This means that if we ignore the effect of $\rho$ the variance of the sample mean would be underestimated. The $t$-statistic thus becomes

$$t = \frac{\sqrt{n}(\bar{y} - \mu)}{s\sqrt{1 + 2\hat{\rho}}} \tag{2.1}$$

To illustrate the unpleasant effect of $\hat{\rho}$ on the $p$-value, let us consider the following hypothetical example. Suppose that we would like to test $H_0 : \mu = 30$ versus $H_1 : y > 30$. The sample information is $y = 32$, $n = 100$, $s = 10$, and $\hat{\rho} = 0.30$. Therefore, from Equation (2.1), $t = 1.58$ and the $p$-value = _0.06. If the effect of $\hat{\rho}$ were ignored, then $t = 2.0$ and the $p$-value = _0.02.

This means that significance will be spuriously declared if the effect of $\hat{\rho}$ is ignored.

**Remark:** It was recommended by Miller (1986) that a preliminary test of $\rho = 0$ is unnecessary, and that it is only the size of $\hat{\rho}$ that is important. See also Barltett (1946).

We now return to the problem of comparing group means. If the observations within each group are collected in time sequence, we compute the serial correlations $\hat{\rho}_i (i = 1, 2, \cdots k)$. Few papers researching the effect of serial correlation on the ANOVA F statistic are available. We shall report on their quantitative merits in Chapter 7, which is devoted to the analysis of repeated measurements and longitudinal data analysis. However, the effect of correlation can be salvaged by using the serial correlation estimates together with one of the available approximate techniques. As an illustration, let $y_{i1}, y_{i2}, \cdots,$ $y_{in_i}$ be the observations from the $i$th group and let $y_i$ and $s_i^2$ be the sample means and variances, respectively. If the data in each sample are taken serially, the serial correlation estimates are $\hat{\rho}_1, \hat{\rho}_2, ..., \hat{\rho}_k$. We can now write the estimated variance of $\bar{y}_i$ as

$$v_i = \hat{var}(\bar{y}_i) = \frac{s_i^2}{n_i} [1 + 2\hat{\rho}_i] \tag{2.2}$$

Therefore, $w_i = v_i^{-1}$ can be used to construct a pooled estimate for the group means under the hypothesis $H_0 : \mu_1 = \cdots = \mu_k$, so that

$$\bar{\bar{y}} = \sum_{i=1}^{k} w_i \bar{y}_i / w, \quad w = \sum_{i=1}^{k} w_i$$

The estimate is

$$\hat{G} = \sum_{i=1}^{k} w_i \left( \bar{y}_i - \bar{\bar{y}} \right)^2 \tag{2.3}$$

The hypothesis is thus rejected for values of exceeding $\chi^2_{k-1}$ at the prescribed $\alpha$-level of significance.

To account for the effect of intraclass correlation, a similar approach may be followed. Procedures for estimating the intraclass correlation will be discussed in Chapter 3.

### Example 2.4

The data presented in Table 2.9 are the average milk production per cow per day, in kilograms, for three herds over a period of 10 months. For illustrative purposes, we would like to use Cochran's statistic G to compare between the herd means. Because the data were collected in serial order, one should account for the effect of the serial correlation.

The estimated variances and serial correlations of the three herds are $s_1^2 = 9.27$, $s_2^2 = 3.41$, $s_3^2 = 6.95$, $\hat{\rho}_1 = 0.275$, $\hat{\rho}_2 = .593$, and $\hat{\rho}_3 = 0.26$. Therefore, $v_1 = 1.36$, $v_2 = 0.75$, and $v_3 = 1.06$. Since $G = 6.13$, the p-value is <0.05; one concludes that the hypothesis of no difference in mean milk production cannot be justified by the data.

**Remarks on SAS programming:** We can test the homogeneity of variances (HOVTEST) of the groups defined by the MEANS effect within PROC GLM. This can be done using either Bartlett's test or Levene's test by adding the following options to the MEANS statement:

MEANS group |HOVTEST = BARTLETT; or

HOVTEST = LEVENE (TYPE = ABS);

**TABLE 2.9**

Milk Production Per-Cow Per-Day in Kilograms for Three Herds over a Period of 10 Months

| Farm | Months | | | | | | | | | |
|------|------|------|------|----|----|------|------|------|------|------|
| | 1 | 2 | 3 | 4 | 5 | 6 | 7 | 8 | 9 | 10 |
| 1 | 25.1 | 23.7 | 24.5 | 21 | 19 | 29.7 | 28.3 | 27.3 | 25.4 | 24.2 |
| 2 | 23.2 | 24.2 | 22.8 | 22.8 | 20.2 | 21.7 | 24.8 | 25.9 | 25.9 | 25.9 |
| 3 | 21.8 | 22.0 | 22.1 | 19 | 18 | 17 | 24 | 24 | 25 | 29 |

## 2.6 Nonparametric Tests

In addition to the aforementioned procedures to deal with situations where the usual assumptions of normality and variance heterogeneity are not tenable, several alternatives are available. Nonparametric procedures are designed to have desirable statistical properties when few assumptions can be made about the underlying distribution of the data. In this section we shall explore:

- Use of nonparametric tests for one- and two-sample location differences
- Testing for dispersion difference
- Testing for one-way layout
- Testing for distribution differences in general

For the "Milk" data we shall illustrate the use of the nonparametric procedures using SAS, while for the "Hospital Admissions" data we shall use R. The results are presented in Table 2.10. The histograms representing the distribution of milk from both farms are shown in Figure 2.3.

**TABLE 2.10**

Basic Summary Statistics for Large Farms (L)

| Moments | | | |
|---|---|---|---|
| N | 60 | Sum weights | 60 |
| Mean | 28.167 | Sum observations | 1690.02 |
| Std deviation | 3.42149601 | Variance | 11.7066349 |
| Skewness | −0.6801441 | Kurtosis | 0.56803226 |
| Uncorrected SS | 48293.4848 | Corrected SS | 690.69146 |
| Coeff variation | 12.1471793 | Std error mean | 0.44171323 |

| Basic Statistical Measures | | | |
|---|---|---|---|
| **Location** | | **Variability** | |
| Mean | 28.16700 | Std deviation | 3.42150 |
| Median | 28.55500 | Variance | 11.70663 |
| Mode | 26.74000 | Range | 16.64000 |
| | | Interquartile range | 4.37000 |

| Tests for Normality | | | |
|---|---|---|---|
| **Test** | | **Statistic** | ***p*-value** |
| Shapiro-Wilk | W | 0.958282 | Pr < W | 0.0388 |
| Kolmogorov-Smirnov | D | 0.106524 | Pr > D | 0.0890 |
| Cramer-von Mises | W-Sq | 0.129199 | Pr > W-Sq | 0.0451 |
| Anderson-Darling | A-Sq | 0.842162 | Pr > A-Sq | 0.0291 |

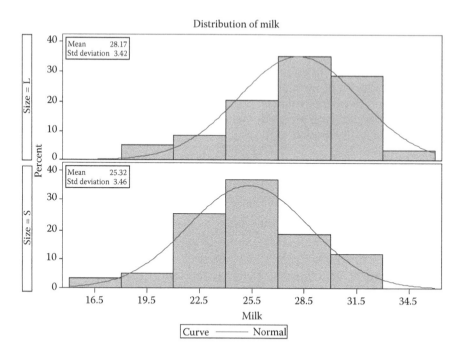

**FIGURE 2.3**
Histogram of the distribution of milk in both types of farms.

### 2.6.1 Nonparametric Analysis of Milk Data Using SAS

The results of the data analyses are shown in Tables 2.10 and 2.11.

The results of the Wilcoxon test (Table 2.12) and the Krusal-Wallis test (Table 2.13) indicate that there is a significant difference in the median levels of milk productions in the two types of farms.

**Example 2.5: Hospital Admissions**

The following data are the average hospital admissions in a certain year classified by the size of the hospital. Here we have three hospital types: small, medium, and large. Similar to the descriptive analyses done to the milk data, we use the R program to conduct similar analyses on the hospital admissions data.

**TABLE 2.11**

Basic Summary Statistics for Small Farms

| Moments | | | |
|---|---|---|---|
| N | 60 | Sum weights | 60 |
| Mean | 25.3241667 | Sum observations | 1519.45 |
| Std deviation | 3.4591908 | Variance | 11.966001 |
| Skewness | -0.6716783 | Kurtosis | 0.56409475 |
| Uncorrected SS | 39184.7991 | Corrected SS | 705.994058 |
| Coeff variation | 13.6596432 | Std error mean | 0.44657961 |

| Basic Statistical Measures | | | |
|---|---|---|---|
| **Location** | | **Variability** | |
| Mean | 25.32417 | Std deviation | 3.45919 |
| Median | 26.03000 | Variance | 11.96600 |
| Mode | 26.07000 | Range | 15.36000 |
| | | Interquartile range | 3.82000 |

| Tests for Normality | | | | |
|---|---|---|---|---|
| **Test** | | | **Statistic** | ***p*-value** |
| Shapiro-Wilk | W | 0.954646 | Pr < W | 0.0259 |
| Kolmogorov-Smirnov | D | 0.116104 | Pr > D | 0.0434 |
| Cramer-von-Mises | W-Sq | 0.110784 | Pr > W-Sq | 0.0825 |
| Anderson-Darling | A-Sq | 0.700621 | Pr > A-Sq | 0.0674 |

**TABLE 2.12**

Testing Equality of Medians Based on Wilcoxon Rank-Sum Test

| Wilcoxon Scores (Rank Sums) for Variable Milk Classified by Variable Size | | | | | |
|---|---|---|---|---|---|
| **Size** | **N** | **Sum of Scores** | **Expected Under H0** | **Std Dev Under H0** | **Mean Score** |
| L | 60 | 4469.0 | 3630.0 | 190.520958 | 74.483333 |
| S | 60 | 2791.0 | 3630.0 | 190.520958 | 46.516667 |
| | | | Average scores were used for ties. | | |

| Wilcoxon Two Sample Test | |
|---|---|
| Statistic | 4469.0000 |
| Normal approximation | |
| Z | 4.4011 |
| One-aided Pr > Z | <.0001 |
| Two-sided Pr > \|Z\| | <.0001 |
| t Approximation | |
| One-sided Pr > Z | <.0001 |
| Two-sided Pr > \|Z\| | <.0001 |
| Z includes a continuity correction of 0.5. | |

**TABLE 2.13**

Results of the SAS Analysis Based
on Wilcoxon Two-Sample Test

| Kruskal-Wallis Test | |
| --- | --- |
| Chi-square | 19.3927 |
| DF | 1 |
| Pr > Chi-square | <.0001 |

Throughout the book we use packages produced by the R Development Core Team (2006, 2010, 2014).

## R Code: Hospital Admissions Data
## R.2.3.1, and R.3.0.3

```
admissions=read.csv(file.choose())
## Have a look at the densities
# We must upload the two functions "car" and "coin".
library(car)
library(coin)
attach(admissions)
names(admissions)
#Simple histogram
hist(admissions$number.Patients)

plot(density(number.Patients))
## Perform the test
shapiro.test(number.Patients)

 ## Plot using a qqplot
qqnorm(number.Patients);qqline(number.Patients, col = 2)
#Adding a normal curve
x=number.Patients
h=hist(x,breaks=10, col="red", xlab="Numner of Patients",
main="Histogram with Normal Curve")
xfit=seq(min(x),max(x),length=40)
yfit=dnorm(xfit,mean=mean(x),sd=sd(x))
yfit=yfit*diff(h$mids[1:2])*length(x)
lines(xfit,yfit,col="blue",lwd=2)
#Boxplot

 boxplot(number.Patients~hospital.Size,data=admissions)
stripchart(number.Patients~hospital.Size,pch=20,vert=TRUE,add=TRUE)
# Levene test
leveneTest(number.Patients~hospital.Size,center=median,data=admissions)

#Applying Nonparametric techniques
#Kruskal Wallis test On Way ANOVA by Ranks
kruskal.test(x~hospital.Size)
#Two samples ttest. We need to compare medium size to large size hospitals)
```

```
twosamples=subset(admissions,hospital.Code >0)
kruskal.test(number.Patients~hospital.Code, data=twosamples)
# I need to construct confidence interval on the difference of two medians.
#The following R code does just that:
a=as.numeric(c(twosamples$number.Patients,hospital.Code="1"))
b=as.numeric(c(twosamples$number.Patients,hospital.Code="2"))
I=wilcox.test(a,b,conf.lev=0.95,conf.int=TRUE,exact=F,correct=T)

    inf=I$conf.int[1]

    sup=I$conf.int[2]
```

## R Output

First, we plot the data histogram, regardless of the hospital size (Figure 2.4). The corresponding density histogram as produced by R is shown in Figure 2.5.

Shapiro-Wilk normality test Data: number .Patients. W = 0.97772, *p*-value = 0.031. We use the Q-Q plot to graphically assess the normality of the data as shown in Figure 2.6.

```
## Plot using a qqplot
```

The box plots for the number of admissions in the three types of hospitals are shown in Figure 2.7.

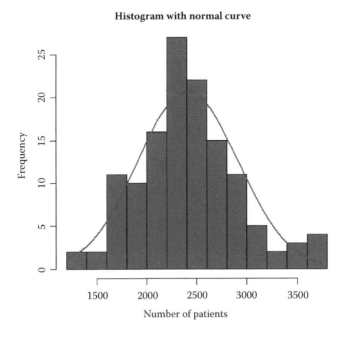

**Histogram with normal curve**

*Number of patients*

**FIGURE 2.4**
The histogram for the number of patients.

**Density.default(x = number.patients)**

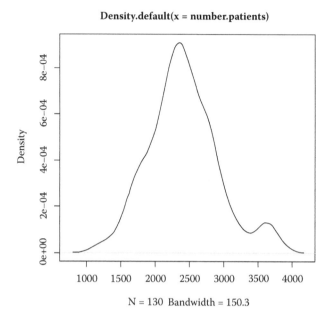

N = 130  Bandwidth = 150.3

**FIGURE 2.5**
Density histogram of the number of patients.

**Normal Q-Q plot**

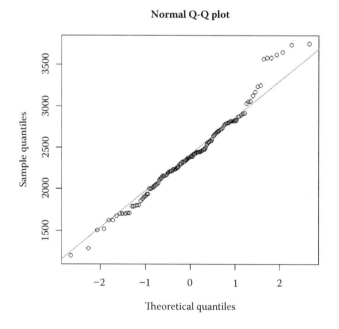

**FIGURE 2.6**
Testing for normality using the Q-Q plot.

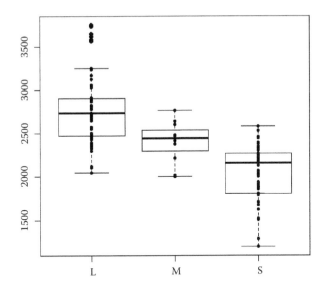

**FIGURE 2.7**
Box plot for comparing the median number of admission in the three types of hospitals.

Testing homogeneity of variances using Levene's test:
Levene's Test for Homogeneity of Variance (center = median)

|                    | Df  | F value | Pr(>F)  |
|--------------------|-----|---------|---------|
| Between hospitals  | 2   | 1.7398  | 0.1797  |
| Within hospitals   | 127 |         |         |

The hypothesis of variance homogeneity is supported by the data.

Now we use the nonparametric Kruskal-Wallis test (one-way ANOVA) to test homogeneity of median number of admissions in the three types of hospitals.

Kruskal-Wallis rank sum test data: x by hospital. Size

```
Kruskal-Wallis chi-squared = 74.666, df = 2, p-value < 2.2e-16
```

The K-W procedure does not do posthoc analysis. Therefore, in case we need to compare medium size to large size hospitals' number of patients admissions, we need to subset the data.

Two samples test:
Kruskal-Wallis rank sum test. data: number. Patients by hospital. Code

```
Kruskal-Wallis chi-squared = 9.1556, df = 1, p-value = 0.00248
```

We may also want to construct 95% confidence interval on the difference between the two medians.

**#Two samples t-test. We need to compare medium size to large size hospitals)**

```
twosamples=subset(admissions,hospital.Code >0)
kruskal.test(number.Patients~hospital.Code, data=twosamples)
# I need to construct confidence interval on the difference of two
medians
a=as.numeric(c(twosamples$number.Patients,hospital.Code="1"))
b=as.numeric(c(twosamples$number.Patients,hospital.Code="2"))
I=wilcox.test(a,b,conf.lev=0.95,conf.int=TRUE,exact=F,
correct=T)

        inf=I$conf.int[1]

        sup=I$conf.int[2]
```

# 3

## Analyzing Clustered Data

### 3.1 Introduction

Clusters are aggregates of individuals or items that are the subject of investigation. A cluster may be a family, school, herd of animals, flock of birds, hospital, medical practice, or an entire community. Data obtained from clusters may be the result of an intervention in a randomized clinical or a field trial. Sometimes interventions in randomized clinical trials are allocated to groups of patients rather than to individual patients. This is called cluster randomization or cluster allocation, and is particularly common in human and animal health research. There are several reasons why investigators wish to randomize clusters rather than the individual study subjects. The first being the intervention may naturally be applicable to clusters. For example, Murray et al. (1992) evaluated the effect of school-based interventions in reducing adolescent tobacco use. A total of 24 schools (of various sizes) were randomized to an intervention condition (SFG = Smoke Free Generation) or to a control condition (EC = Existing Curriculum) are not pair-matched. After two years of follow-up, the number (and proportion) of children in each school who continued to use smokeless tobacco are given in Table 3.1.

It would be impossible to assign students to intervention and control groups because the intervention is through the educational program that is received by all students in a school.

Second, even if individual allocation is possible, there is the possibility of contamination. For example, in a randomized controlled intervention trial the purpose was to reduce the rate of cesarean section. The intervention was that each physician should consult with his/her fellow physician before making the decision to operate, and the control was to allow the physician to make the decision without consulting his/her colleague. Ten hospitals were randomized to receive the intervention, while ten hospitals were kept as controls. In this example, cluster randomization is desired even if the randomization by physician is possible, because of the possibility of significant cross-over contamination. Because the physicians work together, it is likely that a physician in the control group, who did not receive the intervention, might still be affected by it via interactions with colleagues in the intervention group.

**TABLE 3.1**

Smokeless Tobacco Use among Schoolchildren

| Control (EC) | Intervention (SFG) |
|---|---|
| 5/103 | 0/42 |
| 3/174 | 1/84 |
| 6/83 | 9/149 |
| 6/75 | 11/136 |
| 2/152 | 4/58 |
| 7/102 | 1/55 |
| 7/104 | 10/219 |
| 3/74 | 4/160 |
| 1/55 | 2/63 |
| 23/225 | 5/85 |
| 16/125 | 1/96 |
| 12/207 | 10/194 |

Third, cluster allocation is sometimes cheaper or more practical than individual allocation. Many public health interventions are relatively less costly when implemented at an organizational level (e.g., community) than at an individual level.

Similar to cluster randomization, cluster allocation is common in observational studies. In this case, it is sometimes more efficient to gather data from organizational units, such as farms, census tracts, or villages rather than from individuals (see Murray et al., 1992).

## 3.2 The Basic Feature of Cluster Data

When subjects are sampled, randomized, or allocated by clusters, several statistical problems arise. Observations within a cluster tend to be more alike than observations selected at random. If observations within a cluster are correlated, one of the assumptions of estimation and hypothesis testing is violated. Because of this correlation, the analyses must be modified to take into account the cluster design effect. When cluster designs are used, there are two sources of variations in the observations. The first is the between subjects variability within a cluster, and the second is the variability among clusters. These two sources of variation cause the variance to inflate and must be taken into account in the analysis.

The effect of the increased variability due to clustering is to increase the standard error of the effect measure, and thus widen the confidence interval and inflate the type I error rate, compared to a study of the same size using individual randomization. In other words, the effective sample size, and consequently the power are reduced. Conversely, failing to account for clustering in the analysis will result in confidence intervals that are falsely narrow and the *p*-values falsely small. Randomization by cluster, accompanied by an analysis appropriate to randomization by individuals, is an exercise in self-deception (Cornfield, 1978).

Failing to account for clustering in the analysis is similar to another error that relates to the definition of the unit of analysis. The unit of analysis error occurs when several observations are taken on the same subject. For example, a patient may have multiple observations (e.g., systolic blood pressure) repeated over time. In this case the repeated data points cannot be regarded as independent, since measurements are taken on the same subject. For example, if five readings of systolic blood pressure are taken from each of 15 patients, assuming 75 observations to be independent is wrong. Here the patient is the appropriate unit of analysis and is considered as a cluster.

To recognize the problem associated with the unit of analysis, let us assume that we have $k$ clusters each of size $n$ units (the assumption of equal cluster size will be relaxed later on). The data layout may take the general form shown in Table 3.2.

The grand sample mean is denoted by $\bar{y} = \dfrac{1}{nk} \sum_{i=1}^{k} \sum_{j=1}^{n} \bar{y}_{ij}$.

If the observations within a cluster are independent, then the variance of the $i$th cluster mean is

$$V(\bar{y}_i) = \frac{\sigma_y^2}{n} \tag{3.1}$$

**TABLE 3.2**

Data Layout

| Units | Clusters | | | | | |
|---|---|---|---|---|---|---|
| | 1 | 2 | ....... | $i$ | ....... | $K$ |
| 1 | $y_{11}$ | $y_{21}$ | ....... | $y_{i1}$ | ....... | $y_{k1}$ |
| 2 | $y_{12}$ | $y_{22}$ | ....... | $y_{i2}$ | ....... | $y_{k2}$ |
| : | : | : | ....... | : | ....... | : |
| $j$ | $y_{1j}$ | $y_{2j}$ | ....... | $y_{ij}$ | ....... | $y_{kj}$ |
| : | : | : | ....... | : | ....... | : |
| $N$ | $y_{1n}$ | $y_{2n}$ | ....... | $y_{in}$ | ....... | $y_{kn}$ |
| Total | $y_{1.}$ | $y_{2.}$ | ....... | $y_{i.}$ | ....... | $y_{k.}$ |
| Mean | $\bar{y}_1$ | $\bar{y}_2$ | ....... | $\bar{y}_i$ | ....... | $\bar{y}_k$ |

where $\sigma_y^2 = E(y_{ij} - \mu)^2$ and $\mu$ is the population mean. Assuming that the variance is constant across clusters, the variance of the grand mean is

$$V(\bar{y}) = \frac{\sigma_y^2}{nk} \tag{3.2}$$

Now, if $k$ clusters are sampled from a population of clusters, and because members of the same cluster are similar, the variance within cluster would be smaller than would be expected if members were assigned at random. To articulate this concept, we first assume that the $j$th measurement within the $i$th cluster $y_{ij}$ is such that

$$y_{ij} = \mu + \tau_i + e_{ij} \tag{3.3}$$

where $\tau_i$ is the random cluster effect, $e_{ij}$ is the within-cluster deviation from cluster mean, $\mu$ is the grand mean of all measurements in the population, $\tau_i$ reflects the effect of cluster $i$, and $\epsilon_{ij}$ is the error term ($i = 1,2,\ldots, k; j = 1,2,\ldots,n_i$). It is assumed that the cluster effects $\{\tau_i\}$ are normally and identically distributed with mean 0 and variance $\sigma_\tau^2$, the errors $\{e_{ij}\}$ are normally and identically distributed with mean 0 and variance $\sigma_e^2$, and the $\{\tau_i\}$ and $\{e_{ij}\}$ are independent. For this model the intracluster correlation (ICC), which may be interpreted as the correlation $\rho$ between any two members of a cluster, is defined as

$$\rho = \frac{\sigma_\tau^2}{\sigma_\tau^2 + \sigma_e^2} \tag{3.4}$$

It is seen by definition that the ICC is defined as nonnegative in this model, a plausible assumption for the application of interest here. We also note that the variance components $\sigma_\tau^2$ and $\sigma_e^2$ can be estimated from the one-way ANOVA mean squares given in expectation by

$$E(MSB) = \sigma_e^2 + n_0 \sigma_\tau^2 \tag{3.5}$$

$$E(MSW) = \sigma_e^2 \tag{3.6}$$

The ANOVA estimator of $\rho$ is then given by

$$\hat{\rho}_0 = \frac{MSB - MSW}{MSB + (n_0 - 1)MSW} \tag{3.7}$$

where MSB and MSW are obtained from the usual ANOVA table, with corresponding sums of squares:

$$SSB = \sum_{i=1}^{k} n_i (\bar{y}_i - \bar{y})^2$$

$$SSW = \sum_{i=1}^{k} \sum_{j=1}^{n_i} \left( y_{ij} - \bar{y}_i \right)^2$$

where $N = \sum_{i=1}^{k} n_i$ and $n_0 = \frac{1}{k-1}\left[N - \sum_{i=1}^{k} n_i^2 \Big/ N\right]$. Using the delta method, and to the first order of approximation, the variance of $\hat{\rho}_0$ (see Donner, 1986) is given by

$$var(\hat{\rho}_0) = \frac{2(1-\rho)^2(1+(n_0-1)\rho)^2}{n_0^2(k-1)\left(1-\dfrac{k}{N}\right)} \tag{3.8}$$

Note that when $n_i = n$, $i = 1,2,\dots, k$, Equation 3.6 reduces to

$$var(\hat{\rho}_0) = \frac{2(1-\rho)^2(1+(n-1)\rho)^2}{n(n-1)k} \tag{3.9}$$

Note that when $n_i = 2$ (as in twin studies), $var(\hat{\rho}_0) = k^{-1}(1-\rho^2)^2$. An approximate $(1-\alpha)$ 100% confidence interval on $\rho$ may then be constructed as

$$\hat{\rho}_0 \pm z_{1-\alpha/2}\sqrt{var(\hat{\rho}_0)} \tag{3.10}$$

This correlation is known as the ICC. Therefore

$$\sigma_e^2 = \sigma_y^2(1-\rho) \tag{3.11}$$

Equation 3.11 shows that if the observations within cluster are not correlated ($\rho = 0$), then the within-cluster variance is identical to the variance of randomly selected individuals. Since $0 \le \rho < 1$, then the within-cluster variance $\sigma_e^2$ is always less than $\sigma_y^2$.

Simple algebra shows that

$$V(\bar{Y}_i) = \frac{\sigma_y^2}{n}[1+(n-1)\rho] \tag{3.12}$$

and

$$V(\bar{Y}) = \frac{\sigma_y^2}{nk}[1+(n-1)\rho] \tag{3.13}$$

Note that for the binary response, $\sigma^2$ is replaced with $\pi(1-\pi)$. The quantity $[1+(n-1)\rho]$ is called the variance inflation factor (VIF) or the "Design Effect" (DEFF) by Kerry and Bland (1998). It is also interpreted as the relative efficiency of the cluster design relative to the random sampling of subjects and is the ratio

$$DEFF = [1+(n-1)\rho] \tag{3.14}$$

The DEFF represents the factor by which the total sample size must be increased if a cluster design is to have the same statistical power as a design in which individuals are sampled or randomized. If the cluster sizes are not

**TABLE 3.3**

ANOVA Table

| SOV | Df | SOS | MS |
|-----|-----|-----|-----|
| Between clusters | $k - 1$ | BSS | $\text{BMS} = \dfrac{\text{BSS}}{k-1}$ |
| Within clusters | $N - k$ | WSS | $\text{WMS} = \dfrac{\text{WSS}}{N-k}$ |
| Total | $N - 1$ | | |

equal, which is commonly the case, the cluster size $n$ should be replaced with $n' = \sum\limits_{i=1}^{k} n_i^2 / N$, where $N = \sum\limits_{i=1}^{k} n_i$.

Since $\rho$ is unknown, we estimate its value from the one-way ANOVA layout (see Table 3.3), where BSS $= n \sum\limits_{i}^{k} (\bar{y}_i - \hat{\mu})^2$, WSS $= \sum\limits_{i}^{k} \sum\limits_{j}^{n} (y_{ij} - \bar{y}_i)^2$, and the ICC is estimated by

$$\hat{\rho} = \frac{\text{BMS} - \text{WMS}}{\text{BMS} + (n_0 - 1)\text{WMS}} = \frac{\hat{\sigma}_b^2}{\hat{\sigma}_b^2 + \hat{\sigma}_e^2}$$

where $\hat{\sigma}_b^2 = (\text{BMS} - \text{WMS})/n_0$ and $\hat{\sigma}_e^2 = \text{WMS}$ are the sample estimates of $\sigma_b^2$ and $\sigma_e^2$, respectively, where

$$n_0 = \bar{n} - \frac{\sum\limits_{i=1}^{k} (n_i - \bar{n})^2}{k(k-1)\bar{n}} \tag{3.15}$$

Note that when $n_1 = n_2 = \ldots = n_k$, then $n_0 = \bar{n} = n$, where $\bar{n} = \dfrac{N}{k}$.

If $\hat{\rho} > 0$, then the variance in a cluster design will always be greater than a design in which subjects are randomly assigned so that, conditional on the cluster size, the confidence interval will be wider and the $p$-values larger. We note further that, if the cluster size ($n$) is large, the DEFF would be large even for small values of $\hat{\rho}$. For example, an average cluster size 40 and ICC $= 0.02$ would give DEFF of 1.8, implying that the sample size of a cluster randomized design should be 180% as large as the estimated sample size of an individually randomized design in order to achieve the same statistical power.

## 3.3 Effect of One Measured Covariate on Estimation of the Intracluster Correlation

Since in this chapter our focus is on the joint estimation of the regression parameters and the ICC, we shall show how measured covariates affect the point estimation and the precision of the ICC. Let us assume that we have one single covariate. Stanish and Taylor (1983) adjusted Equation 3.3 so that

$$y_{ij} = \mu + \tau_i + \beta\left(x_{ij} - \bar{x}\right) + \epsilon_{ij},\qquad(3.16)$$

where $x_{ij}$ represents a covariate measured without error and $\bar{x} = \dfrac{1}{N}\sum_{i=1}^{k}\sum_{j=1}^{n_i}x_{ij}$.

Again, from Searle et al. (1992) we have E(SSW) = $(N - K - 1)\sigma_\epsilon^2$, E(MSW) = $\sigma_\epsilon^2$, and

$$(\text{MSB}) = \frac{1}{k-1}\text{E(SSB)} = \sigma_\epsilon^2 + n_{01}\sigma_\tau^2 \qquad(3.17)$$

where $n_{01} = \dfrac{1}{k-1}\left[(k-1)n_0 - \dfrac{\sum_{i=1}^{k}n_i^2(\bar{x}_i - \bar{x})^2}{\sum_{i=1}^{k}\sum_{j=1}^{n_i}(x_{ij} - \bar{x})^2}\right]$ and $\bar{x}_i = n_i^{-1}\sum_{j=1}^{n_i}x_{ij}$.

Therefore, the ANCOVA estimator of $\rho$ is given by

$$\hat{\rho}_1 = \frac{\text{MSB} - \text{MSW}}{\text{MSB} + (n_{01} - 1)\text{MSW}}\qquad(3.18)$$

Note that

$$\sum_{i=1}^{k}\sum_{j=1}^{n_i}\left(x_{ij} - \bar{x}\right)^2 = \sum_{i=1}^{k}\sum_{j=1}^{n_i}\left(x_{ij} - \bar{x}_i\right)^2 + \sum_{i=1}^{k}n_i(\bar{x}_i - \bar{x})^2$$

We have three remarks on the preceding setup:

1. If $x_{ij}$ is measured at the cluster level, then $x_{ij} = \bar{x}_i$. Hence

$$n_{01} = n_0 - \frac{1}{k-1}\left[\frac{\sum_{i=1}^{k}n_i^2(\bar{x}_i - \bar{x})^2}{\sum_{i=1}^{k}n_i(\bar{x}_i - \bar{x})^2}\right]$$

   and the expectation of the mean sum of squares between clusters can be written as

$$\text{E(MSB)} = \sigma_\epsilon^2 + n_0\sigma_\tau^2 - \frac{\sigma_\tau^2}{k-1}\left[\frac{\sum_{i=1}^{k}n_i^2(\bar{x}_i - \bar{x})^2}{\sum_{i=1}^{k}n_i(\bar{x}_i - \bar{x})^2}\right]\qquad(3.19)$$

2. It is clear from Equation 3.19 that when one covariate is measured at the cluster level, the expected mean square between clusters is reduced by the amount

$$\frac{\sigma_\tau^2}{k-1}\left[\frac{\sum_{i=1}^{k}n_i^2(\bar{x}_i - \bar{x})^2}{\sum_{i=1}^{k}n_i(\bar{x}_i - \bar{x})^2}\right]$$

The covariate effect

$$\text{CEF} = \sum_{i=1}^{k} n_i^2 (\bar{x}_i - \bar{x})^2 \Big/ \sum_{i=1}^{k} n_i (\bar{x}_i - \bar{x})^2$$

is the ratio of two quantities that measure the extent of the deviation of the cluster mean from the overall mean of the measured covariate. The numerator in this expression is seen to be weighted by the square of the cluster size, and the denominator by the cluster size.

3. We finally note that the degrees of freedom associated with the within-cluster sum of squares is reduced by 1, due to the estimation of the regression coefficient $\beta$.

It is not generally clear how measuring a covariate on the cluster level will affect the estimated value of the ICC. However, in the case $n_i = n$, $i = 1,2,...,k$, $\text{CEF} = n$ and $E(MSB) = \sigma_\epsilon^2 + n\sigma_\tau^2 \left( \dfrac{k-2}{k-1} \right)$. Hence, when we have a large number of clusters, the effect of the measured covariate on the estimated ICC is negligible.

Shoukri et al. (2013) extended the work of Stanish and Taylor (1983) to the case of two covariates and to the case of the binary response variable. They concluded that under the above set up the estimated ICC is mildly affected by inclusion of a covariate measured at the cluster level.

We have demonstrated that applying conventional statistical methods to cluster data assuming independence between the study subjects is wrong, and one has to employ an appropriate approach that duly accounts for the correlated nature of cluster data. The complexity of the approach depends on the complexity of design. For example, the simplest form of cluster data is the one-way layout, where subjects are nested within clusters. Conventionally, this design has two levels of hierarchy: subjects at the first level and clusters at the second level. For example, the sib-ship data which we will analyze in this chapter have two levels of hierarchy; the first is observations from sibs and the second is formed by the family identifiers. Data with multiple levels of hierarchy such as animals within the farms and farms nested within regions may also be available and one must account for the variability at each level of hierarchy.

We shall now review some of the studies reported in the medical literature, which must be included under "cluster randomization trials" where we clearly identify what is meant by "cluster."

- Russell et al. (1983) investigated the effect of nicotine chewing gum as a supplement to general practitioners' advice against smoking. Subjects were assigned by week of attendance (in a balanced design) to one of three groups: (a) nonintervention controls, (b) advice and booklet, and (c) advice and booklet plus the offer of nicotine gum. There were six practices, with recruitment over three weeks, one week to each regime. There were 18 clusters (practices) and 1938 subjects. The unit of analysis will be subject nested within practice.

- In a trial of clinical guidelines to improve general-practice management and referral of infertile couples, Emslie et al. (1993) randomized 82 general practices in the Grampian (Scotland) region and studied 100 couples in each group. The outcome measure was whether the general practitioner had taken a full sexual history, and examined and investigated both partners appropriately, so the general practitioner would be the unit of analysis.

- A third example is the Swedish two-county trial of mammographic screening for breast cancer. In this study, clusters (geographical areas) were randomized within strata, comprising 12 pairs of geographical clusters in Ostergotland County and 7 sets of triplets in Kopperberg County. The strata were designed so that so that clusters within a stratum were similar in socioeconomic terms. It should be noted that for randomization or allocation by cluster, there is a price to be paid at the analysis stage. We can no longer think of our trial subjects as independent individuals, but must do the analysis at the level of the sampling unit. This is because we have two sources of variation that is between subjects within a cluster and between clusters, and the variability between clusters must be taken into account. Clustering leads to a loss of power and a need to increase the sample size to cope with the loss (Murray 1998). The excess variation resulting from randomization being at the cluster level rather than the individual level was neglected by Tabar et al. (1985). They claimed (without supporting evidence) that such excess variation was negligible. This study was later analyzed by Duffy et al. (2003) who used hierarchical modeling to take clustering into account and found evidence for an effect. Taking account of the cluster randomization, there was a significant 30% reduction in breast cancer mortality. They concluded that mammographic screening does indeed reduce mortality from breast cancer, and that the criticisms of the Swedish Two-County Trial were unfounded. The fact is that the criticism was founded, because it was wrong to ignore clustering in such a study. Getting the same answer when we do it correctly is irrelevant.

There are several approaches that can be used to allow for clustering ranging from simple to quite sophisticated:

- Whether the outcome is normally distributed or not, group comparisons may be achieved by adjusting the standard error of the "effect measure" using the design effect (DEFF). These are generally approximate methods, but more realistic than ignoring the within-cluster correlation.

- Multilevel or hierarchical or random effects modeling (Hong and Raudenbush 2006).
- When covariates adjustment are required within the regression analyses, the generalized estimating equation (GEE) approach is used.
- Bayesian hierarchical models

The focus through this book will be on the statistical analysis of correlated data using the first three approaches.

## 3.4 Sampling and Design Issues

As we have seen, the main reasons for adopting trials with clusters as the sampling unit are

- Evaluation of interventions, which by their nature have to be implemented at the community level (e.g., water and sanitation schemes), and some educational interventions (e.g., smoking cessation project).
- Logistical convenience, or to avoid the resentment or contamination that might occur if unblended interventions were provided for some individuals but not others in each cluster.
- It might be desirable to capture the mass effect on disease of applying an intervention to a large proportion of community or cluster members, for example, due to "an overall reduction in the transmission of an infectious agent" (Hayes and Bennett, 1999).

The within cluster correlation can come about by any of several mechanisms, including shared exposure to the same physical or social environment, self selection in belonging to the cluster or the group, or sharing ideas or diseases among members of the group.

As we see from Equation 3.14, the DEFF is a function of the cluster size and the ICC. The values of $\rho$ tend to be larger in small groups, such as a family, and smaller in large groups, such as a county or a village, because the degree of clustering often depends upon the interaction of group members. Family members are usually more alike than individuals in different areas of a large geographic region. Unfortunately, the influence of $\rho$ on study power is directly related to cluster size. Studies with a few large clusters are often very inefficient.

### 3.4.1 Comparison of Means

We assume in this example that $k$ clusters of $n$ individuals are assigned to each of an experimental group (E) and a control group (C). The aim is to test the hypothesis $H_0 : \mu_E = \mu_C$, where $\mu_E$ and $\mu_C$ are the means of the two groups, respectively, of a normally distributed response variable $Y$ having common but unknown variance $\sigma^2$.

Unbiased estimates of $\mu_E$ and $\mu_C$ are given by the usual sample means $\bar{y}_E$ and $\bar{y}_C$, where

$$\bar{y}_E = \frac{1}{nk} \sum_{i=1}^{k} \sum_{j=1}^{n} y_{ij}$$

and

$$V(\bar{y}_E) = \frac{\sigma^2}{nk} [1 + (n-1)\rho] \tag{3.20}$$

with similar expression for $V(\bar{y}_C)$.

For sample size determination, expression 3.20 implies that the usual estimate of the required number of individuals in each group should be multiplied by the inflation factor, or DEFF $= [1 + (n-1)\,\rho]$ to provide the same statistical power as would be obtained by randomizing $nk$ subjects to each group when there is no clustering effect (see Campbell et al. 2004). More formally, let $z_{\alpha/2}$ denote the value of a standardized score cutting of $100\alpha/2\%$ of each tail of a standard normal distribution, and $z_\beta$ denote the value of a standardized sore cutting off the upper $100\,\beta\%$. Then the test $H_0 : \mu_E = \mu_C$ versus $H_1 : \mu_E \neq \mu_C$ has a power of $100\,(1-\beta)\%$ when performed at the $100\alpha$ percent level of significance. The number of individuals $n$ required in each group is given by

$$n' = 2(z_{\alpha/2} + z_\beta)^2 \sigma^2 [1 + (n-1)\rho] / \delta^2 \tag{3.21}$$

where $\delta$ is a "meaningful difference" specified in advance by the investigator. Alternatively Equation 3.21 may be written as

$$n' = (z_{\alpha/2} + z_\beta)^2 [1 + (n-1)\rho] / \Delta^2 \tag{3.22}$$

where $\Delta = \dfrac{\mu_E - \mu_C}{\sigma\sqrt{2}} = \dfrac{\delta}{\sigma\sqrt{2}}$.

At $\rho = 0$ formula 3.21 reduces to the usual sample size specification given by Snedecor and Cochran (1981). When $\rho = 1$, there is variability within cluster, the usual formula applies with $n$ as the number of clusters that must be sampled from each group. In most epidemiologic applications, however, values of $\rho$ tend to be no greater than 0.6, and advance estimates may also be available from previous data. Obtaining an estimate of $\rho$ is no different in principle from the usual requirement imposed on the investigator to supply an advance estimate of the population variance $\sigma^2$. In the case of unequal cluster sizes, we may replace $n$ in the right-hand side of Equation 3.21 or Equation 3.22 by the average cluster size, $\bar{n}$ (or by $n_0$). A conservative approach would be to replace $n$ by $n_{max}$, the largest expected cluster size in the sample.

We now consider a test of significance on $\bar{y}_E - \bar{y}_C$. Note that in most applications, the clusters are likely to be of unequal size $n_i$, $i = 1, 2, ..., k$. In this case formula 3.20 generalizes to

$$V(\bar{y}_E) = \frac{\sigma_e^2}{N}\left[1 + \left(\frac{\sum n_i^2}{N}\right)\frac{\sigma_b^2}{\sigma_e^2}\right] \tag{3.23}$$

An estimate $\hat{V}(\bar{y}_E)$ of $V(\bar{y}_E)$ may be calculated by substituting $\hat{\sigma}_b^2$ and $\hat{\sigma}_e^2$ (defined earlier), the sample estimates of $\sigma_b^2$ and $\sigma_e^2$, respectively, in Equation 3.23.

A large sample test on $H_0 : \mu_E = \mu_C$ may be obtained by calculating

$$Z = \frac{\bar{y}_E - \bar{y}_C}{\left[\hat{V}(\bar{y}_E) + \hat{V}(\bar{y}_C)\right]^{\frac{1}{2}}} \tag{3.24}$$

**Example 3.1: (ch2_milk)**

The milk yield data from 10 Ontario farms, 5 large and 5 small farms, are analyzed. Each farm provided 12 observations representing the average milk yield per cow per day for each month. For the purpose of this example we shall ignore the sampling time as a factor and consider each farm as a cluster of size 12. The following SAS code shows how to read in the data and run the general linear model.

```
data milk;
input farm milk size $;
cards;
1  32.33   L
1  29.47   L
1  30.19   L
...
10 24.12   S
;

proc sort data=milk; by size;

proc glm data = milk;
class farm;
model milk = farm;
run;
```

The ANOVA results from the SAS output are given next:

| Source | DF | Sum of Squares | Mean Square |
|--------|-----|----------------|-------------|
| Model  | 9   | 905.122484     | 100.569165  |
| Error  | 110 | 734.014075     | 6.672855    |

$$\hat{\sigma}_e^2 = 6.67$$

$$\hat{\sigma}_b^2 = \frac{100.57 - 6.67}{12} = 7.83$$

Therefore, the estimated ICC is $\hat{\rho} = \dfrac{7.83}{7.83 + 6.67} = 0.54$.

An important objective of this study was to test whether average milk yield in large farms differs significantly from the average milk yield of small farms. That is, to test $H_0 : \mu_s = \mu_l$ versus $H_1 : \mu_s \neq \mu_l$.

The ANOVA results separately for each farm size (small and large) are now produced. The SAS code is

```
proc glm data = milk;
class farm;
model milk = farm;
by size;
run; quit;
```

Large Farm Size

| Source | DF | Sum of Squares | Mean Square |
|--------|----|----------------|-------------|
| Model | 4 | 283.4424600 | 70.8606150 |
| Error | 55 | 407.2490000 | 7.4045273 |

Small Farm Size

| Source | DF | Sum of Squares | Mean Square |
|--------|----|----------------|-------------|
| Model | 4 | 379.2289833 | 94.8072458 |
| Error | 55 | 326.7650750 | 5.9411832 |

For large farms, since $\bar{y}_l = 28.00$, $k = 5$, $n = 12$, $N = 60$

$$\hat{\sigma}_{el}^2 = 7.40, \quad \hat{\sigma}_{bl}^2 = \frac{70.86 - 7.4}{12} = 5.29$$

$$\hat{\sigma}_{bl}^2 / \hat{\sigma}_{el}^2 = 0.71$$

then $\hat{V}(\bar{y}_l) = \dfrac{7.40}{60}[1 + (12)(0.71)] = 1.17$.

For small farms:

$$\bar{y}_s = 25.32, \quad k = 5, \quad n = 12, \quad N = 60$$

$$\hat{\sigma}_{es}^2 = 5.94, \quad \hat{\sigma}_{bs}^2 = \frac{94.81 - 5.94}{12} = 7.4$$

$$\hat{\sigma}_{bs}^2 / \hat{\sigma}_{es}^2 = 1.24$$

$$\hat{V}(\bar{y}_s) = \frac{5.94}{60}[1 + (12)(1.24)] = 1.57$$

$$Z = \frac{28 - 25.32}{(1.17 + 1.57)^{\frac{1}{2}}} = 1.61$$

$p$-value = 0.10, and there is no sufficient evidence in the data to support $H_O$.

The R code to read in the milk data and produce the ANOVA results is

```
milk <- read.csv(file.choose())

# ANOVA results overall
anova(lm(milk ~ factor(farm), data=milk))

# ANOVA results for large farms
anova(lm(milk~factor(farm), data=milk[milk$size=='L',]))

# ANOVA results for small farms
anova(lm(milk~factor(farm), data=milk[milk$size=='S',]))
```

## 3.5 Regression Analysis for Clustered Data

A fundamental question in many scientific investigations is concerned with how and to what extent a response variable is related to a set of independent variables. For example, a health economist may be interested in the relationship between the effect of intervention and the cost of its administration, a clinical nutritionist may be interested in the relationship between obesity and hypertension, or a radiologist may be interested in the relationship between the ultrasound diagnosis of cancer and the tumor size. The list of situations of this kind in biomedical research is endless, let alone other areas of applications.

Suppose for a given situation, the actual mathematical relationship between the response variable $Y$ and a set of independent variables is known. The investigator is then in a position to understand the factors that control the direction of the response. Unfortunately there are few situations in practice in which the true mathematical model connecting the response to the independent variables is known. Consequently, one is forced to combine practical experience and mathematical theory to develop an approximate model that characterizes the main features of the behavior of the response variable.

Regression analysis is among the most commonly used methods of statistical analysis to model the relationship between variables. Its objective is to describe the relationship of response with independent or explanatory variables. In its very general form, a regression model is written as

$$Y = X\beta + e \qquad (3.25)$$

where $Y$ is the $(n \times 1)$ vector of dependent variable values, $X$ is a $(n \times (p + 1))$ matrix containing the values of the independent variables, $\beta$ is the $((p + 1) \times 1)$ vector of parameters, and $e$ is the $(n \times 1)$ vector of error components. It is well known that the method of least squares is the most preferred method of estimation of the parameter of the regression model. There are fundamental assumptions that should be satisfied to use this method to estimate the parameter vector $\beta$:

    a. The components of $Y$ are uncorrelated with each other.

    b. The error components $e$ are assumed to be uncorrelated with mean zero and common variance $\sigma^2$.

    c. The vector of error components $e$ is uncorrelated with the matrix $X$.

Under these conditions and provided that $(X^TX)^{-1}$ exists, the least squares estimate of $\beta$ is

$$\hat{\beta} = \left(X^TX\right)^{-1}X^TY \tag{3.26}$$

Equation 3.26 is important in regression analysis since it provides the estimates of $\beta$ once we are sure that conditions (a, b, c) are satisfied and the matrix X is specified.

In addition to the linear models (Equation 3.25), regression models include logistic models for binary responses, log-linear models for counts, and survival analysis for time to events. In this chapter we discuss the linear-normal model for continuous responses when the basic assumption that all the observations are independent, or at least uncorrelated, is violated. Recall that the assumption of zero correlation would mean that knowing one subject's response provides no information regarding the status of another subject in the same study. However, the assumption of independence may not hold if the subjects belong to the same cluster as has been already demonstrated. As an example of a regression problem when clusters of subjects are sampled together is Miall and Oldham's arterial blood pressure levels family study. Due to their common household environment and their shared genetic makeup, we would expect a family member to have a greater chance of having elevated blood pressure levels if his/her sibling had. Data from this study can usefully be thought of as being "clustered" into families. Blood pressure levels from different families are likely to be independent; those from the same family are not. This dependence among observations from the same cluster must be accounted for in assessing the relationships between risk factors and health outcomes.

Another example cited by Liang and Zeger (1993) is the growth study of Hmong refugee children. In this example 1000 Hmong refugee children receiving health care at two Minnesota clinics between 1976 and 1985 were examined for their growth patterns. The objective is to study the patterns of growth and its association with age at entry into the United States. It is believed that stature is influenced by both genetic and environmental factors. When the offending environmental factors have been removed, the growth process progresses at a faster rate. To study the growth, repeated measurements of height of each child were recorded. The number of visits per child ranged from 1 to 15 and averaged 5. The correlation between repeated observations on height for each child may be a nuisance but cannot be ignored in regression analysis.

The preceding two examples have common features, although they address questions with different scientific objectives. Data in the preceding two studies are organized in clusters. For family study, the clusters are formed by families, and in the second example a cluster comprises the repeated observations for a child. Another aspect of similarity between the two studies is that one can safely assume that the response variables (blood pressure in the family study and height in the growth study) are normally distributed. The two studies also differ in the structure of the within-cluster correlation. For example, in the family study one may assume that the correlation between the pairs of sibs within the family is equal, that is, we may assume a constant within cluster correlation. For a repeated measures longitudinal study the situation is different. Although the repeated observations are correlated, this correlation may not be constant across time (cluster units). It is common sense to assume that observations taken at adjacent time points are more correlated than observations that are taken at separated time points.

In the remainder of this chapter, we shall focus on regression analysis of clustered data assuming common or fixed within-cluster correlation and the response variable is normally distributed. Other types of response variables and different correlation structures will be discussed in details in subsequent chapters.

Within the framework of linear regression, we illustrate an answer to the question: What happens when the conventional linear regression is used to analyze clustered data?

Let $Y_{ij}$ denote the score on the $j$th offspring in the $i$th family, and $X_i$, the score of the $i$th parent, where $j = 1, 2, ..., n_i$, $i = 1, 2, ..., k$, and $n_i$ is the number of offspring in the $i$th family, and $k$ is the total number of families. We assume that the regression of $Y$ on $X$ is given by

$$Y_{ij} = \mu_y + \beta(X_i - \mu_x) + E_{ij} \qquad (3.27)$$

where $\mu_y = E(Y_{ij})$, $\mu_x = E(X_i)$, $\beta$ is the regression coefficient of $Y$ on $X$, and $E_{ij}$ is the deviation of the $j$th offspring of the $i$th parent. We further assume that

$$\mathrm{Cov}\left(Y_{ij}, Y_{il}\right) = \begin{cases} \rho\sigma^2 & j \neq l \\ \sigma^2 & j = l \end{cases}$$

Under this model, Kempthorne and Tandon (1953) showed that the minimum variance unbiased estimator of $\beta$ is given by

$$b_1 = \sum_{i=1}^{k} w_i(x_i - \hat{x})\bar{y}_i \Big/ \sum w_i(x_i - \hat{x})^2 \qquad (3.28)$$

and $V(b_1) = \sigma^2(1 - \rho) / \sum w_i(x_i - \hat{x})^2$

where $w_i = n_i/(1 + n_i\rho_e)$, $\rho_e = (\rho - \beta^2)/(1 - \beta^2)$, $\bar{y}_i = \sum_j y_{ij}/n_i$, and $\hat{x} = \sum w_i x_i / \sum w_i$.

The most widely used estimator for $\beta$ ignores the ICC $\rho$ and is given by the usual estimator

$$b = \frac{\sum_{i=1}^{k} n_i(\bar{y}_i - \bar{y})(x_i - \bar{x})}{\sum_{i=1}^{k} n_i(x_i - \bar{x})^2} \tag{3.29}$$

where $\bar{y} = \sum n_i\bar{y}_i/N$, $\bar{x} = \sum n_i x_i/N$, and $N = n_1 + n_2 + ...n_k$.

$$V(b) = \frac{\sigma^2(1 - \rho)\sum n_i(1 + n_i\rho_e)(x_i - \bar{x})^2}{\left[\sum n_i(x_i - \bar{x})^2\right]^2} \tag{3.30}$$

If $\rho_e = 0$, then $w_i = n_i$ and $var(b) = var(b_1) = \sigma^2(1 - \rho)/\sum n_i(x_i - \bar{x})^2$, which means that $b_1$ is fully efficient.

Therefore,

$$\frac{V(b)}{V(b_1)} = \left[\frac{\sum n_i(1 + n_i\rho_e)(x_i - \bar{x})^2}{\sum n_i(x_i - \bar{x})^2}\right]$$

or equivalently

$$\frac{V(b)}{V(b_1)} = 1 + \frac{\rho_e \sum n_i^2(x_i - \bar{x})^2}{\sum n_i(x_i - \bar{x})^2} \tag{3.31}$$

Assuming $\rho_e > 0$, $b$ is always less efficient.

The most important message of Equation 3.31 is that ignoring within-cluster correlation can lead to a loss of power when both within-cluster and cluster-level covariate information is being used to estimate the regression coefficient.

To analyze clustered data, one must therefore model both the regression of $Y$ on $X$ and the within-cluster dependence. If the responses are independent of each other, then ordinary least squares can be used, which produces regression estimators that are identical to the maximum likelihood in the case of normally distributed responses. In this chapter we consider two different modeling approaches: marginal and random effects. In marginal models, the regression of $Y$ on $X$ and the within-cluster dependence are modeled separately. The random effects models attempt to address both issues simultaneously through a single model. We shall explore both modeling strategies for a much wider class of distributions named generalized linear models that include the normal distribution as a special case.

## 3.6 Generalized Linear Models

Generalized linear models (GLMs) are a unified class of regression methods for continuous and discrete response variables. There are two components in a GLM: the systematic component and the random component. For the systematic component, one relates $Y$ to $X$ assuming that the mean response $\mu = E(Y)$ satisfies $g(\mu) = X_1\beta_1 + X_2\beta_2 + \ldots + X_p\beta_p$, which may be conveniently written as

$$g(\mu) = X^T\beta \tag{3.32}$$

Here, $g(.)$ is a specified function known as the "link function." The normal regression model for continuous data is a special case of Equation 3.32, where $g(.)$ is the identity link. For binary response variable $Y$, the logistic regression is a special case of Equation 3.32 with the logit transformation as the link. That is,

$$g(\mu) = \log\frac{\mu}{1-\mu} = \log\frac{P(Y=1)}{P(Y=0)} = X^T\beta$$

When the response variable is count, we assume that

$$\log E(Y) = X^T\beta$$

To account for the variability in the response variable that is not accounted for by the systematic component, a GLM assumes that $Y$ has the probability density function given by

$$f(y) = \exp[(y\theta - b(\theta))/\varphi + C(y, \varphi)] \tag{3.33}$$

which is a general form of the exponential family of distributions. This includes among others, the normal, binomial, and the Poisson as special cases. It can be easily shown that

$$E(Y) = b'(\theta)$$

and

$$V(Y) = \varphi b''(\theta)$$

For the normal distribution:

$$\theta = \mu \text{ (identity link)}$$

$$b(\theta) = \mu^2/2, \ \varphi = \sigma^2$$

Hence, $b'(\theta) = \mu$ and $b''(\theta) = 1$, indicating that $\phi$ is the variance. For the Poisson distribution:

$$\theta = \ln\mu \text{ (log–link)}$$

$$b(\theta) = \mu = e^\theta, \text{ and } \varphi = 1$$

$$b'(\theta) = e^{\theta} = \mu$$

$$b''(\theta) = e^{\theta} = \mu$$

Hence, $E(Y) = V(Y) = \phi\mu$.

The scale parameter $\phi$ is called the overdispersion parameter. If $\phi > 1$, then the variance of the counts is larger than the mean.

When the link function in Equation 3.32 and the random component are specified by the GLM, we can estimate the regression parameters $\beta$ by solving the estimating equation:

$$U = \sum_{i=1}^{p} \left(\frac{\partial \mu_i(\beta)}{\partial \beta}\right)^T V^{-1}(Y_i)[Y_i - \mu_i(\beta)] = 0 \tag{3.34}$$

Equation 3.34 provides valid estimates when the responses are independently distributed. For clustered data, the GLM and hence Equation 3.34 are not sufficient, since the issue of within-cluster correlation is not addressed. We now discuss the two modeling approaches commonly used to analyze clustered data.

### 3.6.1 Marginal Models (Population Average Models)

As already mentioned, in a marginal model, the regression of $Y$ on $X$ and the within-cluster dependence are modeled separately. We assume:

- The marginal mean or "population-average" of the response, $\mu_{ij} = E(Y_{ij})$ depends on the explanatory variables $X_{ij}$ through a link function $g(\mu_{ij}) = X_{ij}^T\beta$, where $g$ is a specified link function.

- The marginal variance depends on the marginal mean through $V(Y_{ij}) = \varphi V(\mu_{ij})$, where $V(.)$ is a known variance function, such as $V(\mu_{ij}) = \varphi$ for normally distributed response and $V(\mu_{ij}) = \mu_{ij}$ for count data similar to the GLM setup.

- $Cov(Y_{il}, Y_{il})$, the covariance between pairs within clusters is a function of the marginal means and another additional parameter $\alpha$, that is, $Cov(Y_{ij}, Y_{il}) = \gamma(\mu_{ij}, \mu_{il}, \alpha)$, where $\gamma(.)$ is a known function.

### 3.6.2 Random Effects Models

There are several names given to these types of models; multilevel, hierarchical, random coefficients or mixed effects models. The fundamental feature of these models is the assumption that parameters vary from cluster to cluster, reflecting natural heterogeneity due to unmeasured cluster level covariates.

The general setup for the random effects GLM was described by Zeger and Karim (1991) as follows:

- Conditional on random effects $b_i$, specific to the $i$th cluster, the response variable $Y_{ij}$ follow a GLM with

$$g\left[E\left(Y_{ij}|b_i\right)\right] = X_{ij}^T\beta + Z_{ij}^T b_i \qquad (3.35)$$

  where $Z_{ij}$, a $q \times 1$ vector of covariates, is a subset of $X_{ij}$.
- Conditional on $b_i$, $Y_i = (Y_{i_1}, Y_{i_2}, ...Y_{in})^T$ are statistically independent.
- The $b_i$'s are independent observations from a distribution $F(.,\alpha)$, indexed by some unknown parameter $\alpha$. The term "random effects" was assigned to the random variables $b_i$, because we treat $b_i$ as a random sample from $F$. The random effects $b_i$ are not related to $X_{ij}$.

### 3.6.3 Generalized Estimating Equation (GEE)

The generalized estimating equation (GEE) provides a tool for practical statistical inference on the $\beta$ coefficient under the marginal model when the data are clustered. The regression estimates are obtained by solving the equation

$$U_1(\beta, \alpha) = \sum_{i=1}^{k} \left(\frac{\partial \mu_i}{\partial \beta}\right)^T Cov^{-1}(Y_i, \beta, \alpha)[y_i - \mu_i(\beta)] = 0 \qquad (3.36)$$

where $\mu_i(\beta) = E(Y_i)$, the marginal mean of $Y_i$. One should note that $U_1(.)$ has the same form of $U(.)$ in Equation 3.34, except that $Y_i$ is now $n_i \times 1$ vector, which consists of the $n_i$ observations of the $i$th cluster, and the covariance matrix $Cov(Y_i)$, which depends on $\beta$ and $\alpha$, a parameter that characterizes the within-cluster dependence.

For a given $\alpha$, the solution $\hat{\beta}$ to Equation 3.36 can be obtained by an iteratively reweighted least squares calculations. The solution to these equations is a consistent estimate of $\beta$ provided that $\mu_i(\beta)$ is correctly specified. The consistency property follows because $\left(\dfrac{\partial \mu_i}{\partial \beta}\right)^T Cov^{-1}(Y_i)$ does not depend on the $Y_i$'s, so Equation 3.36 converges to zero and has consistent roots so long as $E(Y_i - \mu_i(\beta)) = 0$ (see Liang and Zeger, 1986, for details).

If an $\sqrt{k}$ consistent estimate of $\alpha$ is available, $\hat{\beta}$ are asymptotically normal, even if the correlation structure is misspecified. Liang and Zeger (1986) proposed a "robust" estimate of the covariance matrix of $\hat{\beta}$ as

$$V\left(\hat{\beta}\right) = A^{-1}MA^{-1}$$

where

$$A = \sum_{i=1}^{k} \tilde{D}_i^T \tilde{V}_i^{-1} \tilde{D}_i$$

$$M = \sum_{i=1}^{k} \tilde{D}_i^T \tilde{V}_i^{-1} Cov(Y_i) \tilde{V}_i^{-1} \tilde{D}_i$$

$$Cov(Y_i) = \left(Y_i - \mu_i\left(\tilde{\beta}\right)\right)\left(Y_i - \mu_i\left(\tilde{\beta}\right)\right)^T$$

(A tilde [~] denotes evaluation at $\tilde{\beta}$ and $\tilde{\alpha}\left(\tilde{\beta}\right)$.)

Liang and Zeger (1986) proposed a simple estimator for $\alpha$ based on Pearson's residuals

$$\hat{r}_{ij} = \frac{y_{ij} - \mu_{ij}(\beta)}{\sqrt{var\left(y_{ij}\right)}}$$

For example, under common correlation structure that $Cor(y_{ij}, y_{il}) = \alpha$ for all $i, j$, and $l$, an estimate of $\alpha$ is

$$\hat{\alpha} = \sum_{i=1}^{k}\sum_{j=1}^{n_i}\sum_{l=j+1}^{n_i-1} \hat{r}_{ij}\hat{r}_{il} \left/ \left[\sum_{i=1}^{k}\binom{n_i}{2} - p\right]\right. \tag{3.37}$$

where $p$ in the denominator is the number of regression parameters.

One of the limitations of the preceding approach is that estimation of $\beta$ and $\alpha$ from Equations 3.36 and 3.37 is done as if $(\beta, \alpha)$ are independent of each other. Consequently, very little information from $\beta$ is used when estimating $\alpha$. This can lead to a significant loss of information on $\alpha$. To improve on the efficiency of estimating $\beta$, Prentice (1988) and Liang et al. (1992) discussed estimating $\theta = (\beta, \alpha)$ jointly by solving

$$U_2(\beta, \alpha) = \sum_{i=1}^{k}\left(\frac{\partial \mu_i^*}{\partial \beta}\right)[Cov(Z_i, \delta)]^{-1}(Z_i - \mu_i^* \theta) = 0$$

where

$$Z_i = \left(Y_{i1}, ... Y_{in_i}, Y_{i1}^2, ... Y_{in_i}^2, Y_{i1}Y_{i2}, Y_{i1}Y_{i3}, ... Y_{in_{i-1}}Y_{in_i}\right)^T$$

where $\mu_i^* = E(Z_i, \theta)$, which is completely specified by the modeling assumptions of the GLM. They called this expanded procedure, which uses both the $Y_{ij}$'s and $Y_{ij}Y_{il}$, the GEE2.

GEE2 seems to improve the efficiency of both $\beta$ and $\alpha$. On the other hand, the robustness property for $\beta$ of GEE is no longer true. Hence, correct inferences about $\beta$ require correct specification of the within-cluster dependence structure. The same authors suggest using a sensitivity analysis when making inference on $\beta$. That is, one may repeat the procedure with different models for the within-cluster dependence structure to examine the sensitivity of $\tilde{\beta}$ to choose the dependence structure.

## 3.7 Fitting Alternative Models for Clustered Data

**Example 3.2: (ch3_family)**

We will use a subset of the data from a survey conducted by Miall and Oldham (1955) to assess the correlations in systolic and diastolic blood pressures among family members living within 25 miles of Rhondaa Fach Valley in South Wales. The purpose of the following analysis is to illustrate the effect of the within-cluster correlation in the case of the normal linear regression model. The part of the data that we use for this illustration consists of the observations made on siblings and their parents. Each observation consists of systolic and diastolic blood pressures to the nearest 5 mmHg or 10 mmHg. In this analysis we will not distinguish among male and female siblings. The following variables will be used to run a few models in this section.

| | |
|---|---|
| familyid | Family ID |
| subjid | Sibling ID |
| sbp | Sibling systolic blood pressure |
| msbp | Mother systolic blood pressure |
| age | Sibling age |
| armgirth | Sibling arm girth |
| cenmsbp | Mother systolic blood pressure centered |
| cenage | Sibling age centered within the family |
| cengirth | Sibling arm girth centered within the family |

The records with missing values of sibling age, mother systolic blood pressure, sibling arm girth, and sibling systolic blood pressure are deleted. The data set consists of 488 observations on 154 families. The family size ranges from 1 to 10. We begin by first analyzing the data using the GEE approach. This is followed by a series of models using the multilevels modeling approach. All models are fitted using the SAS procedures GENMOD for the GEE and the MIXED for the multilevels approach. We also provide an equivalent R code.

The SAS code to read in the data and fit this model is

```
data fam;
input familyid subjid sbp age armgirth msbp;
datalines;
```

```
1  1  85  5 5.75 120
1  2 105 15 8.50 120
.   .    .   .    .   .
200  5 135 40 12.50 255
213  1 120 64 11.00 110
;

* Computing the overall mean for msbp;
proc means data=fam noprint; var msbp; output out=msbp mean=
  mmsbp; run;

* Computing cluster specific means for age and armgirth;
proc means data=fam noprint; class familyid; var age armgirth;
output out=fmeans mean=mage marmgirth; run;
data fmeans; set fmeans; if familyid=. then delete; drop _TYPE_
  _FREQ_;run;

* Centering msbp at overall mean and age and armgirth at cluster
  specific means;
 data family; merge fam fmeans; by familyid; if _n_=1 then set
msbp(drop=_TYPE_ _FREQ_);
cenage=age-mage; cengirth=armgirth-marmgirth; cenmsbp=msbp-
  mmsbp;
keep familyid subjid sbp age armgirth msbp cenage cengirth
  cenmsbp;

proc genmod data=family;
class familyid;
model sbp = msbp cenage cengirth / dist =n lnk=id;
repeated subject = familyid /type = cs corrw;
run;
```

Following is the partial output showing the analysis of the GEE parameter estimates:

| | | | | | | |
|---|---|---|---|---|---|---|
| Analysis of GEE Parameter Estimates | | | | | | |
| Empirical Standard Error Estimates | | | | | | |
| Parameter | Estimate | Standard Error | 95% Confidence Limits | | Z | Pr > \|Z\| |
| Intercept | 119.098 | 0.920 | 117.295 | 120.901 | 129.48 | <.0001 |
| cenmsbp | 0.202 | 0.035 | 0.134 | 0.270 | 5.80 | <.0001 |
| Cenage | 0.180 | 0.198 | −0.209 | 0.569 | 0.91 | 0.3636 |
| cengirth | 3.544 | 0.826 | 1.925 | 5.164 | 4.29 | <.0001 |

This model treats the within-cluster correlation as a nuisance. It is assumed that the within-subject correlation structure is exchangeable or compound symmetry. The estimated working correlation under the common (compound symmetry) structure is 0.328. The analysis indicates that the arm girth and the mother systolic blood pressure are significant predictors of the sibling systolic blood pressure levels.

Now we illustrate the application of the random effects model to analyze clustered data. We followed an informative strategy given by Singer (1998) for fitting multilevel data. Therefore, we shall present four nested random effects models and discuss the relative advantages of each model.

### 3.7.1 Proc Mixed for Clustered Data

Here we illustrate how two-level clustered data are analyzed using PROC MIXED in SAS. By two levels we mean a situation where subjects are nested within organizational units. The subjects in the data set are the siblings and the clusters are the families. We are interested in examining the behavior of a level 1 outcome (siblings outcome) as function of level 1 and level 2 (family) covariates. The siblings level outcome is the systolic blood pressures (sbp), and the covariates measured at the siblings level are age (cenage) and arm girth (cengirth). There are several family level outcomes but we shall restrict to the systolic blood pressure of the mother (cenmsbp).

The first baseline model is called an "unconditional means" model, which examines the variability in sbp across families.

### 3.7.2 Model 1: Unconditional Means Model

Under this model we express sbp ($y_{ij}$) as a one-way random effects model:

$$y_{ij} = \mu + b_i + e_{ij}$$

$$b_i \approx N\left(0, \sigma_b^2\right) \text{ and } e_{ij} \approx N\left(0, \sigma_e^2\right)$$

The model has one fixed effect ($\mu$) and two variance components: one representing the variation between family means ($\sigma_b^2$) and the other variation among siblings within families ($\sigma_e^2$).

The SAS code to fit this model is

```
proc mixed data=family noclprint noitprint;
class familyid;
model sbp= /bw;
random familyid;
run;
```

The selected output is shown next.

Covariance Parameter Estimates

| Cov Parm | Estimate |
|----------|----------|
| Familyid | 106.98 |
| Residual | 166.22 |

Solution for Fixed Effects

| Effect | Estimate | Standard Error | DF | T Value | Pr > \|t\| |
|--------|----------|----------------|-----|---------|-----------|
| Intercept | 118.67 | 1.0554 | 487 | 112.44 | <.0001 |

The mixed procedure produces a set of information; the "familyid" estimate is an estimate of the parameter ($\sigma_b^2$), whereas the "Residual" estimate is an estimate of the parameter ($\sigma_e^2$). The maximum likelihood estimate of the ICC is

$$\hat{\rho} = \hat{\sigma}_b^2/(\sigma_b^2 + \sigma_e^2) = 106.98/(106.98 + 166.22) = 0.39$$

This tells us that there is a great deal of clustering of systolic blood pressure levels of siblings within families.

There is another approach that generalizes more easily to data with multiple levels. This approach expresses the subject-level outcome $y_{ij}$ using a pair of linked models: one at the subject level (level 1) and another at the family level (level 2). At level 1 we express $y_{ij}$ as the sum of an intercept for the subject's family ($\beta_{i0}$) and random error($e_{ij}$):

$$y_{ij} = \beta_{i0} + e_{ij}$$

where $e_{ij} \approx N(0, \sigma_e^2)$.

At the higher level (family level) we express the family level intercept as the sum of an overall mean ($\beta$) and random deviation ($u_{i0}$) so that

$$\beta_{i0} = \beta + u_{i0}$$

where $u_{i0} \approx N(0, \tau_0)$.

Therefore

$$y_{ij} = \beta + u_{i0} + e_{ij}$$

The SAS code for this model is

```
proc mixed data=family noclprint noitprint covtest;
class familyid;
model sbp = / s ddfm=bw;
random intercept/sub = familyid;
run;
```

The parameter estimates under this model are the same as in the previous model. The purpose of the *covtest* option in the "proc mixed" statement is to test the hypothesis for the variance components.

### 3.7.3 Model 2: Including a Family Level Covariate

In this model we include the mother's systolic blood pressure score (msbp) as a predictor of the siblings score. Following Singer (1998), the msbp is centered at the overall mean so that it has mean 0 and allows a meaningful interpretation of the intercept. For this model we write

$$y_{ij} = \beta_{i0} + e_{ij}$$

$$\beta_{i0} = \gamma_{00} + \gamma_{01}x_i + u_{i0}$$

where $e_{ij} \approx N(0, \sigma_e^2)$ and $u_{i0} \approx N(0, \sigma_0^2)$.

Therefore, $y_{ij} = (\gamma_{00} + \gamma_{01}x_i) + (u_{i0} + e_{ij})$, where $x_i = $ msbp − mean(msbp) = cenmsbp.

The preceding model has two components: a fixed part enclosed in the first bracket, and a random part enclosed in the second bracket. The SAS code to fit this model is

```
proc mixed data=family noclprint noitprint;
class familyid;
model sbp = cenmsbp / s ddfm=bw notest;
random intercept / subject = familyid;
run;
```

Note that there is another option in the "model" statement, which is ddfm = bw. This allows SAS to use the "between/within" method for computing the denominator degrees of freedom for tests of fixed effects. See SAS documentation or Littell et al. (1996) for details. The SAS output is

Covariance Parameter
Estimates

| Cov Parm | Subject | Estimate |
|----------|---------|----------|
| Intercept | familyid | 67.4679 |
| Residual | | 163.89 |

Solution for Fixed Effects

| Effect | Estimate | Standard Error | DF | t Value | Pr > \|t\| |
|--------|----------|----------------|----|---------|-----------|
| Intercept | 119.11 | 0.9161 | 152 | 130.01 | <.0001 |
| Cenmsbp | 0.2005 | 0.02748 | 152 | 7.29 | <.0001 |

Note that there two fixed effects to be estimated: the intercept and the covariate (MSBP). The null hypothesis that there is no relationship between mother systolic blood pressure levels and their siblings is not supported by the data. Note also that the variance components estimates are 67.47 and 163.89 for $\tau_0$ and $\sigma_e^2$, respectively. These estimates under the present model have different interpretations. In model 1, there were no covariates, so these were unconditional components. After adding the mother blood pressure as a covariate, these are now conditional components. Note that the conditional within-family component is slightly reduced (from 166.22 to 163.89). The variance component representing variation between families $\tau_0$ or $\sigma_b^2$ has diminished markedly (from 106.98 to 67.47). This tells us that the cluster or family level covariate (mother systolic blood pressure) explains a large percentage of the family-to-family variation. One way of measuring how much variation in family mean blood pressures is explained by mother's blood pressure levels is to compute how much the variance component for this term $\tau_0$ has diminished between the two models. Following Bryk and Raudenbush

(1992) we compute this as $(106.98 - 67.47)/106.98 = 36.9\%$. This is interpreted by saying that about 36% of the explainable variation in the sibling's mean systolic blood pressure levels is explained by the mother's systolic blood pressure levels.

### 3.7.4 Model 3: Including the Sib-Level Covariate

The simplest model may be written as

$$y_{ij} = \beta_{i0} + \beta_{11}Z_{ij} + e_{ij}$$

Here $Z_{ij}$ is the age of the $j$th subject within the $i$th family centered at its mean value. The other terms are defined as before. Let

$$\beta_{i0} = \beta_{00} + u_{i0}$$

$$\beta_{i1} = \beta_{11} + u_{i1}$$

Hence,

$$y_{ij} = \beta_{00} + u_{i0} + (\beta_{11} + u_{i1})Z_{ij} + e_{ij}$$

$$= \left(\beta_{00} + \beta_{11}Z_{ij}\right) + \left(u_{i0} + u_{i1}Z_{ij} + e_{ij}\right)$$

where $e_{ij} \approx N(0, \sigma_e^2)$, $u_i = (u_{i0} \; u_{i1}) \approx BIVN\left(0, \sum\right)$, and $e_{ij}$ is independent of the bivariate normal random vector $u_i$. The $\Sigma$ is a $2 \times 2$ symmetric matrix whose elements are $\delta_{00} = V(u_{i0})$, $\delta_{11} = V(u_{i1})$, and $\delta_{01} = Cov(u_{i0}, u_{i1})$. The SAS code to fit this model is

```
proc mixed data=family noclprint noitprint;
class familyid;
model sbp= cenmsbp/s ddfm=bw notest;
random intercept cenage/ subject=familyid type=un;
run;
```

Notice that the random statement has two random effects: one for intercept and one for the Z-slope. There is also an additional option in the random statement "type=un" indicating that an unstructured specification for the covariance of $u_i$ is assumed. The partial SAS output is shown next.

Covariance Parameter
Estimates

| Cov Parm | Subject | Estimate |
|----------|---------|----------|
| UN(1,1) | familyid | 78.9632 |
| UN(2,1) | familyid | 4.3682 |
| UN(2,2) | familyid | 1.0199 |
| Residual | | 133.16 |

Solution for Fixed Effects

| Effect | Estimate | Standard Error | DF | t Value | Pr > \|t\| |
|--------|----------|----------------|-----|---------|-----------|
| Intercept | 118.43 | 0.8977 | 152 | 131.92 | <.0001 |
| cenmsbp | 0.1973 | 0.02687 | 152 | 7.34 | <.0001 |

We shall start by first interpreting the fixed effects. The estimate of the intercept 118.43 indicates the estimated average sibling systolic blood pressure levels after controlling for the mother systolic blood pressure. The estimate of the cenmsbp indicates that the average slope representing the relationship between siblings' blood pressure and the mother systolic blood pressure is 0.20. The standard errors of these estimates are very small resulting in small $p$-values. We conclude that, on the average, there is a statistically significant relationship between siblings' systolic blood pressures and the mother systolic blood pressure.

The covariance parameter estimates tell us how much these intercepts and slopes vary across families. The $\hat{\delta}_{00} = 78.96$ represents the variability in the intercepts, $\hat{\delta}_{11} = 4.37$ is the variability in the slopes, and $\hat{\delta}_{01} = 1.02$ is the covariance between intercepts and slopes. We can say that the intercepts vary considerably; in other words, families do differ in average systolic blood pressure levels after controlling for the effects of the mother blood pressure levels. We also note that the slopes do not considerably vary between the families and there is no evidence that the effects of mother blood pressure on sibling systolic blood pressure differ between the families.

Finally, we compare the residual error variance of the unconditional model to that of the present model. Recall that the estimated variance of the unconditional model was 166.22. Here we have the conditional estimate of 133.16. Inclusion of the siblings age has therefore explained $(166.22 - 133.16)/166.22 = 20.0\%$ of the explainable variation within families.

### 3.7.5 Model 4: Including One Family Level Covariate and Two Subject Level Covariates

Following Singer (1998), it is always helpful to write the outcome variable as a function of the covariates measured at the lowest (subject) level. Thereafter, we write the slopes and the intercepts as functions of the higher level (in our example a family):

$$y_{ij} = B_{i0} + B_{i1}Z_{ij} + B_{i2}a_{ij} + e_{ij}$$

$$B_{i0} = \gamma_{00} + \gamma_{01}x_i + u_{i0}$$

$$B_{i1} = \gamma_{10} + \gamma_{11}x_i + u_{i1}$$

$$B_{i2} = \gamma_{20} + \gamma_{21}x_i + u_{i2}$$

where $Z_{ij}$ is the centered arm girth of the $j$th subject within the $i$th family, $a_{ij}$ is the centered age of the $j$th subject within the $i$th family, and $x_i$ is the centered mother systolic blood pressure in the $i$th family.

Hence

$$y_{ij} = \gamma_{00} + \gamma_{01}x_i + u_{i0} + Z_{ij}(\gamma_{10} + \gamma_{11}x_i + u_{i1}) + a_{ij}(\gamma_{20} + \gamma_{21}x_i + u_{i2}) + e_{ij}$$

Simplifying we get

$$y_{ij} = \left[\gamma_{00} + \gamma_{01}x_i + \gamma_{10}Z_{ij} + \gamma_{20}a_{ij} + \gamma_{11}x_iZ_{ij} + \gamma_{21}x_ia_{ij}\right] + \left[u_{10} + Z_{ij}u_{i1} + a_{ij}u_{iz} + e_{ij}\right]$$

Terms in the first bracket should appear in the model statement, while in the second bracket they should appear in the random statement.

The SAS code to fit the model is

```
proc mixed data=family covtest noclprint noitprint;
class familyid;
model sbp= cenmsbp cenage cengirth cenmsbp*cenage
  cenmsbp*cengirth /s ddfm=bw notest;
random cenage cengirth/subject=familyid type=un;
run;
```

The variable cengirth has a zero variance component. We fit the model after removing cengirth from the random statement. The results are given in Table 3.4.

Interpretation of the preceding output has been left as an exercise for the reader.

**TABLE 3.4**

Output of the Analysis of the Linear Mixed Model

| Covariance Parameter Estimates | | | | | |
|---|---|---|---|---|---|
| Cov Parm | Subject | Estimate | Standard Error | Z Value | Pr Z |
| UN(1,1) | familyid | 81.8458 | 15.4989 | 5.28 | <.0001 |
| UN(2,1) | familyid | 4.7531 | 1.7062 | 2.79 | 0.0053 |
| UN(2,2) | familyid | 0.6185 | 0.3115 | 1.99 | 0.0235 |
| Residual | | 124.29 | 11.2232 | 11.07 | <.0001 |

| Solution for Fixed Effects | | | | | |
|---|---|---|---|---|---|
| Effect | Estimate | Standard Error | DF | t Value | Pr > \|t\| |
| Intercept | 119.07 | 0.9208 | 152 | 129.32 | <.0001 |
| Cenmsbp | 0.2043 | 0.02777 | 152 | 7.36 | <.0001 |
| Cenage | 0.02669 | 0.1624 | 330 | 0.16 | 0.8695 |
| Cengirth | 4.2129 | 0.7010 | 330 | 6.01 | <.0001 |
| Cenmsbp*cenage | 0.008845 | 0.004130 | 330 | 2.14 | 0.0330 |
| Cenmsbp*cengirth | -0.02641 | 0.01736 | 330 | -1.52 | 0.1292 |

The R code reads the data; computes the centered variables cenage, cengirth, and cenmsbp; and fits the alternative models discussed in this example. Note that the packages "nlme" and "gee" should be installed and loaded for functions "lme" and "gee" to run, respectively.

```
fam <-read.csv(file.choose())
cenage = unlist(tapply(fam[,4], fam[,1], scale, scale=FALSE))
cengirth = unlist(tapply(fam[,5], fam[,1], scale, scale=FALSE))
cenmsbp = scale(fam[,6], center = TRUE, scale = FALSE)
family = data.frame(fam, cenage, cengirth, cenmsbp)

# Generalized Estimating Equations (GEE)
fam.gee <- gee(sbp ~ cenmsbp+cenage+cengirth, familyid,
  data=family, family = gaussian, corstr = "exchangeable")
summary(fam.gee)

# Unconditional mean model - Mixed Model 1
fam.lme1 <- lme(fixed = sbp ~ 1, random= ~ 1 | familyid, data = family)
summary(fam.lme1)

# Mixed model including one cluster-level covariate,
  cenmsbp - Mixed Model 2
fam.lme2 <- lme(fixed = sbp ~ cenmsbp, random= ~ 1 | familyid,
  data = family)
summary(fam.lme2)

# Mixed model including sib-level covariate, cengirth - Mixed Model 3
fam.lme3 <- lme(fixed = sbp ~ cenmsbp, random= ~ cenage | familyid,
  data = family)
summary(fam.lme3)

# Mixed model including one sibling level covariate, cengurth -
  *Mixed Model 4
fam.lme4 <- lme(fixed = sbp ~ cenmsbp + cenage + cengirth +
  cenmsbp*cenage + cenmsbp*cengirth, random= ~ cenage -1| familyid,
  data = family)
summary(fam.lme4)
```

---

# Appendix

## Linear Combinations of Random Variables

Let $x = (x_i, x_2, ..., x_k)$ be a set of random variables such that $E(x_i) = \mu_i$, $V(x_i) = \sigma_i^2$, and $Cov(x_i, x_j) = c_{ij}$. We define a linear combination of the random variable $x$ as

$$y = \sum_{i=1}^{k} w_i x_i$$

where $w_1, w_2, \ldots, w_k$ are constants.

$$E(y) = \sum_{i=1}^{k} w_i \mu_i \qquad (A.1)$$

and

$$V(y) = \sum_{i=1}^{k} w_i^2 \sigma_i^2 + \sum_{\substack{i=1 \\ i \neq j}}^{k} \sum_{j=1}^{k} w_i w_j c_{ij} \qquad (A.2)$$

## Two Linear Combinations

Consider two linear combinations $y_s = \sum_{i=1}^{k} a_{si} x_i$, and $y_t = \sum_{i=1}^{k} a_{ti} x_i$, then the covariance between $y_s$ and $y_t$ is

$$Cov(y_s, y_t) = \sum_{i=1}^{k} \sum_{j=1}^{k} a_{si} a_{tj} Cov\left(x_i, x_j\right) \qquad (A.3)$$

## The Delta Method

Let $g(.)$ be a differentiable function of $x$. Then to the first order of approximation

$$E[g(x)] = g(E(x))$$

and

$$V[g(x)] = \left(\frac{\partial g}{\partial x}\right)^2_{x=\mu} V(x) \qquad (A.4)$$

In general, if $g$ is a differentiable function of $x_1, x_2, \ldots, x_k$, then to the first approximation:

$$V[g(x_1, x_2, \ldots x_k)] \approx \sum_{i=1}^{k} \left(\frac{\partial \dot{g}}{\partial x_i}\right)^2 V(x_i) + \sum_{i=1 \neq j=1}^{k} \left(\frac{\partial \dot{g}}{\partial x_i}\right)\left(\frac{\partial \dot{g}}{\partial x_j}\right) cov\left(x_i, x_j\right) \qquad (A.5)$$

where the $(.)$ on top of $\dfrac{\partial g}{\partial x_i}$ means that they are evaluated at $\mu$.

If we have two differentiable functions $g_1(x_1,...x_k)$ and $g_2(x_1,...,x_k)$, then to the first order of approximation:

$$Cov(g_1, g_2) = \sum_{i=1}^{k} \sum_{j=1}^{k} \left(\frac{\partial \dot{g}_1}{\partial x_i}\right) \left(\frac{\partial \dot{g}_2}{\partial x_j}\right) Cov\left(x_i, x_j\right) \qquad (A.6)$$

**Exercises**

3.1 Suppose that we have $k$ clusters, each of size $n$, and that the model generating the data is the one-way random effects:

$$y_{ij} = \mu + b_i + e_{ij} \qquad b_i \approx N\left(0, \sigma_b^2\right)$$

$$e_{ij} \approx N\left(0, \sigma_e^2\right)$$

Under the same model assumptions, and given that

$$var\left(\hat{\sigma}_e^2\right) = 2\sigma_e^4/k(n-1)$$

$$var\left(\hat{\sigma}_b^2\right) = \frac{2}{n^2}\left[\frac{\left(n\sigma_b^2 + \sigma_e^2\right)^2}{k-1} + \frac{\sigma_e^4}{k(n-1)}\right]$$

and

$$cov\left(\hat{\sigma}_e^2, \hat{\sigma}_b^2\right) = -2\sigma_e^4/kn(n-1)$$

use the delta method to show that

$$var(\hat{\rho}) = \frac{2(1-\rho)^2(1+(n-1)\rho)^2}{kn(n-1)}$$

3.2 Under the same assumptions of Exercise 3.1, it is known that

$$MSB \approx \left(n\sigma_b^2 + \sigma_e^2\right)\chi_{k-1}^2$$

$$MSW \approx \sigma_e^2 \chi_{k(n-1)}^2$$

On defining upper and lower points of the $F$-distribution as $F_u$ and $F_l$ by

$$Pr\left[F_l \le F_{k(n-1)}^{k-1} \le F_u\right] = 1 - \alpha,$$

construct an exact $(1-\alpha)100\%$ confidence interval on $\rho = \sigma_b^2/(\sigma_b^2 + \sigma_e^2)$.

3.3 Suppose that we have a two-arm cluster randomized clinical trials as described in this chapter. Let $\theta = \mu_E/\mu_C$. Use the delta method to find a first-order approximation to the maximum likelihood of $\theta$. Hence, find an approximate $(1 - \alpha)100\%$ confidence interval on assuming that the number of clusters is the same in each arm.

3.4 Suppose that we have $H$ treatment groups and $k$ clusters are randomized in the $h$th group ($h = 1, 2, ..., H$) with $n$ denoting the $j$th cluster size within the $h$th group ($j = 1, 2, ..., k_h$). It is required to test the hypothesis $H_0 : \mu_1 = \mu_2 = ... = \mu_H$. Cochran (1937) suggested that for individual randomization, the statistic

$$G_H^2 = \sum_{h=1}^{H} w_i \bar{y}_i (\bar{y}_{h'} - \bar{\bar{y}})^2$$

has approximately chi-square distribution with $(H - 1)$ degrees of freedom. Here $\bar{y}_{h'}$ is the $h$th group mean

$$\bar{\bar{y}} = \sum_{i=1}^{H} w_i \bar{y}_i \bigg/ \sum_{i=1}^{H} w_i$$

and

$$w_i = [\text{var}(\bar{y}_i)]^{-1}$$

For the case of cluster randomization:

a. What is $w_i$ and var $(\bar{y})$?

b. Show that $G_H$ is equivalent to $Z$ given in Equation 3.15. State your assumptions.

3.5 Under the one-way random effects, define the within-cluster coefficient of variation as $\theta = \sigma_e/\mu$, and its maximum likelihood estimator as $\hat{\theta} = (MSW)^{\frac{1}{2}}/\bar{y}$. Use the delta method (Ox and Hinkley 1974) to derive a first-order approximation to the variance of $\hat{\theta}$. Assume the data are normally distributed.

3.6 The metabolic syndrome is the co-aggregation of hypertension, impaired glucose tolerance, dyslipidemia, and abdominal obesity, and is associated with an increased risk of total and cardiovascular mortality in adults (see Shoukri et al. 2015). Genetics as well as environmental influences have been implicated in obesity and several cardiovascular risk factors. Family is one of the most important factors affecting metabolic risk factors in children, in that family displays an interaction between genetic and shared environmental factors. The data set is given in the Excel sheet. "METBOLIC_SYNDROME_CHAPTER3" has the measured values of the five components of the syndrome for 4000 spousal pairs. The intraclass correlation (ICC) is taken as the measure of familial clustering, where the cluster size $n = 2$. Show that the intraclass correlation for each component is as given in Table 3.5. Let $w_i = v_i^{-1}$ denote the inverse of the variance of the MLE $\hat{\rho}_i$. Following Cochran, the best linear unbiased estimator of the pooled estimator of $\rho_i$ is

$$\bar{\rho} = \sum_{i=1}^{5} w_i \hat{\rho}_i \Big/ \sum_{i=1}^{5} w_i$$

It can be easily shown that the variance of $\bar{\rho}$ is given by

$$\text{var}(\bar{\rho}) \equiv v = \left( \sum_{i=1}^{5} w_i \right)^{-1}$$

Use the preceding results to construct a 95% confidence interval on the overall measure of clustering $\rho$.

In Table 3.6 we show the cut-off points recommended by the Royal Society of Medicine.

**TABLE 3.5**

Mean ± SD and Intraclass Correlation Coefficient

| MS Component | Husband ($\bar{x}_f \pm s_f$) | Wife ($\bar{x}_m \pm s_m$) | | ICCC ± SE |
|---|---|---|---|---|
| A. Central Obesity | 84.3 ± 8.2 | 78.8 ± 8.9 | 0.06 | 0.05 ±.016 |
| B. TG | 1.67 ± 1.5 | 1.3 ± .73 | 0.10 | 0.09 ±.016 |
| C. HDL | 1.19 ± .31 | 1.28 ± .29 | 0.11 | 0.10 ±.016 |
| D. SBP | 127 ± 18.6 | 120.8 ± 19.2 | 0.11 | 0.10 ±.016 |
| DBP | 81.2 ± 11.6 | 75.7 ± 11.2 | 0.13 | 0.12 ±.016 |
| E. Fasting Glucose | 5.6 ± 1.5 | 5.5 ± 1.6 | 0.08 | 0.07 ±.016 |

**TABLE 3.6**

Definition of the Components of Metabolic Syndrome

| Component | IDF | | WHO | | EGIR | | NCEP | |
|---|---|---|---|---|---|---|---|---|
| | M | F | M | F | M | F | M | F |
| Central Obesity | ≥102 | ≥88 | ≥102 | ≥88 | ≥94 | ≥80 | ≥102 | ≥88 |
| Raised TG | ≥1.7 | ≥1.7 | ≥1.7 | ≥1.7 | ≥2.0 | ≥2.0 | ≥1.7 | ≥1.7 |
| Low HDL | <1.03 | <1.29 | ≤.9 | ≤1 | ≤1 | ≤1 | ≤1.03 | ≤1.29 |
| Hypertension | ≥130/85 | ≥130/85 | ≥140/90 | ≥140/90 | ≥140/90 | ≥140/90 | ≥130/85 | ≥130/85 |
| Fasting Glucose | ≥5.6 | ≥5.6 | ≥6.1 | ≥6.1 | ≥6.1 | ≥6.1 | ≥6.1 | ≥6.1 |

*Source:* *Journal of the Royal Society of Medicine*, vol. 99, September 2006.

*Note:* EGIR, European Group for the Study of Insulin Resistance; IDF, International Diabetes Federation; NCEP, National Cholesterol Education Program; WHO, World Health Organization.

3.7 Following standard formulas, for a trial using individual randomization (Armitage et al., 2002) for fixed power $(1 - \beta)$ and fixed sample size $(n)$ per arm, show that the detectable difference between two sample means drawn from two populations with the same variance $d_I$ has variance $\text{var}(d_I) = 2\sigma^2/n$.

3.8 Moreover show that

$$d_I = \sqrt{\frac{2\sigma^2}{n}} \left(z_{\alpha/2} + z_\beta\right)$$

where $z_{\alpha/2}$ denotes the upper $100\alpha/2$ standard normal centile.

3.9 Show that the required sample size per arm for a trial at prespecified power $1 - \beta$ to detect a prespecified difference $d$, is given by

$$n_1 = 2\sigma^2 \left[\frac{\left(z_{\alpha/2} + z_\beta\right)^2}{d^2}\right]$$

3.10 Suppose that each of $k$ clusters is of size $m$, so that we have a total of $n_C = mk$ individuals per arm. Then, by standard results, the variance of the difference to be detected $d_C$ is inflated by the variance inflation factor (VIF). Show that the VIF is given by

$$VIF = [1 + (m - 1)\rho]$$

where $\rho$ is the intracluster correlation (ICC) coefficient, which represents how strongly individuals within clusters are related to each other.

3.11 When the cluster sizes are unequal, show that this variance inflation factor can be approximated by

$$VIF = \left[1 + \left((cv^2 + 1)\bar{m} - 1\right)\rho\right]$$

where $cv$ represents the coefficient of variation of the cluster sizes, and $\bar{m}$ is the average cluster size (Kerry and Bland, 2006).

3.12 Show that the number of individuals per arm is thus given by:

$$n_C = 2\sigma^2 \left[\frac{\left(z_{\alpha/2} + z_\beta\right)^2 [1 + (m - 1)\rho]}{d^2}\right]$$

$$= n_1 [1 + (m - 1)\rho]$$

$$= n_1 \times VIF$$

where $n_1$ is the required sample size per arm using a trial with individual randomization to detect a difference $d$, and VIF can be modified to allow for variation in cluster sizes. This is the standard result that the required sample size for a CRCT is that required under individual randomization, inflated by the VIF.

3.13 Show that, assuming equal cluster sizes, the number of clusters required per arm is then

$$k = \frac{n_1(1 + (m - 1)p)}{m}$$

3.14 The standard sample size formulas for CRCTs assume knowledge of cluster size ($m$) and consequently determine the number of clusters ($k$) required. For a prespecified available number of clusters ($k$), investigators need instead to determine the required cluster size ($m$). Show that for a trial with a fixed number of equal-sized clusters ($k$) the required sample size per arm for a trial with prespecified power $1 - \beta$, to detect a difference of $d$, is $n_C$ such that

$$n_C = n_1[1 + (m - 1)p]$$

$$= n_1\left[1 + \left(\frac{n_C}{k} - 1\right)p\right]$$

$$= \frac{n_1 k[1 - p]}{|k - n_1 p|}$$

Here $n_1$ is the sample size required under individual randomization. This increase in sample size, over that required under individual randomization, is no longer a simple inflation, as the inflation required is now dependent on the sample size required under individual randomization.

3.15 Finally show that the corresponding number of individuals in each of the $k$ equally sized cluster is

$$m = \frac{n_1(1 - p)}{(k - n_1 p)}$$

For questions 3.7 through 3.15, you may read the paper by Hemming et al. (2011).

# 4

## Statistical Analysis of Cross-Classified Data

### 4.1 Introduction

As indicated in Chapter 1, there are two broad categories of investigative studies that produce statistical data: The first is *designed controlled experiments* and the second is *observational studies*. The controlled experiments are conducted to achieve the following two major objectives: (1) to clearly detect with reasonable power the effects of the treatment (or combination of treatments) structure, and (2) to ensure maximum control on the experimental error. There are several advantages of studies run under controlled experiments, the most important being that it permits one to disentangle a complex causal problem by proceeding in a stepwise fashion. As suggested by Schlesselman (1982), within an experimental procedure the researcher can decompose the main problem and explore each component by a series of separate experiments with comparatively simple causal assumptions.

Observational studies are those that are concerned with investigating relationships among characteristics of certain populations where the groups to be compared have not been formed by random allocation. A frequently used example to illustrate this point is that of establishing a link between smoking cigarettes and contracting lung cancer. This cannot be explored using random allocation of patients to smoking and non-smoking conditions for obvious ethical and practical reasons. Therefore, the researcher must rely on observational studies to provide evidence of epidemiological and statistical associations between a possible risk factor and a disease.

One of the most common and important questions in observational investigation involves assessing the occurrence of the disease under study in the presence of a potential risk factor. The most frequently employed means of presenting statistical evidence is the $2 \times 2$ table. The following section is devoted to assessing the significance of association between disease and a risk factor in a $2 \times 2$ table.

## 4.2 Measures of Association in 2 × 2 Tables

In epidemiological research, it is important that the collected data are translated into interpretable results that can be easily communicated to clinicians. The need for "translatable" evidence from research studies is of prime importance in the evaluation of clinical interventions, because they hold the potential to immediately influence the course of patient treatment. When evaluating these studies, the examination of effect size (ES) can be a useful measure of the comparative efficacy of the treatment under investigation. In randomized clinical trials an effect size estimate quantifies the direction and magnitude of an effect of an intervention.

When exposure and disease risk are measured on a binary scale, several measures of effect size are in current use. The odds ratio (OR), the relative risk (RR), and the population attributable risk (AR) are the most commonly used measures of effect size in clinical as well as analytic epidemiology.

The use of OR, RR, and AR as measures of effect size for binary data is ubiquitous in medical and epidemiological research. To illustrate the use of these measures, let us discuss the following hypothetical clinical trial in transplant surgery in which 132 patients with liver cirrhosis were randomly assigned to one of two liver transplant procedures. The number of patients in the first procedure is 70 and in the second procedure is 62. After three weeks, 14 out of the 70 randomized to the first procedure had organ failure (transplant rejection), whereas 8 of the transplanted organs failed in the second group. The results are shown in Table 4.1.

### 4.2.1 Absolute Risk

The absolute risk of an event is the likelihood of occurrence of the event in the population at risk. This is calculated as the number of persons who had the event divided by the total number of individuals exposed to the risk of the event. In the preceding example, the risk of organ failure in the first procedure is 14/70 or 20%. Similarly, the absolute risk of organ failure attributed to the second procedure is 8/62 or 13%. This means that 20% who undergo liver transplant under the first procedure and 13% of those who undergo transplant surgery under the second procedure are expected to suffer from organ failure.

**TABLE 4.1**

Association between Event (Yes/No) and Exposure (Yes/No)

| Transplant Procedure | Organ Failed | Organ Did Not Fail | Total |
|---|---|---|---|
| Procedure 1 | 14 | 56 | 70 |
| Procedure 2 | 8 | 54 | 62 |

## 4.2.2  Risk Difference

Risk difference is the risk of an event that is attributed to the risk factor of interest. By definition, it is the risk of exposed and the risk of unexposed. From the example, the risk difference is 20% – 13% = 7%. This means that 20% of transplant recipients by the first procedure may suffer organ failure, however, only 13% in the organ failure group is attributed to the second procedure. In the remaining 7%, other causes, such as diabetes, obesity, or heritable traits, may be responsible.

## 4.2.3  Attributable Risk

Attributable risk is the percentage of cases in the population at risk that is specifically attributable to the risk factor of interest. It is calculated as

$$(P(D|E) - P(D|E^c))/P(D|E) = (20\% - 13\%)/20\% = 35\%$$

The interpretation is that whatever the actual number of patients who had transplanted and had liver rejection, only 35% of this number would have transplant rejection attributed to the first procedure.

## 4.2.4  Relative Risk

Relative risk is the likelihood of occurrence of an event among those who are exposed to a risk factor relative to the likelihood of occurrence of the same even among those who are not exposed to the same risk factor.

In our example, the risk of organ failure among the patients in procedure 1 is 20%, while the same risk among patients in procedure 2 is 13%. Therefore the RR = 20% / 13% = 1.5. This means relative to procedure 2, procedure 1 is associated with a 1.5-fold increase in organ failure.

Relative risk is a commonly used measure of effect size. An RR of 1.0 is interpreted as an absence of association between the disease and exposure to the risk factor of interest. On the other hand, an RR less than 1.0 means that the risk is lower in the exposed group relative to the unexposed group. An RR that is larger than 1 means an increased risk of disease in the presence of the risk factor.

## 4.2.5  Odds Ratio

The odds of an event are the ratio of the number of times the event occurs to the number of times the ratio does not occur. In our example, the odds of transplant rejection among the procedure 1 patients are $O_1 = 14/56$. Similarly, the odds of transplant rejection among the procedure 2 patients are $O_2 = 8/54$. The odds ratio is the ratio of odds and is given by

$$O_1/O_2 = (14)(54)/(56)(8) = 1.7$$

This means that the odds of having organ failure due to procedure 1 relative to procedure 2 are 1.7:1.

### 4.2.6  Relationship between Odds Ratio and Relative Risk

Unlike risk, there is a problem with the concept of odds as it may be difficult to explain. Odds are easily understood when they are well above one but are not easily visualized when they are less than one. There is a mathematical relationship between OR and RR. Denoting $P(D \mid E) = P_1$ and $P(D \mid E^c) = P_2$, then $RR = P_1/P_2$ and $OR = P_1(1 - P_2)/P_2 (1 - P_1)$.

Simple algebra shows that $OR/RR = (OR - 1) \times P_2 + 1$. OR and RR are numerically different, although both are measures of effect size, as they both quantify the association between disease and exposure to risk factor. When the disease is rare, OR provides a good approximation to the RR.

### 4.2.7  Incidence Rate and Incidence Rate Ratio As a Measure of Effect Size

In all the effect measures that we discussed thus far, the length of follow up from the base line until the event occurs has not been taken into account. The incidence rate is in fact an effect size that overcomes this deficiency. The incidence rate is the number of new cases of disease during a period of time divided by the person-time-at-risk throughout the observation period. The denominator for incidence rate (person-time) is a more exact expression of the population at risk during the period of time when the change from nondisease to disease is being measured. The denominator for incidence rate changes as persons originally at risk develop disease during the observation period and are removed from the denominator.

### 4.2.8  What Is Person-Time?

Person-time is an estimate of the actual time-at-risk contributed to a study. In certain studies, people are followed for different lengths of time, as some will remain disease-free longer than others.

A person remains in the study only so long as that person remains disease-free and, therefore, still at risk of developing the disease of interest. By knowing the number of new cases of disease and the person-time-at-risk contributed to the study, an investigator can calculate the incidence rate of the disease or how quickly people are acquiring the disease.

#### Example 4.1: Calculating Person-Time

Now suppose an investigator is conducting a study of the incidence of redo after a valve stent has been completed. She follows seven patients who have already received a stint, from baseline (first MI) for up to 70 weeks. The results are graphically displayed in Figure 4.1.

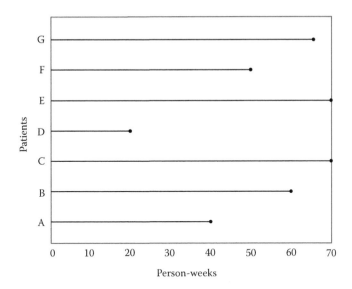

**FIGURE 4.1**
Length of contribution by every participant in the follow-up.

The graph shows the number of weeks each patient remained in the study as a noncase (no redo) from the start of the study. From the graph we can calculate person-time, or person-week. Person-time is the sum of total time contributed by all patients. Therefore, from the graph we have:

Time contributed by subject A = 40 weeks
Time contributed by subject B = 60 weeks
Time contributed by subject C = 70 weeks
Time contributed by subject D = 20 weeks
Time contributed by subject E = 70 weeks
Time contributed by subject F = 50 weeks
Time contributed by subject G = 65 weeks

Therefore, the total person-weeks in this study is T = 40 + 60 + 70 + 20 + 70 + 50 + 65 = 375 person-weeks.

The total number of patients who became cases (A, B, D, F, G) = 5. Therefore, the incidence rate of a stint redo IR1 = R1/T1 = 5/375 = 0.013, or 13 cases per 1000 person-weeks. The sample variance of the logarithm of IR1 is $v1 = var(\log(IR1)) = \dfrac{1}{R1} = 0.2$. Now, suppose that in a similar study conducted in another research facility with the same objectives gave R2 = 35 and T2 = 1060. The incidence rate is IR2 = 35/1060 = 0.033 or 33 cases per 1000 person-weeks, and $v2 = var(\log(IR2)) = \dfrac{1}{R2} = 0.028$. From Rothman and Greenland (2008), we may take the incidence rate ratio (IRR) as a measure of effect size:

$$IRR = IR1/IR2$$

Because of the nonlinearity of this measure, it is better to construct a confidence interval on the natural logarithm of IRR:

$$var\left(\log(IRR)\right) = var\left(\log(IR1)\right) + var\left(\log(IR2)\right) = \frac{1}{R1} + \frac{1}{R2}$$

Hence, IRR = $0.013/0.033$ = 0.393, log(IRR) = $-0.932$, $var(\log(IRR))$ = $\frac{1}{5} + \frac{1}{35} = 0.228$.

The 95% confidence interval on the incidence rate ratio is thus given by

$$\exp(\log(IRR) \pm 1.96\sqrt{var\left(\log(IRR)\right)}$$

$$(0.154, 1.004)$$

## 4.3 Statistical Analysis from the 2 × 2 Classification Data

In this section, an individual classified as *diseased* will be denoted by $D$ and by $\bar{D}$ if *not diseased*. Exposure to the risk factor is denoted by $E$ and $\bar{E}$ for *exposed* and *unexposed*, respectively. Table 4.2 illustrates how a sample of size $n$ is cross-classified according to this notation.

There are, in practice, several methods of sampling by which the table of frequencies (Table 4.2) can be obtained. The three most common methods are the cross-sectional (historical cohort), the cohort study (prospective design), and the case control study (retrospective design). These are described here with regard to the 2 × 2 table analysis.

### 4.3.1 Cross-Sectional Sampling

Cross-sectional sampling calls for the selection of a total of $n$ subjects from a large population after which the presence or absence of disease and the

**TABLE 4.2**

Cross-Classified Data in a 2 × 2 Table

| | Disease | | |
|---|---|---|---|
| Exposure | $D$ | $\bar{D}$ | Total |
| $E$ | $n_{11}$ | $n_{12}$ | $n_{1.}$ |
| $\bar{E}$ | $n_{21}$ | $n_{22}$ | $n_{2.}$ |
| Total | $n_{.1}$ | $n_{.2}$ | $n$ |

presence or absence of exposure is determined for each subject. Only the sample size can be specified prior to the collection of data. With this method of sampling, the issue of association between disease and exposure is the main concern. In the population from which the sample is drawn, the unknown proportions are denoted as in Table 4.3.

The exposure and disease would be independent of each other if and only if $p_{ij} = p_{i.}p_{.j}$ ($i,j = 1,2$). Assessing independence based on the sample outcome is determined by how close the value of $n_{ij}$ is to $e_{ij} = n\hat{p}_{i.}\hat{p}_{.j}$ (the expected frequency under independence), where $\hat{p}_{i.} = \dfrac{n_{i.}}{n}$ and $\hat{p}_{.j} = \dfrac{n_{.j}}{n}$ are the maximum likelihood estimators of $p_{i.}$ and $p_{.j}$, respectively.

There are two commonly used measures of distance between $n_{ij}$ and $e_{ij}$, the chi-square and the Wilks's likelihood ratio test statistics. Historically, emphasis has been placed on large sample chi-square methods for the analysis of contingency tables with an arbitrary number of rows and columns. In more recent years, with the advance of computational power, there has been an increased interest in the exact methods. In choosing an appropriate statistical method for categorical data analysis, one should consider the measurement scale of the response variable as well as the independent variable. As we have already indicated, statistical analyses of contingency tables involve the analysis of two-way tables for the assessment of significance of the association between two variables. We note that statistics exist for situations in which both the exposure variable and the response variable are nominal, when only the response is ordinal, or when both variables are ordinal. We first introduce the Pearson $\chi^2$ statistic:

$$\chi^2 = \sum_i \sum_j \frac{\left(n_{ij} - e_{ij}\right)^2}{e_{ij}}$$

which with Yates's continuity correction becomes

$$\chi^2 = \sum_i \sum_j \frac{\left(\left|n_{ij} - e_{ij}\right| - 0.5\right)^2}{e_{ij}}$$

**TABLE 4.3**

Model 2 × 2 for Cross-Sectional Sampling

|  | Disease | | |
|---|---|---|---|
| Exposure | D | $\bar{D}$ | Total |
| E | $p_{11}$ | $p_{12}$ | $p_{1.}$ |
| $\bar{E}$ | $p_{21}$ | $p_{22}$ | $p_{2.}$ |
| Total | $p_{.1}$ | $p_{.2}$ | 1 |

The hypothesis of independence is rejected for values of $\chi^2$ that exceed $\chi^2_{\alpha,1}$ (the cut-off value of chi-square at $\alpha$-level of significance and 1 degree of freedom). The second is Wilks's statistic:

$$G^2 = 2 \sum_i \sum_j n_{ij}[\ln n_{ij} - \ln e_{ij}]$$

This statistic is called the likelihood-ratio chi-squared statistic. As for the Pearson chi-square statistic, larger values of $G^2$ lead to the rejection of the null hypothesis of independence.

When independence holds, the Pearson chi-square statistic and the likelihood ratio statistic $G^2$ have asymptotic (i.e., in large samples) chi-squared distribution with 1 degree of freedom.

It is not simple to determine the sample size needed for the chi-square distribution to approximate well the exact distributions of $\chi^2$ and $G^2$. For a fixed number of cells (the case being discussed here), $\chi^2$ converges to the chi-square distribution more quickly than the $G^2$ and the approximation is usually poor for $G^2$ when $n < 20$. Further, guidelines regarding sample size considerations and the validity of $\chi^2$ and $G^2$ are given in Agresti (1990, p. 246).

The most commonly used measures of association between disease and exposure are the relative risk and the odds ratio. To explain how these measures are evaluated, some changes in notations in Table 4.2 would be appropriate and are shown in Table 4.4.

The following estimates obtained using the entries in Table 4.4, are important:

- Estimated risk of disease among those exposed to the risk factor:

$$Pr[D/E] = \frac{y_1}{n_1} \equiv \hat{p}_1$$

- Estimated risk of disease among those not exposed to the risk factor:

$$Pr[D/\bar{E}] = \frac{y_2}{n_2} \equiv \hat{p}_2$$

**TABLE 4.4**

Additional Notations for the 2 × 2 Cross-Classified Data

| | Disease | | |
|---|---|---|---|
| Exposure | $D$ | $\bar{D}$ | Total |
| $E$ | $n_{11} = y_1$ | $n_{12} = n_1 - y_1$ | $n_1$ |
| $\bar{E}$ | $n_{21} = y_2$ | $n_{22} = n_2 - y_2$ | $n_2$ |
| Total | $y.$ | $n - y.$ | $N$ |

- The relative risk measured as the risk of disease for those exposed to the risk factor relative to those not exposed:

$$RR = \frac{y_1/n_1}{y_2/n_2} \qquad (4.1)$$

The relative risk represents how much is it more (or less) likely that disease occurs in the exposed group compared to the unexposed group. For $RR > 1$, the association between exposure and disease is positive and negative for $RR < 1$.

- The fourth and probably the most extensively used measure is the odds ratio. Let $T_1$ and $T_2$ denote the occurrence of disease in the exposed and the unexposed groups respectively. The odds of disease in the exposed group may be expressed as

$$\text{odds}(T_1) = \frac{\Pr[T_1]}{1 - \Pr[T_1]} = \frac{\Pr[\text{disease}|\text{exposed}]}{1 - \Pr[\text{disease}|\text{exposed}]} = \frac{y_1/n_1}{1 - y_1/n_1} = \frac{y_1}{n_1 - y_1}$$

where Pr denotes the probability of the event.
Similarly the odds of disease in the unexposed group would be

$$\text{odds}(T_2) = \frac{\Pr[T_2]}{1 - \Pr[T_2]} = \frac{\Pr[\text{disease}|\text{unexposed}]}{1 - \Pr[\text{disease}|\text{unexposed}]} = \frac{y_2/n_2}{1 - y_2/n_2} = \frac{y_2}{n_2 - y_2}$$

The estimated odds ratio usually denoted by RR or $\hat{\psi}$ would then be the ratio of these two odds, thus

$$\hat{\psi} = \frac{\text{odds}(T_1)}{\text{odds}(T_2)} = \frac{y_1(n_2 - y_2)}{y_2(n_1 - y_1)} \qquad (4.2)$$

The odds ratio is an important estimator in at least two contexts. One is that in the situation of rare diseases, the odds ratio approximates relative risk, and second, it can be determined from either cross-sectional, cohort, or case-control studies, as will be illustrated later in this chapter.

- Attributable risk:

$$AR = \frac{P(D|E) - P(D|\overline{E})}{P(D)} \qquad (4.3)$$

### 4.3.2 Cohort and Case-Control Studies

In a cohort study, individuals are selected for observation and followed over time. Selection of subjects may depend on the presence or absence of exposure to risk factors that are believed to influence the development of the disease. This method entails choosing and studying a predetermined number of individuals, $n_1$ and $n_2$, who are exposed and who are not exposed, respectively. This method of sampling forms the basis of prospective or cohort study and the retrospective studies. In prospective studies, $n_1$ individuals with and $n_2$ individuals without a suspected risk factor are followed over time to determine the number of individuals who develop the disease. In retrospective case-control studies, $n_1$ individuals selected due to having the disease (cases) and $n_2$ nondiseased (controls) individuals would be investigated in terms of past exposure to the suspected antecedent risk factor.

The major difference between the cohort and case-control sampling methods is in the selection of study subjects. A prospective cohort study selects individuals who are initially disease-free and follows them over time to determine how many become ill. This would determine the rates of disease in the absence or presence of exposure. By contrast, the case-control method selects individuals on the basis of presence or absence of disease.

Recall that the odds ratio has been defined in terms of the odds of disease in exposed individuals relative to the odds of disease in the unexposed. An equivalent definition can be obtained in terms of the odds of exposure conditional on the disease, so that the odds of exposure among the diseased and not diseased are

$$\text{odds}_D(E) = \frac{\Pr[E|D]}{1 - \Pr[E|D]} = \frac{y_1}{y_2}$$

$$\text{odds}_{\bar{D}}(E) = \frac{\Pr[E|\bar{D}]}{1 - \Pr[E|\bar{D}]} = \frac{n_1 - y_1}{n_2 - y_2}$$

The odds ratio of exposure in diseased individuals relative to the non-diseased is

$$\hat{\psi} = \frac{\text{odds}_D(E)}{\text{odds}_{\bar{D}}(E)} = \frac{y_1(n_2 - y_2)}{y_2(n_1 - y_1)} \tag{4.4}$$

Thus the exposure odds ratio defined by Equation 4.4 is equivalent to the disease odds ratio defined by Equation 4.2. This relationship is quite important in the design of case-control studies.

## 4.4 Statistical Inference on Odds Ratio

Cox (1970) indicated that the statistical advantage of the odds ratio is that it can be estimated from any of the study designs that were outlined in the previous section (cross-sectional survey, prospective cohort study, and the retrospective case-control study).

A problem that is frequently encountered when an estimate of OR is constructed is the situation where $n_{12} n_{21} = 0$, in which case $\psi$ is undefined. To allow for estimation under these conditions, Haldane (1956) suggested adding a correction term $\delta = 0.5$ to all four cells in the $2 \times 2$ tables to modify the estimator proposed earlier by Woolf (1955). The odds ratio estimate is then given by

$$\hat{\psi}_H = \frac{(n_{11} + 0.5)(n_{22} + 0.5)}{(n_{12} + 0.5)(n_{21} + 0.5)}$$

Adding $\delta = 0.5$ to all cells gives a less biased estimate than if it is added only as necessary, such as when a zero cell occurs (Walter, 1985). Another estimator of $\psi$ was given by Jewell (1984, 1986), which is

$$\hat{\psi}_J = \frac{n_{11} n_{22}}{(n_{12} + 1)(n_{21} + 1)}$$

The correction of $\delta = 1$ to the $n_{12}$ and $n_{21}$ cells is intended to reduce the positive bias of the uncorrected estimator $\psi$ and also to make it defined for all possible tables. Walter and Cook (1991) conducted a large scale Monte Carlo simulation study to compare among several point estimators of the OR. Their conclusion was that for sample size $n = 25$, $\psi_J$ has lower bias, mean-square error, and average absolute error than the other estimators included in the study. Approximate variances of $\hat{\psi}$, $\hat{\psi}_H$, and $\hat{\psi}_J$ are given by

$$\text{var}(\hat{\psi}) = \hat{\psi}^2 \left[ \frac{1}{n_{11}} + \frac{1}{n_{12}} + \frac{1}{n_{21}} + \frac{1}{n_{22}} \right]$$

$$\text{var}(\hat{\psi}_H) = \hat{\psi}_H^2 \left[ \frac{1}{n_{11} + 0.5} + \frac{1}{n_{12} + 0.5} + \frac{1}{n_{21} + 0.5} + \frac{1}{n_{22} + 0.5} \right]$$

$$\text{var}(\hat{\psi}_J) = \hat{\psi}_J^2 \left[ \frac{1}{n_{11}} + \frac{1}{n_{12}} + \frac{1}{n_{21} + 1} + \frac{1}{n_{22} + 1} \right]$$

Before we deal with the problem of significance testing of the odds ratio, there are several philosophical points of view concerning the issue of "statistical" significance as opposed to "scientific" significance. It is known that the general approach to test the association between disease and exposure is to contradict the null hypothesis $H_0$: $\psi = 1$. The p-value of this test is a summary measure of the consistency of the data with the null hypothesis. A small p-value provides the evidence that the data are not consistent with the null hypothesis (in this case implying a significant association). As was indicated by Oakes (1986), the p-value should be considered only a guide to interpretation. The argument regarding the role played by the $p$-value in significance tests dates back to Fisher (1932). He indicated that the null hypothesis cannot be affirmed as such but is possibly disproved. On the other hand, scientific inference is concerned with measuring the magnitude of an effect, regardless of whether the data are consistent with the null hypothesis. Therefore, the construction of a confidence interval on $\psi$ is very desirable as an indication of whether the data contain adequate information to be consistent with the $H_0$, or to signal departures from the $H_0$ that are of scientific importance.

### 4.4.1 Significance Tests

The standard chi-square test for association in a 2 × 2 table provides an approximate test on the hypothesis $H_0: \psi = 1$. Referring to the notation in Table 2.3, the $\chi^2$ statistic is given by

$$\chi^2 = \frac{n\left(|y_1(n_2 - y_2) - y_2(n_1 - y_1)| - \frac{n}{2}\right)^2}{n_1 n_2 y(n - y)} \qquad (4.5)$$

Under $H_0$, the above statistic has an approximate chi-square distribution with one degree of freedom and is used in testing the two-sided alternative. However, if there is interest in testing the null hypothesis of no association against a one-sided alternative, the standard normal approximation $Z_\alpha = \pm\sqrt{\chi^2}$ can be used. In this case, the positive value is used for testing the alternative $H_1: \psi > 1$ and the negative values to test $H_1: \psi < 1$. Quite frequently the sample size may not be sufficient for the asymptotic theory of the chi-square statistic to hold. In this case an exact test is recommended.

When large sample methods cannot be justified, owing either to small samples or highly skewed observed table margins, exact methods are employed. These are based on the enumeration of a reference set of tables with the margins fixed to the totals observed in the data. The $p$-values are evaluated by summing probabilities associated with tables from the reference set identified as more extreme that the table observed.

We shall now illustrate the computation of the $p$-value based on Fisher's exact test. First, let $p_{11} \equiv p_1$ and $p_{21} \equiv p_2$ be the proportion of diseased individuals in the population of exposed and unexposed, respectively. Suppose two independent samples of sizes $n_1$ and $n_2$ are taken from the exposed and the unexposed population. Referring to Table 4.4, it is clear that $y_1$ and $y_2$ are independent binomially distributed random variables with parameters $(n_1, p_1)$ and $(n_2, p_2)$. Hence their joint probability distribution is

$$p(y_1, y_2) = \binom{n_1}{y_1}\binom{n_2}{y_2} p_1^{y_1} q_1^{n_1 - y_1} p_2^{y_2} q_2^{n_2 - y_2} \qquad (4.6)$$

Under the transformation $y_. = y_1 + y_2$, Equation 4.4 can be written as

$$p(y, y_1 | \psi) = \binom{n_1}{y_1} \binom{n_2}{y - y_1} \psi^{y_1} q_1^{n_1} p_2^y q_2^{n_2 - y}$$

where $\psi = \dfrac{p_1 q_2}{q_1 p_2}$ is the population odds ratio. Clearly, $\psi = 1$, if and only if $p_1 = p_2$. Conditional on the sum $y_.$, the probability distribution of $y_1$ is

$$p(y_1 | y, \psi) = c(y, n_1, n_2; \psi) \binom{n_1}{y_1} \binom{n_2}{y - y_1} \psi^{y_1} \tag{4.7}$$

where

$$c^{-1}(y, n_1, n_2; \psi) = \sum_x \binom{n_1}{x} \binom{n_2}{y - x} \psi^x$$

Under the hypothesis $\psi = 1$, Equation 4.7 becomes the hypergeometric distribution:

$$p(y_1 | y, 1) = \frac{\binom{n_1}{y_1} \binom{n_2}{y - y_1}}{\binom{n_1 + n_2}{y}}, \quad y_1 = 0, 1, \ldots, y \tag{4.8}$$

The exact $p$-value of the test on the hypothesis $\psi = 1$ is calculated from Equation 4.6 by summing the probabilities of obtaining all tables with the same marginal totals, with observed $y_1$ as extreme as that obtained from the sample.

### Example 4.2

Researchers in veterinary microbiology conducted a clinical trial on two drugs used for the treatment of diarrhea in calves. Colostrum-deprived calves are given a standard dose of an infectious organism (strain B44 E. *coli*) at two days of age and then therapy is instituted as soon as the calves begin to show diarrhea. The data obtained from the trial is shown in Table 4.5.

**TABLE 4.5**

Data in Diarrhea in Calves

|  | Died | Lived | Total |
|---|---|---|---|
| Drug (1) | 7 | 2 | 9 |
| Drug (2) | 3 | 5 | 8 |
| Total | 10 | 7 | 17 |

Since,

$$p(y_1|y,1) = \frac{\binom{9}{y_1}\binom{8}{10-y_1}}{\binom{17}{10}}$$

we have the $p$-values shown in Table 4.6.

The more extreme configurations correspond to $y_1 = 8, 9$. Therefore the one-tailed $p$-value of the Fisher's exact test is $p = .0004 + .013 + .104 = 0.1174$. The two-tailed $p$-value is obtained by adding the probabilities of all the table configurations as probable as the observed one or less.

Therefore the $p$-value is given as

$$p = 0 + .0004 + .013 + .104 + .0345 + .0018 = 0.1537$$

The probabilities of tables corresponding to $y_1 = 4, 5$, and 6 are not included because they are less extreme or more probable than the observed configuration. We conclude that no significant association between treatment and mortality can be justified by the data.

From the properties of the hypergeometric distribution, if $E(y_1.|y,1)$ and $V(y_1|y,1)$ denote the mean and variance of Equation 4.8, they are respectively given by

$$E(y_1|y,\ 1) = \frac{n_1 y}{n_1 + n_2} \tag{4.9}$$

$$V(y_1|y,\ 1) = \frac{n_1 n_2 y(n_1 + n_2 - y)}{(n_1 + n_2)^2(n_1 + n_2 - 1)} \tag{4.10}$$

As an approximation to the tail area required to test $H_0$: $\Psi = 1$, we refer to the table of standard normal distribution:

$$Z = |y_1 - E(y_1|y,\ 1)| \frac{-\frac{1}{2}}{\sqrt{V(y_1|y,\ 1)}} \tag{4.11}$$

**TABLE 4.6**

Calculated $p$-values from All Possible Tables

| $y_1$ | 0 | 1 | 2 | 3 | 4 | 5 | 6 | 7 | 8 | 9 |
|-------|---|---|-----|-----|------|------|------|------|------|------|
| $p(y_1\|y.,1)$ | 0 | 0 | .0018 | .0345 | .1814 | .363 | .302 | .104 | .013 | .0004 |

From the preceding example, $y_1 = 7$ and the estimates of $E(y_.|y_1, 1)$ and $V$ $(y_.|y_1, 1)$ are 5.29 and 1.03, respectively. Hence, $z = 1.19$ and $p = 0.12$ indicating no significant association.

The following SAS code reads the count data and computes the chi-square and the Fisher's exact tests along with other statistics:

```
data calves;
 input drug dead count @@;
cards;
1 1 7 1 2 2 2 1 3 2 2 5
;

proc format;
value drugf 1 = 'Drug 1'
            2 = 'Drug 2';
value deadf 1 = 'Died'
            2 = 'Lived';

proc freq;
weight count;
tables drug*dead/nocol nopercent chisq exact measures;
format drug drugf. dead deadf.; run;
```

The partial SAS output is shown in Table 4.7.

The following R code reads the count data and computes the chi-square (uncorrected) and the Fisher's exact tests.

```
x <- matrix(c(7, 3, 2, 5), nc = 2)
```

**TABLE 4.7**

Partial SAS Output

| Statistic | DF | Value | Prob |
|---|---|---|---|
| Chi-Square | 1 | 2.8367 | 0.0921 |
| Likelihood Ratio Chi-Square | 1 | 2.9151 | 0.0878 |
| Continuity Adj. Chi-Square | 1 | 1.4175 | 0.2338 |
| Mantel-Haenszel Chi-Square | 1 | 2.6698 | 0.1023 |
| Phi Coefficient | | 0.4085 | |
| Contingency Coefficient | | 0.3782 | |
| Cramer's V | | 0.4085 | |
| **Fisher's Exact Test** | | | |
| Cell (1,1) Frequency (F) | 7 | | |
| Left-sided Pr <= F | 0.9866 | | |
| Right-sided Pr >= F | 0.1170 | | |
| Table Probability (P) | 0.1037 | | |
| Two-sided Pr <= P | 0.1534 | | |

```
chisq. test (x, correct=FALSE)
fisher. test (x)
```

   R Output

   Fisher's Exact Test for Count Data

   data: x

   $p$-value = 0.1534. Alternative hypothesis: true odds ratio is not equal to 1

   95 percent confidence interval: 0.492 – 86.117

   sample estimates of odds ratio: 5.187

### 4.4.2 Interval Estimation

To construct an approximate confidence interval on $\psi$, it is assumed that when $n$ is large, $(\hat{\psi} - \psi)$ follows a normal distribution with mean 0 and variance $V(\hat{\psi})$. An approximate $(1 - \alpha)100\%$ confidence interval is

$$\hat{\psi} \pm Z_{\alpha/2}\sqrt{V(\hat{\psi})}$$

To avoid asymmetry, Woolf (1955) proposed constructing confidence limits on $\beta = \ln\Psi$. He showed that

$$V\left(\hat{\beta}\right) = V(\hat{\psi})/\hat{\psi}^2 = \frac{1}{y_1} + \frac{1}{n_1 - y_1} + \frac{1}{y_2} + \frac{1}{n_2 - y_2} \tag{4.12}$$

and the lower and upper confidence limits on $\psi$ are given respectively by

$$\hat{\psi}_L = \hat{\psi}\exp\left[-Z_{\alpha/2}\sqrt{V(\hat{\beta})}\right]$$

$$\hat{\psi}_U = \hat{\psi}\exp\left[+Z_{\alpha/2}\sqrt{V(\hat{\beta})}\right]$$

We refer the reader to Gart and Thomas (1982).

## 4.5 Analysis of Several 2 × 2 Contingency Tables

There are several reasons when we would analyze several cross-classified tables. Confounding and effect modification are the main reasons. These two issues will be explained in details in this section and the next chapter.

   Consider $k$ pairs of mutually independent binomial variables $y_{i1}$ and $y_{i2}$ with corresponding parameters $p_{i1}$ and $p_{i2}$ $p_{i2}$ and sample sizes $n_{i1}$ and $n_{i2}$, where $i = 1, 2, ..., k$. This information has a $k$ 2 × 2 table representation as shown in Table 4.8.

**TABLE 4.8**

2 × 2 Contingency Table for the *i*th Pair

| | | Disease | | |
|---|---|---|---|---|
| | | $D$ | $\overline{D}$ | Total |
| Exposure | $E$ | $y_{i1}$ | $n_{i1} - y_{i1}$ | $n_{i1}$ |
| | $\overline{E}$ | $y_{i2}$ | $n_{i2} - y_{i2}$ | $n_{i2}$ |
| | Total | $y_{i.}$ | $n_i - y_{i.}$ | $n_i$ |

There is considerable literature on the estimation and significance testing of odds ratios in several 2 × 2 tables (Thomas and Gart, 1977; Fleiss, 1979). The main focus of such studies was to address the following questions:

- Does the odds ratio vary considerably from one table to another?
- If no significant variation among the *k* odds ratios is established, is the common odds ratio statistically significant?
- If no significant variation among the *k* odds ratios is established, how can we construct confidence intervals on the common odds ratio after pooling information from all tables?

Before addressing these questions, the circumstances under which several 2 × 2 tables are produced will now be explored in more detail.

One very important consideration is the effect of confounding variables. In a situation where a variable is correlated with both the disease and the exposure factor, confounding is said to occur. Failure to adjust for this effect would bias the estimated odds ratio as a measure of association between the disease and exposure variables.

If we assume for simplicity that the confounding variable has several distinct levels, then one way to control for its confounding effect is to construct a 2 × 2 table for the disease and exposure variable at each level of the confounder. Epidemiologists name this procedure "stratification." Example 4.3 illustrates this idea.

**Example 4.3**

The data in Table 4.9 are from a case-control study on enzootic pneumonia in pigs by Willeberg (1980). Under investigation is the effect of ventilation systems where exposure (*E*) denotes those farms with fans.

$$\hat{\psi} = 2.99$$

$$ln\hat{\psi} = 1.095$$

$$\sqrt{var(\hat{\beta})} = \left[ \frac{1}{91} + \frac{1}{73} + \frac{1}{25} + \frac{1}{60} \right]^{\frac{1}{2}} = 0.285$$

**TABLE 4.9**

Classification of Diseased and Exposed Pigs in Enzootic
Pneumonia Study

| Fan | Disease | | Total |
|---|---|---|---|
| | $D$ | $\bar{D}$ | |
| Yes ($E$) | 91 | 73 | 164 |
| No ($\bar{E}$) | 25 | 60 | 85 |
| Total | 116 | 133 | 249 |

Source: Willeberg, P. (1980). The analysis and interpretation of epidemiological data, Proceedings of the 2nd International Symposium on
Veterinary Epidemiology and Economics, Canberra, Australia.

and the 95% confidence limits are

$$\hat{\psi}_L = 2.99 \, e^{-(1.96)(0.285)} = 1.71$$
$$\hat{\psi}_U = 2.99 \, e^{(1.96)(0.285)} = 5.23$$

A factor that is not taken into account in analyzing the data is the size of
the farms involved in the study. In attempting to filter out the effect of farm
size (if it is present) on the disease, two groups are formed: large and small
farms. By stratifying this possible confounder into large and small farms, the
data are arranged as in Table 4.10 and the odds ratios are given in Table 4.11.

Note: The SAS and R code given in Example 4.3 produce the odds ratio and
the confidence intervals.

**TABLE 4.10**

Stratification by Farm Size

| Farm Size | Fan | Disease | |
|---|---|---|---|
| | | $D$ | $\bar{D}$ |
| Large | Yes ($E$) | 61 | 17 |
| | No ($\bar{E}$) | 6 | 5 |
| Small | Yes ($E$) | 30 | 56 |
| | No ($\bar{E}$) | 19 | 55 |

**TABLE 4.11**

Odds Ratios and Confidence Intervals after Stratification by Farm Size

| Farm Size | $\hat{\psi}_i$ | $\hat{\beta}_i$ | $SE(\hat{\beta}_i)$ | 95% CI |
|---|---|---|---|---|
| Large | 2.99 | 1.095 | 0.665 | (.67, 13.2) |
| Small | 1.55 | 0.438 | 0.349 | (.74, 3.26) |

Now it is evident that the relationship between the disease and the exposure factor is not clear, and this could be due to the possible confounding effect of farm size. For farms of large size, the estimated odds ratio of disease-risk association is $\hat{\psi} = 2.99$, which is identical to the estimate obtained from Table 4.9. However, it is not statistically significant ($p = 0.10$, two-sided). It follows then that pooling the data from large and small farms to form a single 2 × 2 table can produce misleading results. Therefore, the subgroup-specific odds ratio may be regarded as descriptive of the effects. Now, in the context of multiple tables, the three questions posed earlier will be addressed.

### 4.5.1 Test of Homogeneity

This test of homogeneity is a reformulation of the question "Does the odds ratio vary considerably across tables?" This is equivalent to testing the hypothesis $H_0: \psi_1 = \psi_2 = ... = \psi_k = \psi$. Woolf (1955) proposed a test that is based on the estimated log odds ratios ($\hat{\beta}_i$) and their estimated variances. Since the estimated variance of $\hat{\beta}_i$ is

$$V\left(\hat{\beta}_i\right) = w_i^{-1} = \frac{1}{y_{i1}} + \frac{1}{n_{i1} - y_{i1}} + \frac{1}{y_{i2}} + \frac{1}{n_{i2} - y_{i2}}$$

an estimate of $ln\psi$ is constructed as

$$\hat{\beta} = \ln \hat{\psi} = \frac{\sum_{i=1}^{k} w_i \hat{\beta}_i}{\sum_{i=1}^{k} w_i}$$

Furthermore, it can be shown that

$$V(\hat{\beta}) = \frac{1}{\sum_{i=1}^{k} w_i}$$

To assess the homogeneity (i.e., constancy of odds ratios) the statistic

$$\chi_w^2 = \sum_{i=1}^{k} w_i \left(\hat{\beta}_i - \hat{\beta}\right)^2$$

has approximately a chi-square distribution with $k$–1 degrees of freedom. Large values of $\chi_w^2$ provide evidence against the homogeneity hypothesis.

#### Example 4.3 (Continued)

This example applies Woolf's method to the summary estimates from Table 4.9 to test for a difference between the odds ratios of the two groups. The chi-square value is calculated as follows:

$$\hat{\beta}_1 = 1.095 \quad \hat{\beta}_2 = 0.438$$

$$w_1 = 2.26 \quad w_2 = 8.20, \qquad \sum w_1 = 10.46$$

$$\hat{\beta} = \frac{\sum w_i \beta_i}{\sum w_i} = 0.58$$

$$\chi_w^2 = \sum_{i=1}^{k} w_i \left( \hat{\beta}_i - \hat{\beta} \right)^2 = 0.764$$

From the value of the $\chi^2$ with one degree of freedom we can see that there is no significant difference between the odds ratios of the two groups.

If we intend to find a summary odds ratio from several 2 × 2 tables, it is useful to test for interaction. Consider the summary calculations in Table 4.11. The odds ratios for the two strata are 2.99 and 1.55. If the underlying odds ratios for individual strata are really different, it is questionable if a summary estimate is appropriate. In support of this, the implication of a significant value of $\chi_w^2$ should be stressed. A large value of $\chi_w^2$ indicates that the homogeneity of $\psi_i's$ is not supported by the data and that there is an interaction between the exposure and the stratification variable. Thus, in the case of a significant value of $\chi_w^2$ one should not construct a single estimate of the common odds ratio.

### 4.5.2 Significance Test of Common Odds Ratio

Recall that inference on the odds ratio can be based on the total $y_.$, conditional on the marginal totals of the 2 × 2 table. From Cox and Snell (1989), the inference on the common odds ratio $\Psi$ is based on the conditional distribution of $T = \sum_{i=1}^{k} y_{.i}$, given the marginal total of all tables.

Now, we know that under $H_0{:}\psi = 1$, the distribution of $y_{.i}$ (the total of diseased in the $i$th table) for a particular $i$ is given by the hypergeometric distribution (Equation 4.8) and the required distributions of $T$ is the convolution of $k$ of these distributions. It is clear that this is impracticable for exact calculation of confidence limits. However, we can test the null hypothesis that $H_0{:}\psi = 1$ by noting that the mean and variance of $T$ are given respectively as

$$E(T|\psi = 1) = \sum_{i=1}^{k} \frac{n_{i1} y_i}{n_{i1} + n_{i2}} \tag{4.13}$$

$$V(T|\psi = 1) = \sum_{i=1}^{k} \frac{n_{i1} n_{i2} y_i (n_{i1} + n_{i2} - y_i)}{(n_{i1} + n_{i2})^2 (n_{i1} + n_{i2} - 1)} \tag{4.14}$$

where $y_{.i}$ is the observed number of diseased in the $i$th table. A normal approximation, with continuity correction, for the distribution of $T$ will nearly always be adequate. Cox and Snell (1989) indicate that the approximation is good even for a single table and will be improved by convolution. The combined test of significance from several $2 \times 2$ contingency tables is done by referring

$$\chi_1^2 = \frac{\left(|T - E(T|\psi = 1)| - \frac{1}{2}\right)^2}{V(T|\psi = 1)} \tag{4.15}$$

to the chi-square table with one degree of freedom.

Another form of the one degree of freedom chi-square test on $H_0: \psi = 1$ was given by Mantel and Haenszel (1959) as

$$\chi_{mh}^2 = \frac{\left(\sum_{i=1}^{k} \frac{y_{i1}(n_{i2} - y_{i2}) - y_{i2}(n_{i1} - y_{i1})}{n_i}\right)^2}{\sum_{i=1}^{k} \frac{n_{i1}n_{i2}y_{.i}(n_i - y_{.i})}{n_i^2(n_i - 1)}} \tag{4.16}$$

where, $n_i = n_{i1} + n_{i2}$ and $y_{.i} = y_{i1} + y_{i2}$.

In Example 4.3 we consider the data in Table 4.10 and calculate both the Cox and Snell, and Mantel and Haenszel formulations.

### Example 4.3 (Continued)

As a test of significance of the common odds ratio using the data in Table 4.10, we have $T = 91$, $E(T|\psi = 1) = 85.057$, and $V(T|\psi = 1) = 10.3181$. Hence, $\chi_1^2 = 2.872$, and the combined odds ratio is nonsignificant. Using the Mantel and Haenszel $\chi_{mh}^2$ formula we got $\chi_{mh}^2 = 2.882$, which is quite similar to the value obtained using the Cox and Snell method with similar conclusions.

The following SAS code analyzes these data:

```
data willeberg;
 input farms fan disease count @@;
cards;
1 1 1 61 1 1 2 17 1 2 1  6 1 2 2  5
2 1 1 30 2 1 2 56 2 2 1 19 2 2 2 55;

proc format;
value farmsf 1 = 'Large'
             2 = 'Small';
value fanf 1 = 'Yes'
           2 = 'No';
value diseasef 1 = 'Disease'
               2 = 'No Disease';
```

```
proc freq;
weight count;
tables farms*fan*disease/nocol nopercent cmh measures;
format farms farmsf. fan fanf. disease diseasef.;   run;
```

The partial SAS output (Table 4.12) shows the odds ratios stratified by farm size with confidence intervals, the common odds ratios, and confidence intervals, and the Breslow-Day test for the homogeneity of odds ratios.

The following R code reads the grouped data in an array "Willeberg" as required for the Mantel-Haenszel test function in R. Chi-square and the Fisher's exact tests are then computed separately for large and small farms' data followed by the Mantel-Haenszel test and the exact conditional tests for the homogeneity of odds ratios across the strata. The exact conditional test suggests that the odds ratios are homogeneous (p-value = 0.09). However, if odds ratios turn out to be different than the traditional approach, then we would be using the Woolf (1955) test for interaction. The "woolf" function that computes this test is run on the current data set, again indicating that there is

**TABLE 4.12**

SAS Output for the Data in Table 4.10

| Farm Size: Large | | | |
|---|---|---|---|
| Estimates of the Relative Risk (Row1/Row2) | | | |
| Type of Study | Value | 95% Confidence Limits | |
| Case-Control (Odds Ratio) | 2.9902 | 0.8126 | 11.0035 |
| Cohort (Col1 Risk) | 1.4338 | 0.8255 | 2.4901 |
| Cohort (Col2 Risk) | 0.4795 | 0.2216 | 1.0375 |

| Farm Size: Small | | | |
|---|---|---|---|
| Estimates of the Relative Risk (Row1/Row2) | | | |
| Type of Study | Value | 95% Confidence Limit | |
| Case-Control (Odds Ratio) | 1.5508 | 0.7820 | 3.0751 |
| Cohort (Col1 Risk) | 1.3586 | 0.8379 | 2.2031 |
| Cohort (Col2 Risk) | 0.8761 | 0.7140 | 1.0750 |

| Estimates of Common Relative Risk (Row1/Row2) | | | | |
|---|---|---|---|---|
| Type of Study | Method | Value | 95% Confidence Limit | |
| Case-Control | Mantel-Haenszel | 1.7624 | 0.9587 | 3.2396 |
| (Odds Ration) | Logit | 1.7875 | 0.9751 | 3.2767 |

| Breslow-Day Test for Homogeneity of the Odds Ratios | |
|---|---|
| Chi-square | 0.7759 |
| DF | 1 |
| Pr > ChiSq | 0.3784 |

no interaction by the stratifying factor, and therefore the Mantel-Haenszel test is admissible.

```
willeberg=array (c (61, 6, 17, 5, 30, 19, 56, 55),
dim = c(2, 2, 2),
dimnames = list (
Fan = c("Yes", "No"),
Disease = c("Disease", "No Disease"),
Farm.Size = c("Large", "Small")))

apply(willeberg, 3, function(x) chisq.test(x))
apply(willeberg, 3, function(x) fisher.test(x))
mantelhaen. test (willeberg)
mantelhaen. test (willeberg, exact=TRUE)
* * * * * * * * *
woolf <- function(x) {
x <- x + 1 / 2
k <- dim(x)[3]
or <- apply(x, 3, function(x) (x[1,1]*x[2,2])/(x[1,2]*x[2,1]))
w <- apply(x, 3, function(x) 1 / sum(1 / x))
1 - pchisq(sum(w * (log(or) - weighted.mean(log(or), w)) ^ 2), k - 1)
```

The R output for the different tests is given below:

**A.  Testing association stratifying by the farm size.**

**\$Large**

Pearson's Chi-squared test with Yates' continuity correction

data: x

X-squared = 1.7679, df = 1, $p$-value = 0.1836

**\$Small**

Pearson's Chi-squared test with Yates' continuity correction

data: x

X-squared = 1.1835, df = 1, $p$-value = 0.277

**Fisher's Exact Test for Count Data**

**\$Large**

$p$-value = 0.1309. alternative hypothesis: true odds ratio is not equal to 1.

95 percent confidence interval: 0.630, 13.249

sample estimates: odds ratio = 2.946

**\$Small**

$p$-value = 0.2317. alternative hypothesis: true odds ratio is not equal to 1

95 percent confidence interval: 0.743, 3.277

sample estimates: odds ratio = 1.546492

**B.  Mantel-Haenszel chi-squared test with continuity correction**

Mantel-Haenszel X-squared = 2.8717, df = 1, $p$-value = 0.09015. Alternative hypothesis: true common odds ratio is not equal to 1

95 percent confidence interval: 0.959, 3.239

sample estimates: common odds ratio = 1.762

**C.  Exact conditional test of independence in 2 x 2 x k tables**

S = 91, $p$-value = 0.08564. Alternative hypothesis: true common odds ratio is not equal to 1

95 percent confidence interval: 0.922 3.451

sample estimates: common odds ratio = 1.776448

**Woolf's approximation**
The chi-square test yields a $p$-value = 0.364

### 4.5.3  Confidence Interval on the Common Odds Ratio

Mantel and Haenszel (1959) suggested a highly efficient estimate of the common odds ratio from several $2 \times 2$ tables as

$$\hat{\psi}_{mh} = \frac{\sum_{i=1}^{k} \frac{y_{i1}(n_{i2} - y_{i2})}{(n_{i1} + n_{i2})}}{\sum_{i=1}^{k} \frac{y_{i2}(n_{i1} - y_{i1})}{(n_{i2} + n_{i2})}} \qquad (4.17)$$

Note that $\hat{\beta}_{mh} = ln\hat{\psi}_{mh}$ can be regarded as a special form of weighted means, based on the linearizing approximation near $\beta_{mh} = 0$. This equivalence has been used to motivate various estimates of the variance of $\hat{\beta}_{mh}$. Robins et al. (1986) constructed an estimate of the variance as

$$V(\hat{\beta}_{mh}) = \frac{1}{2} \left\{ \sum \frac{A_i C_i}{C^2} + \sum \frac{(A_i D_i + B_i C_i)}{(C\ D\ )} + \sum \frac{B_i D_i}{D^2} \right\} \qquad (4.18)$$

where

$$A_i = (y_{i1} + n_{i2} - y_{i2})/(n_{i1} + n_{i2}), \ B_i = (y_{i2} + n_{i1} - y_{i1})/(n_{i1} + n_{i2})$$

$$C_i = y_{i1}(n_{i2} - y_{i2})/(n_{i1} + n_{i2}), \ D_i = y_{i2}(n_{i1} - y_{i1})/(n_{i1} + n_{i2})$$

and

$$C = \sum C_i \text{ and } D = \sum D_i$$

Using the results from the previous example (Example 4.2), we find that $\hat{\psi}_{mh} = 1.76$ and the 95% confidence limits for $\psi_{mh}$ are (0.96, 3.24).

Hauck (1979) derived another estimator of the large sample variance of $\hat{\psi}_{mh}$ as

$$\hat{V}_{mh} = \frac{\hat{\psi}_{mh}^2 \left( \sum_{i=1}^{k} \hat{w}_i^2 \hat{b}_i \right)}{\left( \sum \hat{w}_i \right)^2} \tag{4.19}$$

where

$$\hat{w}_i = (n_{i1}^{-1} + n_{i2}^{-1})^{-1} \hat{p}_{i2}(1 - \hat{p}_{i1})$$

$$\hat{b}_i = \frac{1}{n_{i1}\hat{p}_{i1}\hat{q}_{i1}} + \frac{1}{n_{i2}\hat{p}_{i2}\hat{q}_{i2}}$$

$$\text{and } \hat{p}_{ij} = \frac{y_{ij}}{n_{ij}} j = 1, 2 \text{ and } \hat{q}_{ij} = 1 - \hat{p}_{ij}.$$

## 4.6 Analysis of Matched Pairs (One Case and One Control)

The *matching* of one or more controls to each case means that the two are paired based on their similarities with respect to some characteristic(s). Pairing individuals can involve attributes such as age, weight, farm, parity, type of hospital of admission and breed. These are just a few examples.

It is important to note that since cases and controls are believed to be similar on the matching variables, their differences with respect to disease may be attributed to different extraneous variables. It was pointed out by Schlesselman (1982, p. 105) that "if the cases and controls differ with respect to some exposure variable, suggesting an association with the study disease, then this association cannot be explained in terms of case-control differences on the matching variables." The main objective of matching is in removing the bias that may affect the comparison between the cases and controls. In other words, matching attempts to achieve comparability on the important potential confounding variables (a confounder is an extraneous variable that is associated with the risk factor and the disease of interest). This strategy of matching is different from what is called "adjusting." Adjusting attempts to correct for differences in the cases and controls during the data analysis step, as opposed to matching which occurs at the design stage.

In this section, we investigate the situation where one case is matched with a single control, where the risk factor is dichotomous. As before, we denote the presence or absence of exposure to the risk factor by $E$ and $\bar{E}$, respectively. In this situation responses are summarized by a $2 \times 2$ table, which exhibits two important features. First, all probabilities or associations may show a symmetric pattern about the main diagonal of the table. Second, the marginal distributions may differ in some systematic way.

The following two sections will address the estimation of the odds ratio and testing the equality of the marginal distributions under the 1:1 matched case-control design.

### 4.6.1 Estimating the Odds Ratio

Suppose that one has matched a single control to each case, and that the exposure under study is dichotomous. Denoting the presence or absence of exposure by $E$ and $\bar{E}$, respectively, there are four possible outcomes for each case-control pair.

To calculate $\psi_{mh}$ for matched pair data, in which each pair is treated as a stratum, we must first change the matched pair table to its unpaired equivalent resulting in four possible outcomes (Tables 4.13 to 4.16).

**TABLE 4.13**

Outcome 1

| Paired | | | | | Unpaired Equivalent | | |
|---|---|---|---|---|---|---|---|
| | | | Control | | | | |
| | | $E$ | $\bar{E}$ | | | Case | Control |
| Case | $E$ | 1 | 0 | $E$ | | $1\ (y_1)$ | $1\ (n_1 - y_1)$ |
| | $\bar{E}$ | 0 | 0 | $\bar{E}$ | | $0\ (y_2)$ | $0\ (n_2 - y_2)$ |

*Note:* $y_1(n_2 - y_2) = y_2(n_1 - y_1) = 0$

**TABLE 4.14**

Outcome 2

| Paired | | | | | Unpaired Equivalent | | |
|---|---|---|---|---|---|---|---|
| | | | Control | | | | |
| | | $E$ | $\bar{E}$ | | | Case | Control |
| Case | $E$ | 0 | 0 | $E$ | | $0\ (y_1)$ | $1\ (n_1 - y_1)$ |
| | $\bar{E}$ | 1 | 0 | $\bar{E}$ | | $1\ (y_2)$ | $0\ (n_2 - y_2)$ |

*Note:* $y_1(n_2 - y_2) = 0$ and $y_2(n_1 - y_1) = 1$

**TABLE 4.15**

Outcome 3

| Paired | | | | Unpaired Equivalent | | |
|--------|---|---|---|---|---|---|
| | | Control | | | Case | Control |
| | | $E$ | $\bar{E}$ | | | |
| Case | $E$ | 0 | 0 | $E$ | 1 ($y_1$) | 0 ($n_1 - y_1$) |
| | $\bar{E}$ | 1 | 0 | $\bar{E}$ | 0 ($y_2$) | 1 ($n_2 - y_2$) |

Note:  $y_1(n_2 - y_2) = 1$ and $y_2(n_1 - y_1) = 0$

**TABLE 4.16**

Outcome 4

| Paired | | | | Unpaired Equivalent | | |
|--------|---|---|---|---|---|---|
| | | Control | | | Case | Control |
| | | $E$ | $\bar{E}$ | | | |
| Case | $E$ | 0 | 0 | $E$ | 0 ($y_1$) | 0 ($n_1 - y_1$) |
| | $\bar{E}$ | 0 | 1 | $\bar{E}$ | 1 ($y_2$) | 1 ($n_2 - y_2$) |

Note:  $y_1(n_2 - y_2) = y_2(n_1 - y_1) = 0$

Since

$$\psi_{mhp} = \frac{\sum y_{i1}(n_{i2} - y_{i2})/n_i}{\sum y_{i2}(n_{i1} - y_{i1})/n_i}$$

and $n_i = 2$, then

$$\hat{\psi}_{mhp} = \frac{\text{number of pairs with } y_{i1}(n_{i2} - y_{i2}) = 1}{\text{number of pairs with } y_{i2}(n_{i1} - y_{i1}) = 1}$$

For the matched pairs Table 4.17, the odds ratio estimate would be $\hat{\psi}_{mhp} = g/h$.

**TABLE 4.17**

Matched Pairs Table

| | | Control | | |
|---|---|---|---|---|
| | | $E$ | $\bar{E}$ | Total |
| Case | $E$ | $f$ ($p_{11}$) | $g$ ($p_{12}$) | $f + g$ ($p_{1.}$) |
| | $\bar{E}$ | $h$ ($p_{21}$) | $i$ ($p_{22}$) | $h + i$ ($p_{2.}$) |
| Total | | $f + h$ ($p_{.1}$) | $g + i$ ($p_{.2}$) | $f + g + h + i$ |

It was shown (see Fleiss, 1981) that for the matched pairs, the variance of the odds ratio estimate is

$$Var\left(\hat{\psi}_{mhp}\right) = \left(\frac{g}{h}\right)^2\left(\frac{1}{g} + \frac{1}{h}\right)$$

For the test of the hypothesis $H_0: \psi_{mhp} = 1$, the statistic

$$\chi^2_{(1)} = \frac{(g - h)^2}{g + h}$$

follows chi-square distribution asymptotically with one degree of freedom. A better approximation includes a correction factor so that the statistic becomes

$$\chi^2_{(1)} = \frac{(|g - h| - 1)^2}{g + h}$$

The implication of a large value of this $\chi^2_{(1)}$ is that the cases and the controls differ with regard to the distribution of the risk factor.

## 4.6.2 Testing the Equality of Marginal Distributions

When $p_{1.} = p_{.1}$, then $p_{2.} = p_{.2}$ and there is marginal homogeneity. When $(p_{1.} - p_{.1}) = (p_{2.} - p_{.2})$, then the marginal homogeneity is equivalent to symmetry of probabilities off the main diagonal, that is, $p_{12} = p_{21}$. Naturally, a test of marginal homogeneity should then be based on

$$D = \hat{p}_1 - \hat{p}_1 = \frac{g - h}{n}$$

From Agresti (1990, p. 348), it can be shown that

$$V(D) = \frac{1}{n}[p_1(1 - p_1) + p_{.1}(1 - p_{.1}) - 2(p_{11}p_{22} - p_{12}p_{21})] \qquad (4.20)$$

Note that the dependence on the sample marginal probabilities resulting from the matching of cases and controls is reflected in the term $2(p_{11}p_{22} - p_{12}p_{21})$. Moreover, dependent samples usually show positive association between responses, that is, $\psi = p_{11}p_{22}/p_{12}p_{21} > 1$, or, equivalently, $p_{11}p_{22} - p_{12}p_{21} > 0$.

From Equation 4.20 this positive association implies that the variance of $D$ is smaller than when the samples are independent. This indicates that matched studies that produce dependent proportions can help improve the precision of the test statistic $D$.

For large samples, $D = \hat{p}_{1.} - \hat{p}_{.1} = (g - h)/n$ has a sampling distribution that is approximately normal. A $(1 - \alpha)100\%$ confidence interval on $p_{1.} - p_{.1}$ is

$$D \pm Z_{\alpha/2}\sqrt{\hat{V}(D)}$$

where

$$\hat{V}(D) = \frac{1}{n}[\hat{p}_1(1 - \hat{p}_1) + \hat{p}_{.1}(1 - \hat{p}_{.1}) - 2(\hat{p}_{11}\hat{p}_{22} - \hat{p}_{12}\hat{p}_{21})]$$

The test statistic

$$Z = \frac{D}{\sqrt{\hat{V}(D)}}$$

is used to test the hypothesis $H_0{:}p_{1.}{=}p_{.1}$. Under $H_0$, it can be shown that the estimated variance of $D$ is

$$\hat{V}(D|H_0) = \frac{g + h}{n^2}$$

The test statistic simplifies to $Z_0 = \dfrac{g - h}{\sqrt{(g + h)}}$. The square of $Z_0$ gives the one degree of freedom chi-square test on $H_0 : \psi = 1$, referred to as McNemar's test.

### Example 4.4

To illustrate McNemar's test, the following hypothetical data (Table 4.18) from a controlled trial with matched pairs is used. The null hypothesis being tested is $H_0 : \psi = 1$.

$$\hat{\psi} = \frac{20}{5} = 4$$

$$SE(\hat{\psi}) = \frac{20}{5}\left[\frac{(20 + 5)}{(20)(5)}\right]^{1/2} = 2$$

$$\chi^2_{(1)} = \frac{(|20 - 5| - 1)^2}{25} = 7.84$$

Based on the calculated value of the $\chi^2$, the data do not support the null hypothesis. Meaning that there is a significant difference ($p < 0.05$) in the rate of exposure among the cases and the controls.

For $2 \times 2$ matched pair data, SAS PROC FREQ computes the uncorrected McNemar test by using the "agree" option in the "table" statement and prints the kappa coefficient, its asymptotic standard error, and the confidence interval.

The following R code prints the $2 \times 2$ table and computes the continuity corrected McNemar test:

```
paired <- matrix(c(15, 5, 20, 85),nr = 2,
dimnames = list("Case" = c("Present", "Absent"),
```

**TABLE 4.18**

Hypothetical Data for Matched-Pair Design

|        |            | Control |         |
|--------|------------|---------|---------|
|        |            | $E$     | $\bar{E}$ |
| Case   | $E$        | 15      | 20      |
|        | $\bar{E}$  | 5       | 85      |

```
"Control" = c("Present", "Absent")))
paired
mcnemar. test (paired, correct=TRUE)
```

The output is

```
McNemar's chi-squared = 7.84, df = 1, p-value = 0.00511
```

## 4.7 Statistical Analysis of Clustered Binary Data

In this section we review methods that have been proposed for comparing the overall proportion of successes among individuals in several groups (treatments) when the clusters are the sampling units. The methods are illustrated by several examples.

### Example 4.5

The data in this example was given by Williams (1975) taken from Weil (1970), and give the results from an experiment comparing two treatments. One group of 16 pregnant female rats was fed a control diet during pregnancy and lactation, while the diet of a second group of 16 pregnant females was treated with a chemical. For each cluster (litter or cluster comprises the pups born to a female rat), the number $n$ of pups alive at four days and the number $y$ of pups that survived at 31-day lactation period were recorded. The data are given as a fraction $y/n$ in Table 4.19.

The purpose of the experiment was to determine if the chemical treatment significantly affects the survival rate among the pups. That is, we need to test $H_0{:}P_1 = P_2$, where $P_i$ is the true proportion of successes in group $i$ ($i = 1, 2$).

The clustering of responses within litters that invalidates the standard chi-square test may be adjusted for, allowing a modified version of this test procedure to be applied (Donner, 1982, 1989; Donald and Donner, 1987). For completeness of the presentation we use the following notations:

We assume that $k_1$ clusters have been randomly assigned to the treatment group and $k_2$ clusters to the control group. The sizes of the clusters in the treatment group denoted by $n_{1j},j = 1, 2, ..., k_1$ and the control group by

**TABLE 4.19**

Weil's Data

| Control | 13/13 | 12/12 | 9/9 | 9/9 | 8/8 | 8/8 | 12/13 | 11/12 |
|---|---|---|---|---|---|---|---|---|
| | 9/10 | 9/10 | 8/9 | 11/13 | 4/5 | 5/7 | 7/10 | 7/10 |
| Treatment | 12/12 | 11/11 | 10/10 | 9/9 | 10/11 | 9/11 | 9/11 | 8/9 |
| | 8/9 | 4/5 | 7/9 | 4/7 | 5/10 | 3/6 | 3/10 | 0/7 |

$n_{2j}, j = 1,2,...k_2$. The total number of individuals in each group is denoted by $N_1 = \sum_{j=1}^{k_1} n_{1j}$ for the treatment, and $N_2 = \sum_{j=1}^{k_2} n_{2j}$ for the control group.

In a completely randomized design with binary response, $\hat{P}_{ij} = Y_{ij}/n_{ij}$ denotes the cluster specific success rate. We further define

$$Y_i = \sum_{j=1}^{k_i} Y_{ij}, \ \hat{P}_i = Y_i/N_i$$

$$\hat{P} = \sum_{i=1}^{2} Y_i/N$$

where $N = N_1 + N_2$ and $k = \sum_{i=1}^{2} k_i$.

The rates $\hat{P}_i$ and $\hat{P}$ denote, respectively, the event rates as computed overall for clusters in group $i$ and the overall event rate observed in the study.

Letting $\bar{n}_{ai} = \sum_{j=1}^{k_i} n_{ij}^2/N_i$, the analysis of variance estimator of the intraclass correlation coefficient $\rho$ is defined as

$$\hat{\rho} = \frac{MSB - MSW}{MSB + (n_o - 1)MSW}$$

where

$$MSB = \sum_{i=1}^{2} \sum_{j=1}^{k_i} n_{ij} \left( \hat{P}_{ij} - \hat{P}_i \right)^2 / (k - 2), \ MSW$$

$$= \sum_{i=1}^{2} \sum_{j=1}^{k_i} n_{ij} \hat{P}_{ij} \left( 1 - \hat{P}_{ij} \right) \Big/ (N - k), \text{ and}$$

$$n_o = \left( N - \sum_{i=1}^{k} \bar{n}_{ai} \right) \Big/ (k - 2)$$

The mean squares *MSB* and *MSW* measure the variation in responses between and within clusters, respectively.

The unadjusted Pearson chi-square statistic for testing $H_0 : P_1 = P_2$ is

$$X_P^2 = \sum_{i=1}^{k} N_i \frac{\left(\hat{P}_i - \hat{P}\right)^2}{\hat{P}\left(1 - \hat{P}\right)} \tag{4.21}$$

### 4.7.1 Approaches to Adjust the Pearson's Chi-Square

The rationale behind these approaches is to directly adjust the standard chi-square test for the clustering of responses within the intact units. These approaches are intuitively attractive since they yield the standard Pearson chi-square statistic if the estimated intraclass correlation is zero.

### 4.7.2 Donner and Donald Adjustment

Donner (1989) discussed an adjustment that depends on clustering correction factors computed separately in each treatment group. The resulting statistic is given by

$$X_A^2 = \sum_{i=1}^{2} \frac{\left(Y_i - N_i\hat{P}\right)^2}{N_i C_i \hat{P}\left(1 - \hat{P}\right)} \tag{4.22}$$

where

$$C_i = 1 + (\bar{n}_{ai} - 1)\hat{\rho} \tag{4.23}$$

is an estimate of the design effect (or "clustering effect") in group $i$ ($i = 1, 2$). Note that in some situations one may assume that $C_1 = C_2$, that is, estimates of population design effects are homogeneous across treatment groups. This assumption will hold under $H_0$ provided the clusters are randomly assigned to treatment groups, but may not hold for observational comparisons, particularly if the mean cluster sizes in the two comparison populations are quite different (Donner et al., 1994). When $\hat{\rho} = 0$, $\chi_A^2$ reduces to the standard Pearson chi-square $\chi_P^2$.

### 4.7.3 Procedures Based on Ratio Estimate Theory

Rao and Scott (1992) proposed a method for testing $H_0$ that regards the observed response rates $\hat{P}_i$ as ratios rather than proportions. The estimated design effects in each group are computed as

$$d_i = \hat{V}_R(\hat{P}_i) / \hat{V}_B(\hat{P}_i)$$

where $\hat{V}_B(\hat{P}_i) = \hat{P}_i(1 - \hat{P}_i)/N_i$ is the estimated binomial variance and $\hat{V}_R(\hat{P}_i)$ is the estimated ratio variance (see Cochran, 1977) given by

$$\hat{V}_R\left(\hat{P}_i\right) = \frac{1}{k_i(k_1-1)N_i^2}\sum_{j=1}^{k_i}\left(y_{ij}-n_{ij}\hat{P}_i\right)^2 = v_i$$

The design effects $d_i$ are used to compute the effective sample size in each group, $\tilde{N}_i = N_i/d_i$ the effective number of successes, and $\tilde{y}_i = y_i/d_i$, and the effective overall response rate

$$\tilde{P} = \sum_{i=1}^{2}\tilde{y}_i \bigg/ \sum_{i=1}^{2}\tilde{N}_i$$

The test statistic that accounts for the design effect is

$$X_{RS}^2 = \sum_{i=1}^{2}\frac{\left(\tilde{y}_i-\tilde{N}_i\tilde{P}\right)^2}{\tilde{N}_i\tilde{P}(1-\tilde{P})} \tag{4.24}$$

The statistic $X_{RS}^2$ does not explicitly involve the concept of within-cluster correlation and does not make any assumptions on the nature of clustering. In particular, it does not assume the homogeneity of design effects. Therefore, the approach is applicable to nonexperimental trials. Donner and Klar (2000) noted, however, that for trials involving random allocation of clusters to treatment groups, the assumption of homogeneous design effect is guaranteed at least under $H_0$. Donner et al. (1994) showed, through extensive Monte Carlo studies, that $X_A^2$ is valid in studies having as few as 10 clusters per group, and the number of clusters per group required to ensure the validity of $X_{RS}^2$ may be substantially more.

### 4.7.4 Confidence Interval Construction

When the effect of intervention is measured by the risk difference $\hat{\delta} = \hat{P}_1 - \hat{P}_2$, one may construct an approximate two-sided $(1-\alpha)100\%$ confidence interval on $P_1-P_2$. The standard error of $\hat{\delta}$ is consistently estimated by

$$\hat{V}\left(\hat{\delta}\right) = \left[\frac{C_1\hat{P}_1\left(1-\hat{P}_1\right)}{N_1} + \frac{C_2\hat{P}_2\left(1-\hat{P}_2\right)}{N_2}\right]$$

where $C_j(j=1,2)$ is given by Equation 4.23.

The proposed confidence interval on $\delta$, the true risk difference, is therefore

$$(\hat{P}_1-\hat{P}_2)\pm z_{\alpha/2}\sqrt{\hat{V}(\hat{\delta})}$$

where $z_{\alpha/2}$ is the $(1-\alpha)100\%$ two-sided critical value of the standard normal distribution.

**Example 4.5 (Continued)**

The summary statistics of Weil's data are

|   | Control | Treatment | Total |
|---|---------|-----------|-------|
| Y | 142     | 112       | 254   |
| N | 158     | 147       | 305   |

Here $\hat{P}_1 = .899,$ $\hat{P}_2 = .760,$ $\hat{P} = .83,$ and $X_P^2 =$

$$\frac{158(.899 - .83)^2 + 147(.760 - .83)^2}{(.83)(.17)} = 10.24.$$

The $p$-value = 0.001, indicating a significant difference in mortality between the two groups.

Now, when we adjust for clustering, the estimated ICC is $\hat{\rho} = 0.25$, $\bar{n}_{a1} = 10.8$, and $\bar{n}_{a2} = 8.92$, so that $C_1 = 3.45$ and $C_2 = 3.20$. The adjusted chi-square statistic is

$$X_A^2 = \frac{(142 - 158(.83))^2}{(158)(3.45)(.83)(.17)} + \frac{(112 - 147(.83))^2}{(147)(3.2)(.83)(.17)} = 3.043$$

and $p$-value = 0.081.

This indicates that the difference in mortality between two groups is not significant at $\alpha = 5\%$ and shows how failure to take into account cluster effects can lead to spurious statistical inference.

Further $\hat{\delta} = .899 - .760 = 0.139$ with an estimated standard error of 0.042, so that the unadjusted 95% confidence limits are (.059, 219).

However, when clustering is adjusted for, the estimated standard error is 0.076 and the confidence interval is (−.010, 0.288), which is much wider than when the clustering is ignored.

The following SAS code reads in the grouped data, generates the ungrouped data, computes the unadjusted chi-square test and runs the linear model to produce statistics for the ICC.

```
* Grouped Data;
data weil;
 input group $ litter n dead @@;
 alive = n-dead;
 cards;
c  1 13 13 c  2 12 12 c  3  9  9 c  4  9  9
c  5  8  8 c  6  8  8 c  7 13 12 c  8 12 11
c  9 10  9 c 10 10  9 c 11  9  8 c 12 13 11
c 13  5  4 c 14  7  5 c 15 10  7 c 16 10  7
t 17 12 12 t 18 11 11 t 19 10 10 t 20  9  9
t 21 11 10 t 22 11  9 t 23 11  9 t 24  9  8
t 25  9  8 t 26  5  4 t 27  9  7 t 28  7  4
```

```
t 29 10  5 t 30  6  3 t 31 10  3 t 32  7  0
;

* Ungrouped Data;
data new; set weil;
  do i=1 to dead; y=1 ; output; end;
  do i=1 to ali ve; y=1 ; output; end;

proc freq data=new;
  tables group*y /nocol nopercent chisq;
run;

* Producing an over all estimate of ICC;
proc glm data=new;
  class litter;
model y = litter;
run;
```

The SAS output is not shown.

The following R code reads in the grouped data, ungroups it to run ANOVA to get statistics for each group for the ICC, and then converts it back into the grouped data format (shown as a technique only) to compute the Rao-Scott and Donner adjusted chi-square tests.

```
# Reading the grouped Weil (1970) data
weiln <- data.frame (group=c (rep ("c",16),rep ("t",16)),
litter=1:32,
n=c (13,12,9,9,8,8,13,12,10,10,9,13,5,7,10,10,12,11,10,9,11,11,
11,9,9,5,9,7,10,6,10,7),
dead=c (13,12,9,9,8,8,12,11,9,9,8,11,4,5,7,7,12,11,10,9,10,9,9,
8,8,4,7,4, 5,3,3,0))

# Un-grouping the data
y1   <-   data.frame (lapply (weiln,   function (x)   rep (x,weiln
$dead)),
y=1) [,c (1,2,5)]
y0   <-   data.frame (lapply (weiln,   function (x)   rep (x,   weiln
$n-
weiln$dead)),y = 0) [,c (1,2,5)]
weil <- rbind (y1,y0)

# Fitting linear model to get statistics by group for ICC
by (weil, weil[,1], function (x) anova (lm (y ~ factor (litter),
data=x)))

# Convering back to grouped data  shown only as a technique
weiln   <-   data.frame (cbind (aggregate (weil[,3],   weil[,1:2],
length), aggregate (weil[,3], weil[,1:2], sum) [,3]))
```

```
names(weiln) <-c("group","litter","n", "dead")

# Computing Rao-Scott Chi-square test adjusted for clusters (litter)
raoscott(cbind(dead, n - dead) ~ group, data = weiln)
# Computing Donner Chi-square test adjusted for clusters (litter)
donner(cbind(dead, n - dead) ~ group, data = weiln)
```

Here is the R output from the Rao-Scott and Donner methods.

```
Test of proportion homogeneity (Rao and Scott, 1993)
- - - - - - - - - - - - - - - - - - - - - - - - - - - - - - - -
raoscott(formula = cbind(dead, n - dead) ~ group, data = weiln)
N = 32 clusters, n = 305 subjects, y = 254 cases, I = 2 groups.

Data and design effects:
 group N   n   y      p      vbin    vratio  deff
1    c 16 158 142 0.8987 0.000576 0.000710 1.232
2    t 16 147 112 0.7619 0.001234 0.004553 3.689

Adjusted chi-squared test:
X2 = 4.9, df = 1, P(> X2) = 0.0266

Test of proportion homogeneity (Donner, 1989)
- - - - - - - - - - - - - - - - - - - - - - - - - - - - - -
donner(formula = cbind(dead, n - dead) ~ group, data = weiln)
N = 32 clusters, n = 305 subjects, y = 254 cases, I = 2 groups.

Data and correction factors:
 group N   n   y      p     C
1    c 16 158 142 0.8987 3.181
2    t 16 147 112 0.7619 3.000
Intra-cluster correlation (anova estimate): 0.2326

Adjusted chi-squared test:
X2 = 3.3, df = 1, P(> X2) = 0.0685
```

### 4.7.5 Adjusted Chi-Square for Studies Involving More than Two Groups

Suppose that it is of interest to compare $H$ groups and each group has $k_h$ clusters ($h = 1, 2, ..., H$). Denote the number of subjects belonging to cluster $i$ in group $h$ by $m_{hi}$. The null hypothesis of interest may be specified as $H_0$: $P_1=P_2=...P_H=P$, where $P_h$ is the probability that the outcome on a randomly selected subject group $h$ is a success.

Let $y_{hi}$ denote the number of successes for cluster $i$ in group $h$. Then the total number of observations in the $h$th group is given by $M_h = \sum_{i=1}^{k_h} m_{hi}$, and the total number of successes by $y_h = \sum_{i=1}^{k_h} m_{hi}$. We denote the number of clusters in

the study by $k = \sum_{h=1}^{H} k_h$.

Let $\hat{P} = \sum_{h=1}^{H} y_h \Big/ \sum_{h=1}^{H} M_h$ denote the overall success rate in the sample, and
$\hat{Q} = 1 - \hat{P}$. If there is no within-cluster effect, $H_0$ may be tested by the standard
Pearson chi-square statistic with $H - 1$ degrees of freedom. Donner et al. (1994)
showed that this statistic may be written as

$$X^2 = \sum_{h=1}^{H} X_h^2$$

where

$$X_h^2 = \frac{\left(y_h - M_h\hat{P}\right)^2}{M_h\hat{P}} + \frac{\left(M_h - y_h - M_h\hat{Q}\right)^2}{M_h}$$

In the presence of clustering, $X^2$ is no longer valid, but may be adjusted to
account for the within-cluster correlation effect. To construct this adjusted
statistic, say, $X_A^2$, let $\hat{P}_{hi}$ denote the proportion of successes in cluster $i$ in group
$h$ $(i=1,2,...k_h, h=1,2,...H)$; $\hat{P}_h = y_h/M_h$, the overall success rate in group $h$; and
$M = \sum_h \sum_i m_{hi}$, the total number of observations in the sample. Then, the
mean square error among clusters within groups is $MSB = \sum_h \sum_i m_{hi}$
$(\hat{P}_{hi} - \hat{P}_h)^2/(k - H)$, and the mean square error within cluster is $MSW = \sum_h \sum_i y_{hi}(1 - \hat{P}_{hi})/(M - k)$.

The clustering effect is measured by the ICC, which is estimated by

$$\hat{\rho} = \frac{MSB - MSW}{MSB + (m_0 - 1)MSW} \tag{4.25}$$

where

$$m_0 = \left[M - \sum_h \sum_i \frac{m_{hi}^2}{m_h}\right] \Big/ (k - H)$$

According to Donner (1982), the adjusted chi-square statistic may now be
computed as

$$\chi_A^2 = \sum_{h=1}^{H} \left(X_h^2/C_h\right)$$

where

$$C_h = \sum_{i=1}^{kh} m_{hi} C_{hi} \bigg/ \sum_{i=1}^{kh} m_{hi}$$

and

$$C_{hi} = 1 + (m_{hi} - 1)\hat{\rho}$$

An approximate test of $H_0$ is obtained by referring $X_A^2$ to tables of chi-square distribution with $H - 1$ degrees of freedom.

### Example 4.6: A Hypothetical Drug Trial

The data are the result of a drug trial aimed at comparing the effect of four antibiotics against "Shipping Fever" in calves. Calves arrive in trucks at the feedlot and are checked upon arrival for signs of the disease. Animals that are confirmed as cases (from the same truck) are randomized as a group to one of the four treatment regimes. Table 4.20 gives the number of treated animals within a two-week period and the number of deaths at the end of the two weeks.

We would like to test the hypothesis that the mortality rates are the same for all treatments. Let $p_i$ denote the mortality rate of the $i$th treatment. Hence the null hypothesis would be $H_0 : p_1 = p_2 = ...p_I$.
Since

$$\hat{p}_i = \sum_{i=1}^{I} y_i \bigg/ \sum_{i=1}^{I} n_i$$

where $n_1 = 144$, $n_2 = 129$, $n_3 = 130$, $n_4 = 139$, and $\hat{p}(1 - \hat{p}) = 0.1236(0.8764) = 0.1083$, then the uncorrected $\chi^2$ is given by

**TABLE 4.20**

Hypothetical Drug Trial to Compare the Effect of Four Antibiotics on Shipping Fever in Calves

| Drug 1 | Treated animals | 30 | 25 | 25 | 32 | 12 | 10 | 10 |
|--------|-----------------|----|----|----|----|----|----|----|
|        | Deaths          | 3  | 0  | 4  | 10 | 0  | 0  | 1  |
| Drug 2 | Treated animals | 30 | 30 | 15 | 15 | 20 | 19 |    |
|        | Deaths          | 1  | 1  | 1  | 1  | 4  | 0  |    |
| Drug 3 | Treated animals | 30 | 35 | 30 | 10 | 25 |    |    |
|        | Deaths          | 5  | 7  | 9  | 1  | 2  |    |    |
| Drug 4 | Treated animals | 40 | 10 | 25 | 20 | 19 | 25 |    |
|        | Deaths          | 10 | 1  | 1  | 2  | 1  | 2  |    |

$$\chi^2 = 0.002606 + 4.520 + 4.47 + 0.00217 = 8.995$$

Comparing this to a $\chi^2_3$ value of 7.81, the null hypothesis is rejected. The $p$-value is 0.029. This implies that there is a significant difference in the proportions of dead animals for the four different antibiotics.

To estimate the intracluster correlation, the ANOVA results obtained from SAS are summarized in Table 4.21.

$MSB = 0.208$ and $MSW = .103$, $n_0 = 21.934$ and thus $\hat{\rho} = 0.044$.

Since $\bar{n}_{a1} = 24.43$, $\bar{n}_{a2} = 23.34$, $\bar{n}_{a3} = 28.85$, and $\bar{n}_{a4} = 26.708$, then $C_1 = 2.023$, $C_2 = 1.98$, $C_3 = 2.22$, and $C_4 = 2.13$.

To facilitate the computations, we summarize the results in Table 4.22.

Hence $\chi^2_D = 4.29$, which indicates that there is no significant difference between the mortality rates.

**Rao and Scott's Method**

The computations are summarized as follows (also see Table 4.23):

Drug 1: $\hat{P}_i = \dfrac{18}{144} = 0.125$

$v_1 = 0.003$, $d_1 = 3.775$,

$\tilde{Y}_1 = 4.768$, $\tilde{N}_1 = 38.141$

Drug 2: $\hat{P}_2 = \dfrac{8}{129} = 0.062$

**TABLE 4.21**

ANOVA Summary Statistics

| Treatment | SSB | DF | SSW | DF | $N_i$ |
|-----------|-------|----|--------|-----|-----|
| 1 | 1.915 | 6 | 13.835 | 137 | 144 |
| 2 | .504 | 5 | 7.00 | 123 | 129 |
| 3 | .763 | 4 | 18.807 | 125 | 130 |
| 4 | .974 | 5 | 13.947 | 133 | 139 |

**TABLE 4.22**

Detailed Computations

| Treatment | $C_i$ | $\hat{P}_i$ | $N_i$ | Components of $\chi^2_A$ |
|-----------|-------|-------------|-------|--------------------------|
| 1 | 2.023 | 0.13 | 144 | 0.0013 |
| 2 | 1.98 | 0.06 | 129 | 2.28 |
| 3 | 2.22 | 0.18 | 130 | 2.008 |
| 4 | 2.13 | 0.12 | 139 | 0.001 |

**TABLE 4.23**

Summary Calculations for RS

| Cluster | $y_{1j}$ | $n_{1j}$ | $(y_{1j} - n_{1j}\hat{P}_i)$ |
|---------|----------|----------|------------------------------|
| 1 | 3 | 30 | 0.563 |
| 2 | 0 | 25 | 9.766 |
| 3 | 4 | 25 | 0.766 |
| 4 | 10 | 32 | 36.00 |
| 5 | 0 | 12 | 2.25 |
| 6 | 0 | 10 | 1.563 |
| 7 | 1 | 10 | 0.063 |
| Total | 18 | 144 | 50.969 |

| Cluster | $y_{2j}$ | $n_{2j}$ | $(y_{2j} - n_{2j}\hat{P}_i)$ |
|---------|----------|----------|------------------------------|
| 1 | 1 | 30 | 0.74 |
| 2 | 1 | 30 | 0.74 |
| 3 | 1 | 15 | 0.005 |
| 4 | 1 | 15 | 0.005 |
| 5 | 4 | 20 | 7.616 |
| 6 | 0 | 19 | 1.388 |
| Total | 8 | 129 | 10.495 |

$v_2 = 0.0008$, $d_2 = 1.678$, $\tilde{Y}_2 = 4.767$, $\tilde{N}_2 = 76.863$

Drug 3: $\hat{P}_3 = \dfrac{24}{130} = 0.184$

| Cluster | $y_{3j}$ | $n_{3j}$ | $(y_{3j} - n_{3j}\hat{P}_i)$ |
|---------|----------|----------|------------------------------|
| 1 | 5 | 30 | 0.27 |
| 2 | 7 | 35 | 0.31 |
| 3 | 9 | 30 | 12.11 |
| 4 | 1 | 10 | 0.716 |
| 5 | 2 | 25 | 6.76 |
| Total | 24 | 130 | 20.17 |

$v_3 = 0.0015$, $d_3 = 1.285$, $\tilde{Y}_3 = 18.676$, $\tilde{N}_3 = 101.161$

Drug 4: $\hat{P}_4 = \dfrac{17}{139} = 0.122$

| Cluster | $y_{4j}$ | $n_{4j}$ | $(y_{4j} - n_{4j}\hat{P}_i)$ |
|---------|----------|----------|------------------------------|
| 1 | 10 | 40 | 26.09 |
| 2 | 7 | 35 | 0.05 |
| 3 | 9 | 30 | 4.234 |
| 4 | 1 | 10 | 0.198 |
| 5 | 2 | 25 | 1.752 |
| 6 | 2 | 25 | 1.118 |
| Total | 17 | 139 | 33.442 |

$v_4 = 0.0021$, $d_4 = 2.69$, $\tilde{Y}_4 = 6.32$, $\tilde{N}_4 = 51.68$

Hence, $\chi^2_{RS} = 5.883$, which again does not exceed the tabulated $\chi^2_{3,0.05} = 7.81$, so we have no reason to reject the null hypothesis. This implies that there is no significant difference between the mortality rates.

From this example we can see the importance of accounting for intracluster correlations as the resulting conclusion about the equality of the proportions is different when the correlation is taken into account.

The following SAS program reads the grouped data, generates the ungrouped data, and runs the linear models to compute statistics for the ICC for each drug level.

```
* Grouped Data;
data drug;
  input drug truck n dead @@;
  alive = n-dead;
  cards;
1  1 30  3 1   2 25  0 1   3 25   4 1   4 32 10
1  5 12  0 1   6 10  0 1   7 10   1 2   8 30   1
2  9 30  1 2 10 15  1 2 11 15   1 2 12 20   4
2 13 19  0 3 14 30  5 3 15 35   7 3 16 30   9
3 17 10  1 3 18 25  2 4 19 40 10 4 20 10   1
4 21 25  1 4 22 20  2 4 23 19   1 4 24 25   2
;

* Un-grouped Data;
data new; set drug;
  do i=1 to dead; y=1 ; output; end;
  do i=1 to alive; y=1 ; output; end;

* Producing statistics to compute ICC for each drug level;
proc glm data=new;
  class truck;
model y = truck;
by drug;
run; quit;
```

The SAS output is not shown.

The following R code reads in the grouped data, ungroups it to run ANOVA to get statistics for each drug level for the ICC, and then uses the grouped data to compute the Rao-Scott and Donner adjusted chi-square tests.

```
# Reading the grouped 'drug' data
drugn <- data.frame(drug=c(rep(1,7),rep(2,6),rep(3,5),rep(4,6)),
truck=1:24,
n=c(30,25,25,32,12,10,10,30,30,15,15,20,19,30,35,30,10,25,40,
   10,25,20,19,25),
dead=c(3,0,4,10,0,0,1,1,1,1,1,4,0,5,7,9,1,2,10,1,1,2,1,2))

# Un-grouping the data
y1 <- data.frame(lapply(drugn, function(x) rep(x,drugn$dead)),
y=1)[,c(1,2,5)]
y0 <- data.frame(lapply(drugn, function(x) rep(x, drugn$n-
drugn$dead)),y = 0)[,c(1,2,5)]
drug <- rbind(y1,y0)

# Fiting linear model to get statistics by group for ICC
by(drug, drug[,1], function(x) anova(lm(y ~ factor(truck), data=x)))

# Computing Rao-Scott Chi-square test adjusted for clusters (litter)
raoscott(cbind(dead, n - dead) ~ drug, data = drugn)
# Computing Donner Chi-square test adjusted for clusters (litter)
donner(cbind(dead, n - dead) ~ drug, data = drugn)
```

Here is the output from the Rao-Scott and Donner methods.

```
Test of proportion homogeneity (Rao and Scott, 1993)
- - - - - - - - - - - - - - - - - - - - - - - - - - - - - - -
raoscott(formula = cbind(dead, n - dead) ~ drug, data = drugn)
N = 24 clusters, n = 542 subjects, y = 67 cases, I = 4 groups.
```

```
Data and design effects:
 drug N   n  y      p      vbin      vratio  deff
1    1 7 144 18 0.12500 0.0007595 0.0028676 3.775
2    2 6 129  8 0.06202 0.0004509 0.0007568 1.678
3    3 5 130 24 0.18462 0.0011579 0.0014880 1.285
4    4 6 139 17 0.12230 0.0007723 0.0020771 2.690
```

```
Adjusted chi-squared test:
X2 = 5.9, df = 3, P(> X2) = 0.1174
Test of proportion homogeneity (Donner, 1989)
- - - - - - - - - - - - - - - - - - - - - - - - - - -
donner(formula = cbind(dead, n - dead) ~ drug, data = drugn)
N = 24 clusters, n = 542 subjects, y = 67 cases, I = 4 groups.
```

```
Data and correction factors:
 drug  N    n   y        p     C
1     1  7  144  18  0.12500  2.030
2     2  6  129   8  0.06202  1.982
3     3  5  130  24  0.18462  2.224
4   4 6 139 17 0.12230 2.129
```

```
Intra-cluster correlation (anova estimate): 0.0439
```

```
Adjusted chi-squared test:
X2 = 4.3, df = 3, P(> X2) = 0.2318
```

## 4.8 Inference on the Common Odds Ratio

Consider a series of $2 \times 2$ tables wherein the notation has been changed to accommodate multiple tables, groups, and clusters. Let $n_{ijt}$ be the size of the $i$th cluster in the $j$th group ($j = 1$ for *exposed* and $j = 2$ for *unexposed*) from the $i$th table, and $k_{ij}$ is the number of cluster in the $j$th group from the $i$th table. To further clarify this altered notation, Table 4.24 describes the data layout for the $i$th stratum.

The summed values are shown in Table 4.25.

$$\text{where } y_{ij} = \sum_{t=1}^{k_{ij}} y_{ijt} \text{ and } n_{ij} = \sum_{t=1}^{k_{ij}} n_{ijt} \ (j = 1, 2; i = 1, 2,..., k).$$

Now, because the sampling units are clusters of individuals, the $\chi^2_{mh}$ statistic used to test $H_0 : \psi = 1$ would not be appropriate. To adjust this statistic for the clustering effect we introduce two procedures by Donald and Donner (1987) and by Rao and Scott (1992).

### 4.8.1 Donald and Donner's Adjustment

Because we have $2k$ rows in Table 2.24, an intraclass correlation $\rho$ is first estimated from each row. Let $\hat{\rho}_{ij}$ be the estimate of $\rho_{ij}$ from the $j$th row in the

**TABLE 4.24**

Data Layout for Stratified Analysis

|  |  | Cluster |  |  |  |
|---|---|---|---|---|---|
| **Stratum** |  | 1 | 2 | ... | $k_{ij}$ |
| Exp. $(E^+)j = 1$ | Cluster size | $n_{i11}$ | $n_{i12}$ | ... | $n_{i1kij}$ |
|  | Number of deaths | $y_{i11}$ | $y_{i12}$ | ... | $y_{i1kij}$ |
| Unexp. $(E^-)$ | Cluster size | $n_{i21}$ | $n_{i22}$ | ... | $n_{i2kij}$ |
| $j = 2$ | Number of deaths | $y_{i21}$ | $y_{i22}$ | ... | $y_{i2kij}$ |

**TABLE 4.25**

Summed Values of the Response Variables

|           |         | $D^+$            | $D^-$                  |          |
|-----------|---------|------------------|------------------------|----------|
| $j = 1$   | $E^+$   | $y_{i1}$         | $y_{i1} - n_{i1}$      | $n_{i1}$ |
| $j = 2$   | $E^-$   | $y_{i2}$         | $y_{i2} - n_{i2}$      | $n_{i2}$ |

$i$th table. Under a common correlation assumption, it is reasonable to construct an overall estimate of $\rho$ as

$$\hat{\rho}_A = \frac{1}{2k} \sum_{i=1}^{k} \sum_{j=1}^{2} \hat{\rho}_{ij}$$

Let $\hat{D}_{ijt} = 1 + (n_{ijt} - 1)\hat{\rho}_A$ be the correction factor for each cluster, and let $B_{ij}$ be the weighted average of such correction factors where the weights are the cluster sizes themselves. Therefore, we write

$$B_{ij} = \frac{\sum_{t=1}^{k_{ij}} n_{ijt} \hat{D}_{ijt}}{\sum_{t=1}^{k_{ij}} n_{ijt}}$$

Donald and Donner (1987) suggested that the Mantel Haenszel test statistic on $H_0 : \psi = 1$, adjusted for the clustering effect, is given by

$$\chi_{mhc}^2 = \frac{\left( \sum_{i=1}^{k} \dfrac{y_{i1}(n_{i2} - y_{i2}) - y_{i2}(n_{i1} - y_{i1})}{n_{i1}B_{i1} + n_{i2}B_{i2}} \right)^2}{\sum_{i=1}^{k} \dfrac{n_{i1}n_{i2}y_i(n_i - y_i)}{(n_{i1}B_{i1} + n_{i2}B_{i2})n_i^2}}$$

Since cluster sampling affects the variance of the estimated parameters, the variance $\hat{V}_{mh}$ of $\hat{\psi}_{mh}$ is no longer valid. Donald and Donner defined the cluster variant $\hat{b}_i^*$ of the $b_i$ contained in Hauck's formula:

$$\hat{b}_i^* = \frac{B_{i1}}{n_{i1}\hat{P}_{i1}\hat{q}_{i1}} + \frac{B_{i2}}{n_{i2}\hat{P}_{i2}\hat{q}_{i2}}$$

The corrected variance $\hat{V}_{mhc}$ is

$$\hat{V}_{mhc} = \frac{\hat{\psi}_{mh}^2 \sum_{i=1}^{k} \hat{w}_i^2 \hat{b}_i^*}{\left( \sum_{i=1}^{k} \hat{w}_i \right)^2}$$

Hence an $(1 - \alpha)100\%$ confidence limits on $\psi_{mh}$ after adjusting for clustering is $\hat{\psi}_{mh} \pm Z_{\alpha/2}\sqrt{\hat{V}_{mhc}}$.

### 4.8.2 Rao and Scott's Adjustment

This adjustment requires the computation of the variance inflation factors $d_{ij}$ using the cluster-level data $(y_{ijt}, n_{ijt})$, where $t = 1, 2, \ldots, k_{ij}$. The inflation factor is computed in two steps. First,

$$v_{ij} = \frac{k_{ij}}{(k_{ij} - 1)n_{ij}^2} \sum_{t=1}^{k_{ij}} \left( y_{ijt} - n_{ijt}\hat{p}_{ij} \right)^2$$

where

$$\hat{p}_{ij} = \frac{y_{ij}}{n_{ij}}$$

and then

$$d_{ij} = \frac{n_{ij} v_{ij}}{\hat{p}_{ij}\hat{q}_{ij}}$$

An asymptotically (that is for large $k_{ij}$ for each $i$ and $j$) valid test of $H_0 : \psi = 1$ is obtained by replacing $(y_{ij}, n_{ij})$ with $(_{ij}, _{ij})$, where $_{ij} = y_{ij}/d_{ij}$ and $_{ij} = n_{ij}/d_{ij}$.

To construct an asymptotic confidence interval on $\psi_{mh}$, $(y_{ij}, n_{ij})$ is replaced with $(_{ij}, _{ij})$ in $\hat{\psi}_{mh}$ to get

$$\tilde{\psi}_{mhc} = \frac{\displaystyle\sum_{i=1}^{k} \frac{\tilde{y}_{i1}(\tilde{n}_{i2} - \tilde{y}_{i2})}{(\tilde{n}_{i1} + \tilde{n}_{i2})}}{\displaystyle\sum_{i=1}^{k} \frac{\tilde{y}_{i2}(\tilde{n}_{i1} - \tilde{y}_{i1})}{(\tilde{n}_{i1} + \tilde{n}_{i2})}}$$

as the clustered adjusted point-estimator of $\psi_{mh}$. Similarly Hauck's variance estimator for $\tilde{\psi}_{mhc}$ becomes

$$\tilde{V}_{mhc} = \frac{\tilde{\psi}_{mhc}^2 \left( \displaystyle\sum_{i=1}^{k} \tilde{w}_i^2 \tilde{b}_i \right)}{\left( \displaystyle\sum_{i=1}^{k} \tilde{w}_i \right)^2}$$

where

$$\tilde{w}_i = \left( \tilde{n}_{i1}^{-1} + \tilde{n}_{i2}^{-1} \right)^{-1} \tilde{p}_{i2}(1 - \tilde{p}_{i1}), \quad \tilde{b}_i = \frac{d_{i1}}{n_{i1}\tilde{p}_{i1}(1 - \tilde{p}_{i1})} + \frac{d_{i2}}{n_{i2}\tilde{p}_{i2}(1 - \tilde{p}_{i2})}$$

and

$$\tilde{P}_{ij} = \frac{\tilde{y}_{ij}}{\tilde{n}_{ij}} \, j = 1, 2$$

**Example 4.7**

The data in Table 4.26 (summarized in Table 4.27) is the result of a case-control study to investigate the association between bovine leukemia virus (BLV) infection and bovine immunodeficiency virus (BIV) infection. Each BLV$^+$ cow was matched on sex and age (within two years) with a BLV$^-$ cow from a different herd. The pedigree relatives of a BLV$^+$ cow constituted clusters of BIV$^+$ or BIV$^-$ while the pedigree relatives of a BLV$^-$ cow constituted clusters of BIV$^+$ or BIV$^-$.

**TABLE 4.26**

Case Control Study Results for Investigation of Association between Bovine Leukemia Virus Infection and Bovine Immunodeficiency Infection

| Region 1 | | | | Region 2 | | | |
|---|---|---|---|---|---|---|---|
| BLV$^+$ Cows | | Controls | | BLV$^+$ Cows | | Controls | |
| BIV$^+$ | BIV$^-$ | BIV$^+$ | BIV$^-$ | BIV$^+$ | BIV$^-$ | BIV$^+$ | BIV$^-$ |
| 1 | 4 | 0 | 4 | 7 | 1 | 0 | 2 |
| 1 | 5 | 0 | 4 | 6 | 1 | 0 | 0 |
| 1 | 2 | 0 | 3 | 0 | 0 | 1 | 6 |
| 2 | 4 | 0 | 2 | 1 | 1 | 0 | 6 |
| 0 | 1 | 0 | 4 | 0 | 3 | 0 | 2 |
| 2 | 0 | 0 | 7 | 0 | 1 | 1 | 0 |
| 1 | 1 | 0 | 3 | 1 | 1 | 1 | 0 |
| 1 | 2 | 1 | 1 | | | | |
| 2 | 1 | 2 | 5 | | | | |
| 3 | 0 | 0 | 3 | | | | |
| 1 | 1 | 1 | 2 | | | | |
| 2 | 4 | 0 | 6 | | | | |
| 1 | 1 | 0 | 4 | | | | |

**TABLE 4.27**

Collapsed Data

| | Region 1 | | Region 2 | |
|---|---|---|---|---|
| | BIV$^+$ | BIV$^-$ | BIV$^+$ | BIV$^-$ |
| BLV$^+$ | 18 | 26 | 15 | 8 |
| BLV$^-$ | 4 | 48 | 3 | 16 |

A region-stratified (unmatched) analysis is conducted to test the above hypothesis using

a. The Mantel-Haenszel one-degree-of-freedom chi-square test on the significance of the common odds ratio and a 95% confidence interval using Hauck's variance formula
b. The adjusted chi-square test and the variance expression for clustering using Donald and Donner's procedure and Rao and Scott's procedure.

The MH common odds ratio is given by

$$\hat{\psi}_{MH} = \frac{\dfrac{(18)(48)}{96} + \dfrac{(15)(16)}{42}}{\dfrac{(4)(26)}{96} + \dfrac{(3)(8)}{42}} = 8.89$$

and Hauck's variance is 17.85.

Now, the one degree of freedom chi-squared test of the common odds ratio will be shown using three methods:

a. Uncorrected
b. Donald and Donner's adjustment
c. Rao and Scott's adjustment

**Uncorrected Chi-Squared Test of Common Odds Ratio**
$H_0$: No association between BIV and BLV status in pedigree.

$$\chi^2 = \frac{\left(|33 - 19.94| - \frac{1}{2}\right)^2}{6.86} = 24.863$$

Thus, since 24.863 exceeds the $\chi^2_{(1,0.05)}$ value of 3.84, the null hypothesis is rejected implying that there is a strong association between BIV and BLV status.

**Donald and Donner's Adjustment**
$H_0$: No association between BIV and BLV status in pedigree.
The $\chi^2_{mhc}$ for Donald and Donner is

$$\chi^2_{mhc} = \frac{\left(\displaystyle\sum_{i=1}^{k} \frac{y_{i1}(n_{i2} - y_{i2}) - y_{i2}(n_{i1} - y_{i1})}{n_{i1}B_{i1} + n_{i2}B_{i2}}\right)^2}{\displaystyle\sum_{i=1}^{k} \frac{n_{i1}n_{i2}y_i(n_i - y_i)}{(n_{i1}B_{i1} + n_{i2}B_{i2})n_i^2}}$$

The estimated common intraclass correlation is given by

$$\hat{\rho}_A = \frac{1}{2k}\sum\sum \hat{\rho}_{ij} = \frac{1}{2(2)}(-0.0186 + 0.08655 + 0.5920 + 0.3786) = 0.26$$

Now, the correction factor for each cluster is

$$\hat{D}_{ijt} = 1 + (n_{ijt} - 1)\hat{\rho}_A$$

and the weighted average of the correction factors is

$$B_{ij} = \frac{\sum n_{ijt} D_{ij}}{\sum n_{ijt}}$$

So that

$$B_{11} = \text{Region 1 BLV}^- = 1.870026 \qquad B_{12} = \text{Region 1 BLV}^+ = 1.768454$$

$$B_{21} = \text{Region 2 BLV}^- = 1.952450 \qquad B_{22} = \text{Region 2 BLV}^+ = 2.047695$$

and the $\chi^2_{mhc}$ using Donald and Donner's adjustment is

$$\chi^2_{mhc} = \left\{ \frac{\dfrac{18(52-4)-4(44-18)}{44(1.7685)+52(1.87)} + \dfrac{15(19-3)-3(23-15)}{23(2.048)+19(1.952)}}{\dfrac{52(44)(22)(96-22)}{(96)(96)(44(1.7685)+52(1.87))} + \dfrac{23(19)(18)(42-18)}{(42)(42)(23(2.048)+19(1.952))}} \right\}^2$$

$$= 13.33$$

Because 13.33 is larger than the $\chi^2_{(1,0.05)}$ value of 3.84 the null hypothesis is rejected implying that when we adjust for the intracluster correlation, there is a significant association between the BLV and the BIV status.

The value of the variance is now

$$\hat{V}_{mhc} = \frac{\hat{\psi}^2_{mh}\displaystyle\sum_{i=1}^{k}\hat{w}_i^2 \hat{b}_i^*}{\left(\displaystyle\sum_{i=1}^{k}\hat{w}_i\right)^2}$$

where

$$\hat{b}_i^* = \frac{B_{i1}}{n_{i1}\hat{P}_{i1}\hat{q}_{i1}} + \frac{B_{i2}}{n_{i2}\hat{P}_{i2}\hat{q}_{i2}},$$

$$\hat{w}_i = \left(\frac{1}{n_{i1}} + \frac{1}{n_{12}}\right)^{-1}\hat{P}_{i2}q_{i1}$$

and $\hat{q}_{ij} = 1 - p_{ij}$, $j = 1, 2$.

The results of these computations show that $\hat{w}_1 = 1.085, \hat{w}_2 = 0.572$, $\hat{b}_1^* = 0.673$, and $\hat{b}_2^* = 1.165$. So the variance is equal to 33.791.

### Rao and Scott's Adjustment

The null hypothesis of no association between BIV and BLV status is tested using the following steps. The chi-square is adjusted by the variance inflation factor $d_{ij}$. First, we compute

$$v_{ij} = \frac{k_{ij}}{(k_{ij} - 1)n_{ij}^2} \sum_{t=1}^{k_{ij}} \left(y_{ijt} - n_{ijt}\hat{p}_{ij}\right)^2$$

where $\hat{p}_{ij} = \dfrac{y_{ij}}{n_{ij}}$ The inflation factor $d_{ij}$ is calculated using

$$d_{ij} = \frac{n_{ij}v_{ij}}{\hat{p}_{ij}\hat{q}_{ij}}, \quad \tilde{y}_{ij} = \frac{y_{ij}}{d_{ij}}, \quad \tilde{n}_{ij} = \frac{n_{ij}}{d_{ij}}$$

Here $d_{11} = 0.93$, $d_{12} = 1.32$, $d_{21} = 2.23$, and $d_{22} = 1.70$. Calculation of the adjusted chi-square, $\chi_{mhc}^2$, is therefore given by $\chi_{hmc}^2 = 18.69$.

Once again we reject the null hypothesis, which means that there is a significant association between the BLV and the BIV status. With the Rao and Scott adjustment, the variance estimate is now

$$\tilde{V}_{mhc} = \frac{\tilde{\psi}_{mhc}^2 \left(\displaystyle\sum_{i=1}^{k} \tilde{w}_i^2 \hat{b}_i\right)}{\left(\displaystyle\sum_{i=1}^{k} \tilde{w}_i\right)^2} = 29.851$$

The following SAS program sets up the data, computes the Mantel-Haenszel common odds ratio, and tests for the equality of the odds ratio across the strata. The SAS PROC GLM produces the ANOVA results for four combinations of Region and Bovine Leukemia Virus status from where the overall common intracluster correlation and adjusted tests of the significance of the common odds ratio may be computed. The SAS output is not shown.

```
data virus;
input region pedigree BLV$ BIV count @@;
cards;
1 1 y 1 1 1 1 y 0 4 1 2 y 1 1 1 2 y 0 5
1 3 y 1 1 1 3 y 0 2 1 4 y 1 2 1 4 y 0 4
1 5 y 1 0 1 5 y 0 1 1 6 y 1 2 1 6 y 0 0
1 7 y 1 1 1 7 y 0 1 1 8 y 1 1 1 8 y 0 2
1 9 y 1 2 1 9 y 0 1 1 10 y 1 3 1 10 y 0 0
1 11 y 1 1 1 11 y 0 1 1 12 y 1 2 1 12 y 0 4
1 13 y 1 1 1 13 y 0 1 1 14 n 1 0 1 14 n 0 4
1 15 n 1 0 1 15 n 0 4 1 16 n 1 0 1 16 n 0 3
1 17 n 1 0 1 17 n 0 2 1 18 n 1 0 1 18 n 0 4
```

```
1 19 n 1 0 1 19 n 0 7 1 20 n 1 0 1 20 n 0 3
1 21 n 1 1 1 21 n 0 1 1 22 n 1 2 1 22 n 0 5
1 23 n 1 0 1 23 n 0 3 1 24 n 1 1 1 24 n 0 2
1 25 n 1 0 1 25 n 0 6 1 26 n 1 0 1 26 n 0 4
2 27 y 1 7 2 27 y 0 1 2 28 y 1 6 2 28 y 0 1
2 29 y 1 0 2 29 y 0 0 2 30 y 1 1 2 30 y 0 1
2 31 y 1 0 2 31 y 0 3 2 32 y 1 0 2 32 y 0 1
2 33 y 1 1 2 33 y 0 1 2 34 n 1 0 2 34 n 0 2
2 35 n 1 0 2 35 n 0 0 2 36 n 1 1 2 36 n 0 6
2 37 n 1 0 2 37 n 0 6 2 38 n 1 0 2 38 n 0 2
2 39 n 1 1 2 39 n 0 0 2 40 n 1 1 2 40 n 0 0
;

proc format;
value $blvf y='BLV+'
           n='BLV-';
value bivf 1='BIV+'
           0='BIV-';

* Odds ratios of the two regions and the common odds ratio;
proc freq data=virus order=formatted;
     weight count;
     tables region*blv*biv / norow nocol nopercent measures cmh
noprint;
format blv $blvf. biv bivf.; run;

* Computing the intraclass correlations;
proc sort data=virus; by region blv;

proc glm; by region blv;
   class pedigree;
   freq count;
 model biv = pedigree;
run; quit;
```

The following R code reads the data in grouped form, ungroups it to run ANOVA to get statistics for each drug level to compute the overall ICC, and then uses the grouped data to compute the Rao-Scott and Donner adjusted chi-square tests.

```
# Reading the virus data in the grouped (y/n) form
virusn <- data.frame(region=c(rep(1,26),rep(2,12)),
blv=c(rep("y",13),rep("n",13),rep("y",6),rep("n",6)),
pedigree=1:38,
```

```
n=c(5,6,3,6,1,2,2,3,3,3,2,6,2,4,4,3,2,4,7,3,2,7,3,3,6,4,8,7,2,
3,1,2,2,7,6,2,1,1),
biv =c(1,1,1,2,0,2,1,1,2,3,1,2,1,0,0,0,0,0,0,0,1,2,0,1,0,0,7,6,
1,0,0,1,0,1,0,0,1,1))
# Un-grouping the data
y1 <- data.frame(lapply(virusn, function(x) rep(x,virusn$biv)),
 y=1)[,c(1:3,6)]
y0 <- data.frame(lapply(virusn, function(x) rep(x, virusn$n-
virusn$biv)),y = 0)[,c(1:3,6)]
virus <- rbind(y1,y0)

# Fitting linear models by clusters (pedigree) to get statistics for
 ICC
by(virus, virus[,c(1,2)], function(x) anova(lm(y ~ factor(pedi-
gree), data=x)))
```

## R Output

---

**region: 1**
**blv: n**

**Analysis of Variance Table**

**Response: y**

|                  | Df | Sum Sq | Mean Sq | F value | Pr(>F) |
|------------------|----|--------|---------|---------|--------|
| factor(pedigree) | 12 | 1.0971 | 0.091   | 1.374   | 0.219  |
| Resudials        | 39 | 2.5952 | 0.066   |         |        |

---

**region: 2**
**blv: n**

**Analysis of Variance Table**

**Response: y**

|                  | Df | Sum Sq | Mean Sq | F value | Pr(>F) |
|------------------|----|--------|---------|---------|--------|
| factor(pedigree) | 5  | 1.669  | 0.334   | 5.063   | 0.0085 |
| Residuals        | 13 | 0.857  | 0.066   |         |        |

---

**region: 1**
**blv: y**

**Analysis of Variance Table**

**Response: y**

|                  | Df | Sum Sq | Mean Sq | F value | Pr(>F) |
|------------------|----|--------|---------|---------|--------|
| factor(pedigree) | 12 | 2.836  | 0.236   | 0.939   | 0.522  |
| Residuals        | 31 | 7.800  | 0.25161 |         |        |

---

region: 2
blv: y

**Analysis of Variance Table**

Response: y

|                 | Df       | Sum Sq | Mean Sq | F value | Pr(>F) |
|-----------------|----------|--------|---------|---------|--------|
| factor(pedigree) | 5        | 2.485  | 0.497   | 3.093   | 0.036  |
| Residuals       | 17 2.732 | 0.160  |         |         |        |

## 4.9 Calculations of Relative and Attributable Risks from Clustered Binary Data

The concept of AR introduced by Levin (1953) is a widely used measure of the amount of disease that can be attributed to a specific risk factor. The AR combines the RR and the prevalence of exposure, P(E), to measure the public health burden of a risk factor by estimating the proportion of cases of a disease that would not have occurred if we remove the risk factor. The concept of AR and its statistical characteristics have been reviewed in Fleiss (1981). Statistical inferences on AR require the availability of data from subjects randomly assigned to intervention groups.

Levin (1953) defined the AR as "the amount of disease that can be attributed to a specific risk factor." From Fleiss (1981) the AR estimator may be written as

$$\widehat{AR} = \frac{Y_1(N_2 - Y_2) - Y_2(N_1 - Y_1)}{(Y_1 + Y_2)N_2}$$

The asymptotic variance of $\hat{\theta} = ln(1 - \widehat{AR}\ )$ is shown to be

$$\text{Var}\left(\hat{\theta}\right) = \frac{N_1(1 - P_e)C_1}{P_e(N_1 + N_2\Psi)^2} + \frac{N_1^2(1 - P_e\Psi)C_2}{N_2 P_e\Psi(N_1 + N_2\Psi)^2}$$

Here $P_e = \dfrac{N_1}{N_1 + N_2}$, and $\Psi = (1 - \widehat{AR})/(1 + \widehat{AR})$.

A consistent estimator of Var($\hat{\theta}$) may be obtained on replacing the parameters by their moment estimators. A $(1 - \alpha)100\%$ confidence interval on AR is thus given as

$$\left(1 - \exp\left(\hat{\theta} + z_{\alpha/2}\sqrt{\text{var}}\left(\hat{\theta}\right)\right),\quad 1 - \exp\left(\hat{\theta} - z_{\alpha/2}\sqrt{\text{var}}\left(\hat{\theta}\right)\right)\right)$$

For Weil's data given in Table 4.19, we can show that the point estimate of AR is

$$AR = \frac{(35)(142) - (16)(112)}{(51)(158)} = 39.4\%$$

Direct application gives, $se(\hat{\theta}) = 0.0555$, and the 95% confidence interval on AR is $0.33 < AR < 0.45$.

We shall leave it as an exercise to calculate the variance of the log-relative risk under cluster randomization.

## 4.10 Sample Size Requirements for Clustered Binary Data

### 4.10.1 Paired-Sample Design

Determination of the sample size in a study that aims at testing for difference in proportions is an important issue. For the matched-pair design where the McNemar test is used, Miettinen (1968) and Connor (1987) obtained several approximations to sample size. Specifically, the basic model used by both authors is represented in Table 4.28.

Here $P_{ij}$ is the probability that a matched pair has response $Y_1 = i$ and $Y_2 = j$, $(i, j = 0, 1)$. Also $P_1 = P_{11} + P_{10}$, $P_2 = P_{11} + P_{01}$, and $\eta = P_{10} + P_{01}$ is the probability that the response of a case and its matched control disagree. Formally, we need to determine the number of pairs $n$ necessary to have power $1 - \beta$ of detecting a difference $\delta = P_1 - P_2 > 0$ when a one-sided test of size $\alpha$ is to be used. By an argument similar to that of Snedecor and Cochran (1980), Connor showed that

$$n = \frac{\left[ Z_\alpha n^{1/2} + Z_\beta (\eta - \delta^2)^{1/2} \right]^2}{\delta^2}$$

where $Z_\alpha$ and $Z_\beta$ are the percentiles of the standard normal distribution as defined in Chapter 1.

**TABLE 4.28**

Paired Dichotomous Data

| | | $Y_2$ | | |
|---|---|---|---|---|
| | | 1 | 0 | |
| $Y_1$ | 1 | $P_{11}$ | $P_{10}$ | $P_1$ |
| | 0 | $P_{01}$ | $P_{00}$ | $1-P_1$ |
| | | $P_2$ | $1-P_2$ | 1 |

Note that the preceding equation requires $\eta$ to be known for the sample size to be determined. Since one can write $\eta = P_{10} + P_{01} = P_1 + P_2 - 2P_2 Pr(Y_1 = 1 \mid Y_2 = 1)$, then $\eta$ is minimum when the conditional probability $P_{1\mid 1} = Pr(Y_1 = 1 \mid Y_2 = 1)$ is maximum, and $\eta$ is a maximum when $P_{1\mid 1}$ is minimum. Because $P_{1\mid 1} = P_{11}/P_2$, it is maximized or minimized when $P_{11}$ is maximized or minimized. Clearly the issue of providing a plausible value for $\eta$ is quite complicated. However, Connor (1987) argued that sample size formula for the independent outcome situation provides a conservative estimate of the sample size. This is given by

$$n' = (Z_\alpha \sigma_0 + Z_\beta \sigma_A)^2 / \delta^2$$

where

$$\sigma_0^2 = 2P_1(1 - P_1)$$

and

$$\sigma_A^2 = P_1(1 - P_1) + (P_1 - \delta)(1 - P_1 + \delta)$$

### 4.10.2 Comparative Studies for Cluster Sizes Greater or Equal to Two

Recall that for continuous responses, the introduction of the parameter $\rho$ allows the formalization of the within-cluster dependence. In the case of a dichotomous response, an analogue to $\rho$ has been developed, as seen in the previous sections. Donner et al. (1981) developed a sample size expression required for testing, $H_0 : P_1 = P_2$, where $P_1$ and $P_2$ are the underlying success probabilities, given by

$$N' = (Z_a + Z_\beta)^2 \sigma_d^2 (1 + (n - 1)\rho) / \delta^2$$

where $n$ is the cluster size, $N' = nk$, and $\sigma_d^2 = P_1(1 - P_1) + P_2(1 - P_2)$.
    In case exact cluster size $n$ is not known, it should be replaced by the largest expected cluster size.

#### Example 4.8

Consider a study in which the outcome of interest is hypertension status and the unit of randomization is a spouse pair. Previous studies estimated the within-pair correlation by $\hat{\rho} = 0.25$. Suppose the epidemiologist would like a 90% probability at $\alpha = .05$ (two-tailed) of detecting a significant difference between two ethnic groups (e.g., whites versus blacks), where it is believed that the percentage of hypertension among whites is 15% and 20% among blacks. Using the expression given earlier we have

$$N' = \frac{(1.96 + 1.28)^2[(0.15)(0.85) + (0.20)(0.80)][1 + 0.25]}{(0.05)^2}$$

$$= 1509$$

Thus the study must be designed to allocate $1509/2 \approx 755$ couples from each group. If no clustering effect were present ($\rho = 0$), $N' = 1207$, and hence 604 would be required in each group to provide the same statistical power.

## 4.11 Discussion

An extensive literature has been developed on the methodological issues that arise when the risk of disease is compared across treatment groups when the sampling units are clusters of individuals. The presence of so-called cluster effects invalidates the standard Pearson's chi-square test of homogeneity and makes the reported confidence interval on the common odds ratio unrealistically narrow. In the previous section we introduced the most commonly used techniques of adjusting clustering effects. It is debatable as to which technique is most appropriate. Extensive simulation studies conducted by Ahn and Maryon (1995) and Donner et al. (1994) showed that each approach has its advantages. Ahn and Maryon preferred Rao and Scott's adjustment on the length of the confidence interval of the Mantel-Haenszel common odds ratio. The reason was that the ANOVA estimator of the intraclass correlation $\rho$, needed for Donner's adjustment, is positively biased, and hence the confidence interval would be spuriously wide when $\rho = 0$. On the other hand, Donner's et al. (1994) simulations showed that the adjustment to Pearson's chi-square test of homogeneity based on the pooled estimate of the intraclass correlations performs better than methods that separately estimate design effects from each group. The reason was that under randomization, the assumption of common population design effect is more likely to hold, at least under the null hypothesis. They also stressed the need for more research in this area to investigate the finite-sample properties of methods that do not require the assumption of a common design effect. The A-Cluster 2.0 software (http://www.who.int/reproductive-health/hrp/research/acluster.en.html) by Pinol and Piaggio (2000) implements the analysis of cluster randomized trials with continuous as well as the binary outcomes discussed in this chapter. The program incorporates the analysis of data and the estimation of sample size for parallel group, matched pair, and the stratified designs using Donald and Donner (1987) and Rao and Scott (1992) adjustments for clustering to the usual chi-square test statistics.

## Exercises

4.1. The following table summarizes the result of a case-control study. The main objective of the study was to investigate risk factors associated with the incidence of tick-borne diseases in "small-holder" dairy farms in one of the northern districts near Nairobi, Kenya. A case farm is a farm where animals are sprayed regularly, while a control farm is a farm where animals are not sprayed at all. The following table gives the distribution of the number of farms classified by size (i.e., number of animals on the farm) and the number of animals confirmed to have a tick borne disease.

   a. Separately, for the case and the control farms, test whether the number of sick animals follows a binomial distribution.

   b. Is the rate of infection in control farms different from the case farms? State your hypotheses and assumptions. Explain each step in the analyses and write your conclusions clearly.

| | Herd Size | | | | | | | |
|---|---|---|---|---|---|---|---|---|
| | 1 | | 2 | | 3 | | 4 | |
| **Number of Sick Animals** | ca | ct | ca | ct | ca | ct | ca | ct |
| 0 | 10 | 2 | 20 | 2 | 10 | 1 | 5 | 1 |
| 1 | 8 | 1 | 10 | 2 | 5 | 0 | 1 | 1 |
| 2 | | | 2 | 10 | 0 | 4 | 0 | 2 |
| 3 | | | | | 0 | 4 | 0 | 5 |
| 4 | | | | | | | 0 | 5 |

*Note:* ca, case; ct, control.

4.2. The data in the following table are from a study on the teratogenic effects of a certain chemical (tetrabromodibenzo-p-dioxin or TBDD) in C57BL/6N mice. The responses are the proportions $Y/n$ of cleft palate incidence in each litter for pregnant dams treated on gestation day 10 and examined on gestation day 18.

   a. Is there a significant difference between dose levels on the incidence of cleft palate?

   b. Test the above hypothesis by accounting for the dose structure in a logistic regression model (Take $X = \log_{10}$ (Dose) as a covariate.) What do you conclude?

| Dose | $Y_{ij}$ | $n_{ij}$ | Dose | $Y_{ij}$ | $n_{ij}$ | Dose | $Y_{ij}$ | $n_{ij}$ | Dose | $Y_{ij}$ | $n_{ij}$ |
|---|---|---|---|---|---|---|---|---|---|---|---|
| 3 | 0 | 7 | 6 | 0 | 8 | 24 | 0 | 9 | 96 | 6 | 7 |
| 3 | 0 | 11 | 6 | 0 | 9 | 24 | 0 | 9 | 96 | 7 | 7 |
| 3 | 0 | 10 | 6 | 0 | 9 | 24 | 0 | 7 | 96 | 3 | 3 |
| 3 | 0 | 9 | 6 | 0 | 10 | 24 | 0 | 9 | 96 | 9 | 9 |
| 3 | 0 | 10 | 6 | 0 | 8 | 24 | 0 | 5 | 96 | 10 | 10 |
| 3 | 0 | 8 | 6 | 0 | 8 | 24 | 0 | 9 | 96 | 2 | 3 |
| 3 | 0 | 7 | 6 | 0 | 10 | 24 | 0 | 8 | 96 | 7 | 8 |
| 3 | 0 | 10 | 12 | 0 | 3 | 24 | 1 | 9 | 96 | 1 | 3 |
| 3 | 0 | 9 | 12 | 0 | 9 | 24 | 0 | 11 | 96 | 9 | 9 |
| 3 | 0 | 10 | 12 | 0 | 7 | 24 | 0 | 6 | 96 | 8 | 9 |
| 3 | 0 | 2 | 12 | 0 | 8 | 24 | 0 | 9 | 96 | 8 | 8 |
| 3 | 0 | 9 | 12 | 0 | 9 | 24 | 0 | 9 | 192 | 6 | 6 |
| 3 | 0 | 9 | 12 | 0 | 5 | 24 | 0 | 8 | 192 | 9 | 9 |
| 3 | 0 | 10 | 12 | 0 | 6 | 24 | 0 | 6 | 192 | 4 | 4 |
| 3 | 0 | 9 | 12 | 0 | 8 | 24 | 0 | 9 | 192 | 6 | 6 |
| 6 | 0 | 11 | 12 | 0 | 8 | 48 | 3 | 5 | 192 | 7 | 7 |
| 6 | 0 | 6 | 12 | 0 | 9 | 48 | 2 | 9 | 192 | 10 | 10 |
| 6 | 0 | 3 | 12 | 0 | 10 | 48 | 0 | 8 | 192 | 7 | 7 |
| 6 | 0 | 7 | 12 | 0 | 6 | 48 | 0 | 8 | 192 | 5 | 5 |
| 6 | 0 | 3 | 12 | 0 | 8 | 48 | 0 | 10 | 192 | 9 | 9 |
| 6 | 0 | 9 | 12 | 0 | 9 | 48 | 0 | 5 | 192 | 4 | 4 |
| 6 | 0 | 10 | 12 | 0 | 10 | 48 | 0 | 8 | 192 | 7 | 7 |
| 6 | 0 | 9 | 12 | 0 | 9 | 48 | 0 | 3 | 192 | 8 | 8 |
| 6 | 0 | 3 | 12 | 0 | 7 | 48 | 4 | 9 | 192 | 9 | 9 |
| 6 | 0 | 9 | 24 | 0 | 8 | 48 | 0 | 9 | 192 | 10 | 10 |
| 6 | 0 | 11 | 24 | 0 | 11 | 48 | 0 | 8 | | | |

4.3. Run the SAS and R programs given in Example 4.7 to compute the common odds ratio and test the significance of the common odds ratio using unadjusted and adjusted chi-square tests and verify the calculations in the example.

4.4. An investigator is interested in comparing prevalence of mastitis in four breeds of dairy cattle. He selects herds of the four breeds (exactly one breed in each herd) to obtain the following data. The table shows herd size, followed by the number of cases in brackets.

| | | | | | |
|---|---|---|---|---|---|
| Breed 1 | 30 (10) | 47 (2) | 48 (2) | 24 (0) | |
| Breed 2 | 21 (0) | 31 (0) | 46 (11) | | |
| Breed 3 | 10 (5) | 43 (20) | 32 (9) | 43 (0) | 37 (22) |
| Breed 4 | 17 (1) | 35 (10) | 41 (3) | | |

a. Verify the following results: pooled prevalence $\hat{p} = 0.188$ and mean herd size $m = 33.667$.

b. The usual chi-square, ignoring herds, is

$$\sum \frac{(\text{observed} - \text{expected})^2}{\text{expected}} = 37.926$$

Compare with the chi-square tables with 3 df. Verify that the $p$-value is $< .001$.

c. To account for clustering, we need to calculate the intracluster correlation coefficient. Write a SAS program to calculate the following quantities needed to evaluate the pooled intraclass correlation. For each breed calculate $SSB$ and $SSW$ and verify the following results.

| Breed | SSB | SSW | Number of herds ($n_h$) |
|-------|------|--------|-------------------------|
| 1 | 2.186 | 10.498 | 4 |
| 2 | 1.396 | 8.370 | 3 |
| 3 | 8.409 | 28.585 | 5 |
| 4 | 1.028 | 10.865 | 3 |

$$MSB = \frac{\sum SSB}{\sum (n_h - 1)}, \quad MSW = \frac{\sum SSW}{N - \sum n_h}$$

where $N$ is the total number of animals.

d. Show that $MSB = 1.184$. $MSW = 0.119$, and hence

$$\hat{\rho} = \frac{MSB - MSW}{MSB + (M - 1)MSW} = 0.210$$

correction factor $C = 1 + (m - 1)\hat{\rho} = 7.857$, and corrected chi-square is $37.926/7.857 = 4.827$.

e. What is the number of appropriate degrees of freedom associated with the above corrected chi-square value?

4.5. Stratified data, adjusting the Mantel-Haenszel test: This exercise is an illustration on the use of stratified analysis when the randomization unit is a cluster (litter). Use the chi-square test to compare the two treatments with respect to death rate.

| Stratum I | Treatment I | Litter size | 15 | 13 | 12 | 15 | 10 | 9 | |
|---|---|---|---|---|---|---|---|---|---|
| | | Mortality | 2 | 0 | 2 | 5 | 0 | 0 | $\hat{p} = .122$ |
| | Treatment II | Litter size | 16 | 15 | 13 | 13 | 11 | 12 | |
| | | Mortality | 1 | 0 | 1 | 0 | 3 | 0 | $\hat{p} = .063$ |
| Stratum II | Treatment I | Litter size | 14 | 16 | 15 | 9 | 14 | | |
| | | Mortality | 4 | 5 | 3 | 0 | 3 | | $\hat{p} = .122$ |
| | Treatment II | Litter size | 16 | 7 | 12 | 15 | 11 | | |
| | | Mortality | 4 | 0 | 0 | 3 | 0 | | $\hat{p} = .122$ |

| | Stratum I | | | Stratum II | | |
|---|---|---|---|---|---|---|
| | D $(x)$ | A $(y)$ | Total | D $(x)$ | A $(y)$ | Total |
| Treatment I | 9 | 65 | $74 = n_{11}$ | 15 | 53 | $68 = n_{21}$ |
| Treatment II | 5 | 75 | $80 = n_{12}$ | 7 | 54 | $61 = n_{22}$ |
| Total | $14 = x_1$ | $140 = y_1$ | $154 = n_1$ | $122 = x_2$ | $107 = y_2$ | $129 = n_2$ |

If the odds ratio ($\psi$) is assumed to be the same in both strata, the Mantel-Haenszel procedure tests the hypothesis $\psi = 1$. (When $\psi = 1$, the relative risk is 1.)

$$X_{MH}^2 = \frac{\left[ \sum_t \dfrac{x_{1t}y_{2t} - x_{2t}y_{1t}}{n_t} \right]^2}{\sum_t \dfrac{n_{1t}n_{2t}x_t y_t}{(n_t - 1)n_t^2}}$$

a. Show that $X_{MH}^2 = 414$. What is the $p$-value associated with the test?

b. To account for clustering, compute *SSB* and *SSW* for each of the four treatment groups (rows) using PROC GLM in SAS. Verify the following results:

| | SSB | SSW | Litters |
|---|---|---|---|
| Stratum 1, Row 1 | 1.172 | 6.733 | 6 |
| Stratum 1, Row 2 | 0.645 | 4.042 | 6 |
| Stratum 2, Row 1 | 0.639 | 11.052 | 5 |
| Stratum 2, Row 2 | 0.792 | 5.400 | 5 |

$$MSB = \frac{\sum SSB}{\sum (n_h - 1)} = 0.181, \quad MSW = \frac{\sum SSW}{N- \sum n_h} = 0.104$$

$$\hat{\rho} = \frac{MSB - MSW}{MSB + (m-1)MSW} = 0.054$$

c. Compute the correction factor for each cluster size, $mc_m = 1 + (m-1)(0.054)$ and for each row compute the weighted mean, $B_{jt}$.

For example: $_B11 = \dfrac{15C_{15} + 13C_{13} + \dots + 9C_9}{15 + 13 + \dots + 9} = 1.633.$

d. Show that the corrected Mantel-Haenszel chi-square statistic is as given next:

$$X^2_{MH} = \frac{\left( \sum_t \dfrac{x_{1t}y_{2t} - x_{2t}y_{1t}}{n_{1t}B_{1t} + n_{2t}B_{2t}} \right)^2}{\sum_t \dfrac{n_{1t}n_{2t}x_t y_t}{(n_{1t}B_{1t} + n_{2t}B_{2t})n_t^2}}$$

$$= \frac{(1.373 + 2.03)^2}{3.198 + 4.548}$$

$$= 2.489$$

4.6. Analyze the smoking cessation data of Chapter 3 to compare the two proportions of correlated responses. Consider each school a cluster.

4.7. Run the SAS and R programs given in Example 4.6. Analyze Weil's data and verify the results reported in the example.

4.8. Use the delta method to construct a large sample variance of the logarithm of the relative risk. Use a similar procedure to that used in the construction of a confidence interval on AR to construct a 95% CI on RR. Use Weil's data in Table 4.19 to illustrate the methodology.

4.9. Berg (1981) reported data consisting of a sample of the same-sex Norwegian twin pairs collected as part of a large study of coronary heart disease. One of the major objectives of the study was to evaluate the degree of association within a twin-pair with regard to the presence ($y = 1$) or absence of atherogenic lipoprotein LP(A) ($y = 0$). The data layout for a $2 \times 2$ table is shown next. We use the intraclass correlation as a measure of within-twin pair similarity. We also have data on trouble concentrating (TC) and the presence or absence of atherogenic lipoprotein LP(a), the latter being much smaller in size. These data are displayed in the following table.

|  | | Index Twin ($Y_1$) | |
|---|---|---|---|
|  | | 1 | 0 |
| Co-Twin | 1 | $n_{11}$ | $n_{10}$ |
| ($Y_2$) | 0 | $n_{01}$ | $n_{00}$ |

The TC and the LP(a) Twin Data Sets

|  |  | Index Twin | | |
|---|---|---|---|---|
| Data | Co-Twin | y = 1 | y = 0 | Total |
| TC | y = 1 | 1275 | 323 | 1598 |
|  | y = 0 | 310 | 203 | 513 |
|  | Total | 1585 | 526 | 2111 |
| LP(a) | y =1 | 14 | 12 | 26 |
|  | y = 0 | 11 | 70 | 81 |
|  | Total | 25 | 82 | 107 |

Clearly, the data are special case when the cluster size $n_i = 2$.

a. Estimate the prevalence of the condition for both TC and LP(a), using the expression $\hat{\pi} = \dfrac{2n_{11} + n_{10} + n_{01}}{n}$ Also, calculate the intraclass correlation for both TC and LP(a):

$$\hat{\rho} = \frac{4(n_{11}n_{22} - n_{10}n_{01}) - (n_{10} - n_{01})^2}{(2n_{11} + n_{10} + n_{01})(2n_{22} + n_{10} + n_{01})} \qquad (4.24)$$

$$\text{var}(\hat{\rho}) = \frac{1-\rho}{N}\left[(1-\rho)(1-2\rho) + \frac{\rho(2-\rho)}{2\pi(1-\pi)}\right] \qquad (4.25)$$

b. Construct an 95% confidence interval on the prevalence and the intraclass correlation for both TC and LP(a).

c. Test the null hypothesis of equality of the intraclass correlations of both TC and LP(a).

4.10. Consider the matched case control data in the following table.

|  |  | Control | |
|---|---|---|---|
|  |  | $E$ | $\bar{E}$ |
| Case | $E$ | 15 | 20 |
|  | $\bar{E}$ | 5 | 85 |

In terms of proportions, the table becomes:

|      |         | Control |         |
|------|---------|---------|---------|
|      |         | $E = 1$ | $\bar{E} = 0$ |
| Case | $E = 1$ | .12     | .16     |
|      | $\bar{E} = 0$ | .05 | .68   |

Show that the odds ratio estimate $\hat{\psi}$ is 4, and that the Pearson's correlation coefficient $\hat{\varphi}$ is 0.867.

4.11. Suppose that the preceding table is written with the general formats as

|      |         | Control   |          |
|------|---------|-----------|----------|
|      |         | $E = 1$   | $\bar{E} = 0$ |
| Case | $E = 1$ | $p_{11}$  | $p_{10}$ |
|      | $\bar{E} = 0$ | $p_{01}$ | $p_{00}$ |

Using the notations $p_1 = p_{11} + p_{10}$, $q_1 = 1 - p_1$, $p_0 = p_{11} + p_{01}$, and $q_0 = 1 - p_0$, show that Pearson's correlation coefficient $\varphi$ may be written as $\Phi = (p_{11}p_{00} - p_{10}p_{01})/(p_1 q_1 \cdot p_0 q_0)^{1/2}$.

4.12. Show that all the cell probabilities can be written as functions of the sums and the odds ratio. For example, it can be shown that $p_{10} = \psi(p_1 - p_0)/(\psi - 1)$, and $p_{01} = (p_1 - p_0)/(\psi - 1)$, and so on.

4.13. Derive an expression for the Pearson's correlation coefficient as a function of the marginal sums $(p_1, q_1, p_0, q_0)$ and the odds ratio $\psi$. Let $pdisc = p_{10} + p_{01}$ and $pdiff = p_{10} - p_{01}$.

4.14. Connor (1987) showed that the sample size needed to test the departure of population odds ratio from unity base on McNemar's test is given by $n = \left( \dfrac{Az_{1-\alpha} + B\, z_{1-\beta}}{pdiff} \right)^2$ where $A = (pdisc)^{1/2}$ and $B = (pdisc - pdiff^2)^{1/2}$.

a. Run the following R code to estimate the sample size:
   **library(MESS)#**
   paid = $p_{01}$ = Probability that a case is exposed and the matched control is not exposed.
   power_mcnemar_test(n=NULL, paid=.04, psi=4, power=.8, method="normal")
   power_mcnemar_test(n=NULL, paid=.04, psi=4, power=.8)

b. Run the following SAS code to calculate the required number of matched case control pairs:

```
proc power;
    pairedfreq dist=normal method=connor
        discproportions = 0.04 | 0.16
        npairs = .
        power = .8;
run;
```

# 5

# Modeling Binary Outcome Data

## 5.1 Introduction

In many research problems, it is of interest to study the effects that some variables exert on others. One sensible way to describe this relationship is to relate the variables by some sort of mathematical equation. In most applications, statistical models are mostly mathematical equations constructed to approximately relate the response (dependent) variables to a group of predictor (independent) variables.

When the response variable, denoted by $y$, is continuous and believed to depend linearly on $k$ variables $x_1, x_2, \ldots, x_k$ through unknown parameters $\beta_0, \beta_1, \ldots, \beta_k$, then this linear (where "linear" is used to indicate linearity in the unknown parameters) relationship is given as

$$y_i = \sum_{j=0}^{k} \beta_j x_{ji} + \varepsilon_i, \tag{5.1}$$

where $x_{0i} = 1$ for all $i = 1, 2, \ldots, n$.

The term $\varepsilon_i$ is an unobservable random error representing the residual variation and is assumed to be independent of the systematic component $\sum_{j=0}^{k} \beta_j x_{ji}$. It is also assumed that $E(\varepsilon_i) = 0$ and $V(\varepsilon_i) = \sigma^2$; hence, $E(y_i) = \sum_{j=0}^{k} \beta_j x_{ji}$ and $V(y_i) = \sigma^2$.

To fit the model 5.1 to the data $(y_i, x_i)$, one has to estimate the parameters $\beta_0, \beta_1, \ldots, \beta_k$. The most commonly used methods of estimation are the method of least squares and the method of maximum likelihood.

Applications of these methods of estimation to the linear regression model 5.1 are extensively discussed in Mosteller and Tukey (1977), Draper and Smith (1981), and many other sources. It should be noted that no assumptions are needed on the distribution of the response variable $y$ (except the independence of $(y_1, y_2, \ldots, y_n)$) to estimate the parameters by the method of least squares. However, the maximum likelihood requires that the sample $y = (y_1,$

$y_2,\ldots,y_n)$ is randomly drawn from a distribution where the specified structure of that distribution in most applications is

$$N\left(\sum_{j=0}^{k}\beta_j x_{ji},\, \sigma^2\right)$$

The least squares estimates of the regression parameters will then coincide with those obtained by the method of maximum likelihood. Another remark that should be made here is that there is nothing in the theory of least squares that restricts the distribution of the response variable to be continuous, discrete, or of bounded range. For example, suppose that we would like to model the proportion $\hat{p}_i = \frac{y_i}{n_i}$ $(i = 1,2,\ldots,m)$ of individuals suffering from some respiratory illness, observed over several geographical regions, as a function of $k$ covariates, where $y_i = \sum_{j=1}^{n_i} y_{ij}$, where $y_{ij} = 1(0)$ if disease is present (absent). We further assume, conditional on the observed covariates, that $y_i$ is binomially distributed with mean $p_i$. That is, we assume the relationship between $E(y_i \mid x_{ji}) = p_i$ and the covariates to be

$$p_i = \sum_{j=0}^{k}\beta_j x_{ji} \tag{5.2}$$

The least squares estimates are obtained by minimizing

$$s^2 = \sum_{i=1}^{n}\left(\hat{p}_i - \sum_{j=0}^{k}\beta_j x_{ji}\right)^2$$

Several problems are encountered when the least squares method is used to fit model 5.2. One of the assumptions of the method of least squares is the homogeneity of variance that is, $V(y_i) = \sigma^2$; the variance does not vary from one observation to another. For binary data, $y_i$ follows a binomial distribution with mean $n_i p_i$ and variance $n_i p_i (1 - p_i)$ so that $V(\hat{p}_i) = \frac{p_i(1-p_i)}{n_i}$.

As proposed by Cox and Snell (1989), one can deal with the variance heterogeneity by applying the method of weighted least squares using the reciprocal variance as a weight. The weighted least squares estimates are thus obtained by minimizing:

$$s_w^2 = \sum_{i=1}^{n} w_i \left(\hat{p}_i - \sum_{j=0}^{k}\beta_j x_{ji}\right)^2 \tag{5.3}$$

where $w_i = [V(\hat{p}_i)]^{-1}$ depends on $p_i$, which in turn depends on the unknown parameters $\beta_0, \beta_1, \ldots, \beta_k$ through the relationship 5.2. The model 5.2 can be fitted quite easily using PROC REG in SAS together with the WEIGHT statement, where $\hat{w}_i$ are the specified weights. Note that $0 \le p_i \le 1$, and the estimates $\hat{\beta}_0, \hat{\beta}_1, \ldots, \hat{\beta}_k$ of the regression parameters are not constrained. That is, they are permitted to attain any value in the interval $(-\infty,\infty)$. Since the fitted values are obtained by substituting $\hat{\beta}_0, \hat{\beta}_1, \ldots, \hat{\beta}_k$ in Equation 5.2:

$$\hat{p}_i = \sum_{j=0}^{k} \hat{\beta}_j x_{ji}$$

Thus, there is no guarantee that the fitted values should fall in the interval [0,1].

To overcome the difficulties of using the method of least squares to fit a model where the response variable has a restricted range, it is suggested that a suitable transformation be employed so that the fitted values of the trans-formed parameter vary over the interval $(-\infty,\infty)$.

This chapter is devoted to the analysis of binomially distributed data where the binary responses are independent, with special attention given to situa-tions where the binary responses are obtained from clusters and hence cannot be assumed to be independent. The chapter is organized as follows: In Section 5.2 we introduce the logistic transformation and establish the relationship between the regression coefficient and the odds ratio parameter. Model building strategies are discussed in Sections 5.3 and 5.4, and measures of goodness of fit are introduced in Section 5.5. Approaches to modeling cor-related binary data are discussed in Section 5.6. In this section, population average and cluster specific models are investigated with examples.

## 5.2 The Logistic Regression Model

Let us consider the following example that may help the reader understand the motives of the logistic transformation. Suppose that we have a binary variable $y$ that takes the value 1 if a sampled subject is diseased and zero otherwise. Let $P(D) = P[y = 1] = \pi$ and $P(\bar{D}) = P[y = 0] = 1 - \pi$. Moreover, suppose that $X$ is a risk factor that has normal distribution with mean $\mu_1$ and variance $\sigma^2$ in the population of diseased, that is

$$X|D \equiv N(\mu_1, \sigma^2)$$

Meaning that, given the information that the individual is diseased, the conditional distribution of $X$ is $N(\mu_1,\sigma^2)$. Hence

$$f(X|D) = \frac{1}{\sigma\sqrt{2\pi}} \exp\left[-(x - \mu_1)^2/2\sigma^2\right]$$

In a similar manner, we assume that the distribution of the risk factor $X$, in the population of nondiseased, has a mean $\mu_2$ and variance $\sigma^2$, that is

$$f(X|\bar{D}) = \frac{1}{\sigma\sqrt{2\pi}} \exp\left[-(x-\mu_2)^2/2\sigma^2\right]$$

Since $p = P[y = 1|X = x] = \dfrac{p(y = 1, X = x)}{f(x)}$, then from Bayes's theorem:

$$p = \frac{f(X|D)P(D)}{f(X|D)P(D) + f(X|\bar{D})P(\bar{D})}$$

$$= \frac{\dfrac{\pi}{\sigma\sqrt{2\pi}} \exp\left[-(x-\mu_1)^2/2\sigma^2\right]}{\dfrac{\pi}{\sigma\sqrt{2\pi}} \exp\left[-(x-\mu_1)^2/2\sigma^2\right] + \dfrac{1-\pi}{\sigma\sqrt{2\pi}} \exp\left[-(x-\mu_2)^2/2\sigma^2\right]}$$

Therefore,

$$p = P[y = 1|X = x] = \frac{e^{\beta_0+\beta_1 x}}{1 + e^{\beta_0+\beta_1 x}} \tag{5.4}$$

In Equation 5.4, we have $\beta_0 = -\ln\dfrac{1-\pi}{\pi} - \dfrac{1}{2\sigma^2}(\mu_1 - \mu_2)(\mu_1 + \mu_2)$, $\beta_1 = \dfrac{\mu_1-\mu_2}{\sigma^2}$, and

$$\ln\left(\frac{p}{1-p}\right) = \beta_0 + \beta_1 x \tag{5.5}$$

Therefore, the log-odds is a linear function of the explanatory variable $x$ (here, the risk factor).

**Remarks:** The regression parameter $\beta_1$ has log-odds ratio interpretation in epidemiologic studies. To show this, suppose that the exposure variable has two levels (exposed, not exposed). Let us define a dummy variable $x$ that takes the value 1 if the individual is exposed to the risk factor, and 0 if not exposed.

From Equation (5.4) we have

$$P_{11} = P(y = 1|X = 1) = \frac{e^{\beta_0+\beta_1}}{1 + e^{\beta_0+\beta_1}}$$

$$P_{01} = P(y = 0|X = 1) = 1 - P_{11}$$

$$P_{10} = P(y = 1|X = 0) = \frac{e^{\beta_0}}{1 + e^{\beta_0}}$$

$$P_{00} = P(y = 0|X = 0) = 1 - P_{10}$$

The odds ratio is $\psi = \dfrac{P_{11}P_{00}}{P_{10}P_{01}} = e^{\beta_1}$, and it follows that $\ln(\psi) = \beta_1$.

The representation can be extended such that logit($p$) is a function of more than just one explanatory variable.

Let $y_1, y_2, \ldots, y_n$ be a random sample of $n$ successes out of $n_1, n_2, \ldots, n_n$ trials and let the corresponding probabilities of success be $p_1, p_2, \ldots, p_n$. If we wish to express the probability $p_i$ as a function of the explanatory variables $x_{1i}, \ldots, x_{ki}$, then the generalization of Equation 5.5 is

$$\text{logit}(p_i) = \log\left(\frac{p_i}{1 - p_i}\right) = \sum_{j=0}^{k} \beta_j x_{ji}, \quad x_{0i} = 1 \text{ for all } i = 1, 2, \ldots, n$$

We shall denote the linear function $\sum_{j=0}^{k} \beta_j x_{ji}$ by $\eta_i$, which is usually known as the *link function*. Hence, $p_i = e^{\eta_i} / (1 + e^{\eta_i})$.

The binomially distributed random variable $y_i$, $i = 1, 2, \ldots, n$ has mean $\mu_i = n_i p_i$ and variance, $n_i p_i q_i$. Fitting the model to the data is achieved by estimating the model parameters $\hat{\beta}_0, \hat{\beta}_1, \ldots, \hat{\beta}_k$.

The maximum likelihood method is used to estimate the parameters where the likelihood function is given by

$$L(\beta) = \prod_{i=1}^{n} \binom{n_i}{y_i} p_i^{y_i} q_i^{n_i - y_i}$$

$$= \prod_{i=1}^{n} \binom{n_i}{y_i} \left(e^{\eta_i}\right)^{y_i} \left(\frac{1}{1 + e^{\eta_i}}\right)^{n_i}$$

The logarithm of the likelihood function is thus

$$l(\beta) = \sum_{i=1}^{n} \left[y_i \eta_i - n_i \ln\left(1 + e^{\eta_i}\right)\right] \tag{5.6}$$

Differentiating $l(\beta)$ with respect to $\beta_r$, we have

$$\ell_r = \frac{\partial l(\beta)}{\partial \beta_r} = \sum_{i=1}^{n} \left[y_j x_{ri} - n_i x_{ri} e^{\eta_i} (1 + e^{\eta_i})^{-1}\right]$$

$$= \sum_{i=1}^{n} x_{ri}(y_i - n_i p_i) \quad r = 0, 1, 2, \ldots, k. \tag{5.7}$$

The $k+1$ equations in 5.7 can be solved numerically. The large sample variance–covariance matrix of the maximum likelihood estimators is the inverse of Fisher's information matrix $I^{-1}$ (see Cox and Snell, 1989), where

$$I = -E(\ell_{rs}) = -E\left[\frac{\partial^2 l(\beta)}{\partial \beta_r \partial \beta_s}\right] = \sum_{i=1}^{n} n_i x_{ri} x_{si} p_i (1 - p_i)$$

The $r$th diagonal element in $I^{-1}$ is $\hat{V}(\hat{\beta}_r) = V_{rr}$.

Once the parameters have been estimated, the predicted probabilities of success are given by

$$\hat{p}_i = e^{\hat{\eta}_i}/1 + e^{\hat{\eta}_i}, \text{ where } \hat{\eta}_i = \sum_{j=0}^{k}\hat{\beta}_j x_{ij}$$

The maximum likelihood based $(1 - \alpha)100\%$ confidence interval on $\beta_r$ is given approximately by $\hat{\beta}_r \pm z_{\alpha/2}(V_{rr})^{1/2}$. The Wald's test on the null hypothesis $H_0 : \beta_r = 0$ is given by referring $z = \dfrac{\hat{\beta}}{(V_{rr})^{1/2}}$ to the tables of standard normal distribution.

## 5.3 Coding Categorical Explanatory Variables and Interpretation of Coefficients

Recall that when Equation 5.4 is applied to the simple case of one independent variable that has two levels (exposed, not exposed), we defined a dummy variable X such that $X_i = 1$ if the $i$th individual is exposed, and $X_i = 0$ if the $i$th individual is not exposed.

Suppose that the exposure variable X has $m > 2$ categories. For example, X may be the strain or the breed of the animal, or it may be the ecological zone from which the sample is collected. Each of these variables (strain, breed, zone, etc.) is *qualitative* or a factor variable that can take a finite number of values known as the levels of the factor. To see how qualitative independent variables or factors are included in a logistic regression model, suppose that F is a factor with $m$ distinct levels. There are various methods in which the indicator variables or the dummy variables can be defined. The choice of a particular method will depend on the goals of the analysis. One way to represent the $m$ levels of the factor variable F is to define $m - 1$ dummy variables $f_1, f_2, \ldots, f_{m-1}$ such that the portion of the design matrix corresponding to those variables looks like Table 5.1.

**TABLE 5.1**

Dummy Variables for the Factor Variable F with $m$ Levels

| Level | $f_1$ | $f_2$ | ... | $f_{m-1}$ |
|-------|-------|-------|-----|-----------|
| 1 | 0 | 0 | | 0 |
| 2 | 1 | 0 | ... | 0 |
| 3 | 0 | 1 | ... | 0 |
| . | . | . | ... | . |
| . | . | . | ... | . |
| . | . | . | ... | . |
| $m$ | 0 | 0 | ... | 1 |

**TABLE 5.2**

Hypothetical Results of a Carcinogenic Experiment:
Counts of Rats with Tumors

|           | Strain 1 | Strain 2 | Strain 3 | Strain 4 |
|-----------|----------|----------|----------|----------|
| $y_i$     | 10       | 20       | 5        | 2        |
| $n_i - y_i$ | 45     | 20       | 20       | 43       |
| $n_i$     | 55       | 40       | 25       | 45       |
| $\hat{\psi}$ | 4.780 | 21.500   | 5.375    |          |

**Example 5.1**

The following data are the results of a carcinogenic experiment. Different strains of rats have been injected with carcinogen in their foot pad. They were then put on a high fat diet and at the end of week 21 the number with tumors ($y$) were counted.

The last row of Table 5.2 gives the odds ratio for each strain level using strain 4 as the reference level. For example, for strain 1 the estimated odds ratio is

$$\hat{\psi}[\text{strain 1; strain 4}] = \frac{(10)(43)}{(45)(2)} = 4.780$$

Using strain 4 as a reference group we define the three dummy variables:

$$X_j = \begin{cases} 1 & \text{if rat is from strain } j \quad j = 1, 2, 3 \\ 0 & \text{otherwise} \end{cases}$$

The results of fitting the logistic regression model to the data in Table 5.2 are given in Table 5.3.

Note that $\ln\hat{\psi}$ [strain 1; strain 4] = ln (4.78) = 1.56 = $\hat{\beta}_1$. The SE column has the square roots of the diagonal elements of $I^{-1}$. Hence, as mentioned earlier, the estimated parameters maintain their log-odds ratio interpretation even if we have more than one explanatory variable in the logistic

**TABLE 5.3**

Maximum Likelihood Analysis

| Variable  | Estimates | SE | Wald Chi-Square | $\hat{\psi}$ |
|-----------|-----------|-----|----------------|--------------|
| Intercept | $\hat{\beta}_0 = -3.068$ | .723 | 17.990 | |
| Strain 1  | $\hat{\beta}_1 = 1.564$ | .803 | 3.790 | 4.780 |
| Strain 2  | $\hat{\beta}_2 = 3.068$ | .790 | 15.103 | 21.500 |
| Strain 3  | $\hat{\beta}_3 = 1.682$ | .879 | 3.658 | 5.375 |

regression function. One should also note that the estimated standard error of the odds ratio estimate from a univariate analysis is identical to the standard error of the corresponding parameter estimate obtained from the logistic regression analysis. For example,

$$SE[\ln \hat{\psi}(strain3; strain4)] = \left( \frac{1}{5} + \frac{1}{20} + \frac{1}{2} + \frac{1}{43} \right)^{1/2} = .879$$

which is identical to $SE(\hat{\beta}_3)$ as shown in Table 5.3. Since the approximate $(1 - \alpha)100\%$ confidence limits for the parameter $\beta_j$ are $\hat{\beta}_j \pm Z_{\alpha/2} SE(\hat{\beta}_j)$. The limits for the corresponding odds ratio are obtained by taking the antilog transformation giving the limits on the parameter $\beta_j$ as $\exp[\hat{\beta}_j \pm Z_{\alpha/2} SE(\hat{\beta}_j)]$.

In studying the association between a response variable and categorical covariates and possible confounders, one may be interested in comparing the odds ratio of the confounder. Basing this comparison on the *p*-values may be problematic due to what is known as the multiplicity problem. In this case, a graphical comparison among the odds ratio may be informative. We illustrate this situation in Example 5.2

### Example 5.2: ch5_calcium: What Is Coronary Artery Calcium Scoring (CS)?

The coronary artery calcium score (CS) is a measurement of the amount of calcium in the walls of the arteries that supply the heart muscle using a special computed tomography (CT) scan of the heart. It shows the amount of hardening of the artery wall (a disease called atherosclerosis) that a person has. It tells physicians about patient's risk of a heart attack or stroke (brain attack) in the next 5 to 10 years. The more calcium (and therefore the more atherosclerosis), the higher the risk of having a heart attack or stroke. A high calcium score does not mean that you will have a heart attack, only that you are much more likely to have one than someone with a low score. Even a person with a score of zero could have a heart attack. Quite often the CS is categorized into two levels: 0 and above 0. Age and many other risk factors such as hypertension, diabetes, smoking, and family history are believed to be associated with the CS. In this example, we present data depicting the relationship between CS (high versus low), age, and hypertension (yes versus no).

The SAS input data statement shows the age groups, hypertension status, calcium score, and the number of subjects in each cell. The code produces graphical representation of the odds ratio and Figure 5.1.

```
data calcium;
input age_group $      hypertension $      cal_score $  count;
cards;
25_30 y     h      150
25_30 y     l      120
25_30 n     h      100
25_30 n     l      100
30_40 y     h      170
30_40 y     l      125
30_40 n     h      300
30_40 n     l      200
40_50 y     h      600
40_50 y     l      370
40_50 n     h      600
40_50 n     l      405
```

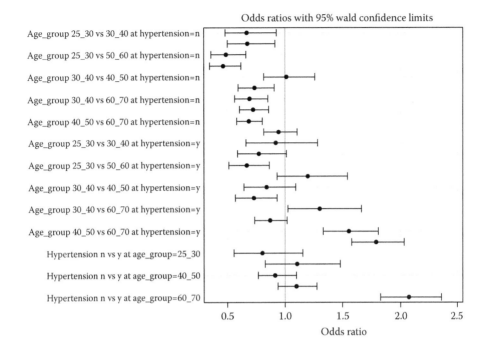

**FIGURE 5.1**
Graph of multiple comparisons of odds ratio measuring the association between calcium scoring and age groups, conditional on the hypertension status.

```
50_60 y   h   1160
50_60 y   l   620
50_60 n   h   820
50_60 n   l   400
60_70 y   h   1200
60_70 y   l   1150
60_70 n   h   1300
60_70 n   l   600
;
ODS RTF;
ODS GRAPHICS ON;
/* graph of odds ratio*/
proc logistic data=calcium plots(only)=oddsratio;
weight count;
class hypertension cal_score age_group /param=glm;
model    cal_score(event="h")=    hypertension    age_group
hypertension*age_group;
oddsratio age_group;
oddsratio hypertension;
ods output oddsRatiosWald =ORPlot;
run; quit;
ODS GRAPHICS OFF;ODS RTF CLOSE;
```

Figure 5.1 is the graphical representation of the 95% confidence odds ratio. Confidence intervals that cross the vertical line of unity are deemed nonsignificant. Odds ratio intervals that are to the right of the line of unity indicate that the odds of the disease are increased with the exposure. On the other hand, when the odds of disease are decreased with the presence of exposure, the confidence intervals fall to the left of the vertical line of unity.

Note that Figure 5.1 is the Forest plot that is frequently used in meta-analysis. Here it will be used for descriptive purposes, and no $p$-values will be attached.

---

## 5.4 Interaction and Confounding in Logistic Regression

In Example 5.1, we showed how the logistic regression model can be used to model the relationship between the proportion or the probability of developing tumors and the strain. Other variables could also be included in the model such as sex, the initial weight of each mouse, age, or other relevant variables. The main goal, of what then becomes a rather comprehensive model, is to adjust for the effect of all other variables and the effect of the differences in their distributions on the estimated odds ratios. It should be pointed out that the effectiveness of the adjustment (measured by the reduction in the bias of the estimated coefficient) depends on the appropriateness of the log-odds transformation and the assumption of constant slopes across the levels of a factor variable (Hosmer and Lemeshow, 1989). Departure from the constancy of slopes is explained by the existence of interaction. For example, if we have two factor variables, $F_1$ and $F_2$, where the response at a particular level of $F_1$ varies over the levels of $F_2$, then $F_1$ and $F_2$ are said to interact. To model the interaction effect, one should include terms representing the main effects of the two factors as well as terms representing the two-factor interaction, which is represented by a product of two dummy variables. Interaction is known to epidemiologists as "effect modification." Thus, a variable that interacts with a risk factor of exposure variable is termed an "effect modifier." To illustrate, suppose that factor $F_1$ has three levels $(a_1, a_2, a_3)$, and factor $F_2$ has two levels $(b_1, b_2)$. To model the main effects and their interactions, we first define the dummy variables:

$$A_i = \begin{cases} 1 & \text{if an observation belongs to the } i^{\text{th}} \text{ level of factor } F_1 (i = 1, 2), \\ 0 & \text{Otherwise} \end{cases}$$

which means that $a_3$ is the referent group. Similarly, we define a dummy variable $B = 1$ if the individual belongs to the second level of factor $F_2$. Suppose that we have the five data points, as shown in Table 5.4.

To model the interaction effect of $F_1$ and $F_2$, the data layout would be as in Table 5.5.

**TABLE 5.4**

Hypothetical Data of a Two-Factor Experiment

| Observation | $y$ | $F_1$ | $F_2$ |
|---|---|---|---|
| 1 | 1 | $a_1$ | $b_1$ |
| 2 | 1 | $a_1$ | $b_2$ |
| 3 | 0 | $a_2$ | $b_2$ |
| 4 | 0 | $a_3$ | $b_2$ |
| 5 | 1 | $a_3$ | $b_1$ |

**TABLE 5.5**

Dummy Variables for Modeling Interaction of Two Factors

| Observation | $Y$ | $A_1$ | $A_2$ | $B$ | $A_1B$ | $A_2B$ |
|---|---|---|---|---|---|---|
| 1 | 1 | 1 | 0 | 0 | 0 | 0 |
| 2 | 1 | 1 | 0 | 1 | 1 | 0 |
| 3 | 0 | 0 | 1 | 1 | 0 | 1 |
| 4 | 0 | 0 | 0 | 1 | 0 | 0 |
| 5 | 1 | 0 | 0 | 0 | 0 | 0 |

Another important concept to epidemiologists is confounding. A confounder is a covariate that is associated with both the outcome variable (disease, say) and a risk factor. When both associations are detected, then the relationship between the risk factor and the outcome variable is said to be confounded.

Kleinbaum et al. (1988) recommended the following approach (which we extend to logistic regression) to detect confounding in the context of multiple linear regression. Suppose that we would like to describe the relationship between the outcome variable $y$ and a risk factor $x$, adjusting for the effect of other covariates $x_1, x_2, ..., x_{k-1}$ (assuming no interactions are involved) so that $\eta = \beta_0 + \beta x + \beta_1 x_1 + ... \beta_{k-1} x_{k-1}$.

Kleinbaum et al. (1988) argued that confounding is present if the estimated regression coefficient $\beta$ of the risk factor $x$, ignoring the effects of $x_1, x_2, ..., x_{k-1}$, is meaningfully different from the estimate of $\beta$ based on the linear combination, which adjusts for the effect of $x_1, x_2, ..., x_{k-1}$.

**Example 5.3: ch5_mycotoxin**

To determine the disease status of a herd with respect to listeriosis (a disease caused by bacterial infection), fecal samples are collected from selected animals, and a "group testing" technique is adopted to detect the agent responsible for the occurrence. One positive sample means that there is at least one affected animal in a herd, hence the entire herd is identified as a "case." In this example, we will consider possible risk factors, such as herd size, type of feed, and level of mycotoxin in the feed.

Herd sizes are defined as small (25–50), medium (50–150), and large (150–300). The two types of feed are dry and not dry; the three levels of

**TABLE 5.6**

Occurrence of Listeriosis in Herds of Different Sizes with Different Levels of Mycotoxin

| | | Herd Size | | | | | |
|---|---|---|---|---|---|---|---|
| | | Small | | Medium | | Large | |
| Level of Mycotoxin | Type of Feed | Case | Control | Case | Control | Case | Control |
| Low | Dry | 2 | 200 | 15 | 405 | 40 | 300 |
| | Not dry | 1 | 45 | 5 | 60 | 5 | 4 |
| Medium | Dry | 5 | 160 | 1 | 280 | 40 | 100 |
| | Not dry | 4 | 55 | 3 | 32 | 0 | 42 |
| High | Dry | 2 | 75 | 60 | 155 | 40 | 58 |
| | Not dry | 8 | 10 | 10 | 20 | 4 | 4 |

mycotoxin are low, medium, and high. Table 5.6 indicates the levels of mycotoxin found in different farms and the associated disease occurrences.

Before analyzing this data using multiple logistic regression, we should calculate some basic results. The study looks at the type of feed as the risk factor of primary interest. Suppose that we would like to look at the crude association between the occurrence of the disease (case/control) and the level of mycotoxin (low, medium, and high). Table 5.7 summarizes the relationship between the disease and levels of mycotoxin, ignoring the effect of herd size and the type of feed.

The estimated odds of disease for a farm with medium level of mycotoxin relative to a farm with low level is

$$\hat{\psi} = \frac{(1014)(67)}{(68)(629)} = 1.59$$

This suggests that herds with medium levels are about 1.5 times more likely to have diseased animals than herds with low levels. On the other hand, the likelihood of disease of a farm with high levels is about 6 times relative to a farm with low levels.

$$\hat{\psi} = \frac{(1014)(124)}{(322)(68)} = 5.74$$

**TABLE 5.7**

Disease Occurrence at Different Levels of Mycotoxin

| | Disease | | |
|---|---|---|---|
| Mycotoxin | $D^+$ | $D^-$ | $\hat{\psi}$ |
| Low | 68 | 1014 | 1.00 |
| Medium | 67 | 629 | 1.59 |
| High | 124 | 322 | 5.74 |

*Note:* "Low" is the referent level.

**TABLE 5.8**

Listeriosis Occurrence When Considering Herd Size and Type of Feed

| | Herd Size | | | | | |
|---|---|---|---|---|---|---|
| | Small | | Medium | | Large | |
| Type of Feed | Case | Control | Case | Control | Case | Control |
| Dry | 9 | 435 | 90 | 840 | 120 | 458 |
| Not dry | 13 | 110 | 18 | 112 | 9 | 10 |
| $n$ | | 567 | | 1060 | | 597 |
| $\hat{\psi}$ | | 0.175 | | 0.667 | | 0.291 |

Now we reorganize the data (Table 5.8) so as to look at the association between disease and the type of feed while adjusting for the herd size.

When considering each herd size separately we have the following results:

**Small**

| | Case | Control | Total |
|---|---|---|---|
| Dry | 9 | 435 | 444 |
| Not dry | 13 | 110 | 123 |
| Total | 22 | 545 | 567 |

$$\hat{\psi}_1 = \frac{990}{5655} = 0.175$$

$$\ln \hat{\psi}_1 = -1.743$$

$$SE(\ln \hat{\psi}_1) = \left( \frac{1}{9} + \frac{1}{435} + \frac{1}{13} + \frac{1}{110} \right)^{\frac{1}{2}}$$

$$= 0.446$$

**Medium**

| | Case | Control | Total |
|---|---|---|---|
| Dry | 90 | 840 | 930 |
| Not dry | 18 | 112 | 130 |
| Total | 108 | 952 | 1060 |

$$\hat{\psi}_2 = 0.667$$

$$\ln \hat{\psi}_2 = -0.405$$

$$SE(\ln \hat{\psi}_2) = 0.277$$

**Large**

|          | Case | Control | Total |
|----------|------|---------|-------|
| Dry      | 120  | 458     | 578   |
| Not dry  | 9    | 10      | 19    |
| Total    | 129  | 468     | 597   |

$$\hat{\psi}_3 = 0.291$$

$$\ln \hat{\psi}_3 = -1.234$$

$$SE(\ln \hat{\psi}_3) = 0.471$$

The odds ratio of the association of feed type and diseased farm adjusting for the herd size is now computed using Woolf's method (1955) as

$$
\ln \hat{\psi} = \frac{\sum \dfrac{\ln\left(\hat{\psi}_j\right)}{v\left(\ln \hat{\psi}_j\right)}}{\sum \dfrac{1}{v\left(\ln \hat{\psi}_j\right)}}
$$

$$
= \frac{\left(\dfrac{-1.743}{0.1991}\right) + \left(\dfrac{-0.405}{0.0767}\right) + \left(\dfrac{-1.234}{0.2216}\right)}{\left(\dfrac{1}{0.1991}\right) + \left(\dfrac{1}{0.0767}\right) + \left(\dfrac{1}{0.2216}\right)}
$$

$$= -0.867.$$

Thus, $\hat{\psi}_W = e^{-0.867} = 0.420$.
The test for interaction using the Woolf's $\chi^2$ is obtained as follows:

$$
V(\ln \hat{\psi}_W) = \frac{1}{\sum \dfrac{1}{V\left(\ln \hat{\psi}_j\right)}}
$$

$$= 0.0443$$

$$
\chi_w^2 = \sum \frac{1}{V\left(\ln \hat{\psi}_j\right)} \left[\ln \hat{\psi}_j - \ln \hat{\psi}\right]^2
$$

$$
= \frac{(-1.74 + 0.867)^2}{0.1991} + \frac{(-0.405 + 0.867)^2}{0.0767} + \frac{(-1.234 + 0.867)^2}{0.2216} = 7.22
$$

Since $\chi_w^2$ is greater than $\chi_{(0.05,2)}^2$ there is evidence of interaction between the herd size and the type of feed used on the farm. We investigate this further in the following section.

## 5.5 The Goodness of Fit and Model Comparisons

Measures of goodness of fit are statistical tools used to explore the extent to which the fitted responses obtained from the postulated model compare with the observed data. Clearly, the fit is good if there is a good agreement between the fitted and the observed data. The Pearson's chi-square ($\chi^2$) and the likelihood ratio test (LRT) are the most commonly used measures of goodness of fit for categorical data. The following sections will give a brief discussion on how each of the chi-square and the LRT criteria can be used as measures of goodness of fit of a logistic model.

### 5.5.1 The Pearson's $\chi^2$ Statistic

The Pearson $\chi^2$ statistic is defined by

$$\chi^2 = \sum_{i=1}^{n} \frac{(y_i - n_i \hat{p}_i)^2}{n_i \hat{p}_i \hat{q}_i} \tag{5.8}$$

For the logistic regression model, $\hat{p}_i = \dfrac{e^{\hat{\eta}_i}}{1+e^{\hat{\eta}_i}}$, and $\hat{\eta}_i = \sum_{j=1}^{k} x_{ji}\hat{\beta}_j$ is the linear predictor, obtained by substituting the MLE of $\beta_j$ in $\eta_i$. The distribution of $\chi^2$ is asymptotically that of a chi-square with $(n - k - 1)$ degrees of freedom (see Collett, 2003). Large values of $\chi^2$ can be taken as evidence that the model does not adequately fit the data. Because the model parameters are estimated by the method of maximum likelihood, it is recommended that one uses the LRT statistic as a criterion for goodness of fit of the logistic regression model.

### 5.5.2 The Likelihood Ratio Criterion (Deviance)

Suppose that the model we would like to fit a model (called current model) with $k + 1$ parameters, and that the log-likelihood of this model given by Equation 5.6 is denoted by $l_c$. That is

$$l_c = \sum_{i=1}^{n} [y_i \ln p_i + (n_i - y_i) \ln(1 - p_i)]$$

Let $\hat{p}_i$ be the maximum likelihood estimator of $p_i$ under the current model. Therefore, the maximized log-likelihood function under the current model is given by

$$\hat{l}_c = \sum_{i=1}^{n} [y_i \ln \hat{p}_i + (n_i - y_i) \ln(1 - \hat{p}_i)] \tag{5.9}$$

McCullagh and Nelder (1989) indicated that in order to assess the goodness of fit of the current model, $\hat{l}_c$ should be compared with another log likelihood of a model where the fitted responses coincide with the observed responses. Such a model has as many parameters as the number of data points and is thus called the *full* or *saturated* model and is denoted by $\hat{l}_s$. Since under the saturated model, the fitted $p_i$ are the same as the observed proportions $\tilde{p}_i = y_i/n_i$, the maximized log-likelihood function under the saturated model is

$$\tilde{l}_s = \sum_{i=1}^{n} [y_i \ln \tilde{p}_i + (n_i - y_i) \ln(1 - \tilde{p}_i)]$$

The metric $D = -2[\hat{l}_c - \tilde{l}_s]$, which is called the *deviance*, was suggested by McCullagh and Nelder (1989) as a measure of goodness of fit of the current model.

As can be seen, the deviance is in fact the likelihood ratio criterion for comparing the current model with the saturated model. Now, since the two models are trivially nested, it is tempting to conclude from the large sample likelihood theory that the deviance is distributed as a chi-square with $(n - k - 1)$ degrees of freedom if the current model holds. However, from the standard theory leading to the chi-square approximation for the null distribution of the likelihood ratio statistic we find that if model A has $p_A$ parameters and model B (nested in model A) has $p_B$ parameters with $p_B < p_A$, then the likelihood ratio statistic that compares the two models has chi-square distribution with degrees of freedom $p_B - p_A$ as n $\rightarrow \infty$ (with $p_A$ and $p_B$ both fixed). If A is the saturated model, $p_A = n$, so the standard theory does not hold. In contrast to what has been reported in the literature, Collett (2003) pointed out that the deviance does not, in general, have an asymptotic chi-square distribution in the limit as the number of data points increase. Consequently, the distribution of the deviance may be far from chi-square, even if $n$ is large.

There are situations, however, when the distribution of the deviance can be reasonably approximated by a chi-square. The binomial model with large $n_i$ is an example. In this situation, a binomial observation $y_i$ may be considered a sufficient statistic for a sample of $n_i$ independent binary observations each with the same mean, so that $n_i \rightarrow \infty$ ($i = 1,2,...n$) plays the role in asymptotic computations as the usual assumption $n \rightarrow \infty$. In other words, the validity of the large-sample approximation to the distribution of the deviance, in the logistic regression model fit, depends on the total number of individual binary observations, $\sum n_i$ rather than on $n$, the actual number of data points $y_i$. Therefore, even if the number of binomial observations is small, the chi-square approximation to the distribution of the deviance can be used so long as $\sum n_i$ is reasonably large.

More fundamental problems with the use of the deviance for measuring goodness of fit arise in the important special case of binary data, where $n_i = 1$ ($i = 1, 2, ..., n$). The likelihood function for this case is

$$L(\beta) = \prod_{i=1}^{n} p_i^{y_i}(1 - p_i)^{1 - y_i}$$

For the saturated model $\hat{p}_i = y_i$ and

$$\hat{l}_c = \sum_{i=1}^{n}[y_i \ln \hat{p}_i + (1 - y_i)\ln(1 - \hat{p}_i)]$$

Now since $y_i = 0$ or $1$, $\ln y_i = (1 - y_i)\ln(1 - y_i) = 0$, the deviance is

$$D = -2\sum_i \left\{ y_i \ln\left(\frac{\hat{p}_i}{1 - \hat{p}_i}\right) + \ln(1 - \hat{p}_i) \right\} \tag{5.10}$$

We now show that $D$ depends only on the fitted values $\hat{p}_i$ and so is uninformative about the goodness of fit of the model. To see this, using Equation 5.7 where $n_i = 1$, we have

$$\frac{\partial l(\beta)}{\partial \beta_r} = \sum_i x_{ri}(y_i - \hat{p}_i) = 0 \tag{5.11}$$

Multiplying both sides of Equation 5.11 by $\hat{\beta}_r$ and summing over $r$

$$\sum (y_i - \hat{p}_i) \sum_r \beta_r x_{ri} = 0$$

$$= \sum (y_i - \hat{p}_i) \ln\left(\frac{\hat{p}_i}{1 - \hat{p}_i}\right)$$

from which

$$\sum y_i \ln\left(\frac{\hat{p}_i}{1 - \hat{p}_i}\right) = \sum \hat{p}_i \ln\left(\frac{\hat{p}_i}{1 - \hat{p}_i}\right)$$

Substituting the last equation into Equation 5.10 we get:

$$D = -2\sum_i \left\{ \hat{p}_i \ln\left(\frac{\hat{p}_i}{1 - \hat{p}_i}\right) + \ln(1 - \hat{p}_i) \right\} \tag{5.12}$$

Therefore, $D$ is completely determined by $\hat{p}_i$ and hence useless as a measure of goodness of fit of the current model. However, in situations where model A is nested in another model B, the difference in the deviances of the two models can be used to test the importance of those additional parameters in model B.

The Pearson's chi-square $\chi^2 = \sum_{i=1}^{n}\frac{(y_i - \hat{p}_i)^2}{\hat{p}_i \hat{q}_i}$ encounters similar difficulties. It can be verified that, for binary data ($n_i = 1$ for all $i$), $\chi^2$ always takes the value $n$

and is therefore completely uninformative. Moreover, the $\chi^2$ statistic, unlike the difference in deviance, cannot be used to judge the importance of additional parameters in a model that contains the parameters of another model.

We now explain how the difference in deviance can be used to compare models. Suppose that we have two models with the following link functions:

| Model | Link ($\eta$) |
|-------|---------------|
| A | $\sum_{j=0}^{p_1} \beta_j x_{ji}$ |
| B | $\sum_{j=0}^{p_1+p_2} \beta_j x_{ji}$ |

Model A contains $p_1 + 1$ parameters and is therefore nested in model B, which contains $p_1 + p_2 + 1$ parameters. The deviance under model A denoted by, $D_A$ carries $(n - p_1 - 1)$ degrees of freedom, while that of model B denoted by $D_B$ carries $(n - p_1 - p_2 - 1)$ degrees of freedom. The difference $D_A - D_B$ has an approximate chi-square distribution with $(n - p_1 - 1) - (n - p_1 - p_2 - 1) = p_2$ degrees of freedom. This chi-square approximation to the difference between two deviances can be used to assess the combined effect of the $p_2$ covariates in model B on the response variable $y$. The model comparisons based on the difference between deviances is equivalent to the analysis based on the likelihood ratio test. In the following example, we illustrate how to compare models using the LRT.

### Example 5.3: ch5_mycotoxin (Continued)

Here we apply the concept of difference between deviances or LRT to test for the significance of the added variables. First, we fit two logistic regression models: one with main effects (A), and another with main effects and interactions (B). The data structure needed for the analysis is given in Table 5.9.

```
* A- Model with main effects only;
proc logistic descending data=toxin;
weight count;
class toxin (ref='low') feed (ref='dry') hsize (ref='small')/
param=ref;
model case= toxin feed hsize;
run;

*Model with main effects and mycotoxin by feed type interaction;
proc logistic descending data=toxin;
weight count;
class toxin (ref='low') feed (ref='dry') hsize (ref='small')/
param=ref;
model case= toxin feed hsize toxin*feed;
run;
```

**TABLE 5.9**

Data of the Previous Example Printed from SAS

| Obs | record | myco | feed | size | case | n |
|-----|--------|------|------|------|------|-----|
| 1 | 1 | low | 1 | s | 2 | 202 |
| 2 | 2 | low | 0 | s | 1 | 46 |
| 3 | 3 | med | 1 | s | 5 | 165 |
| 4 | 4 | med | 0 | s | 4 | 59 |
| 5 | 5 | high | 1 | s | 2 | 77 |
| 6 | 6 | high | 0 | s | 8 | 18 |
| 7 | 7 | low | 1 | mid | 15 | 420 |
| 8 | 8 | low | 0 | mid | 5 | 65 |
| 9 | 9 | med | 1 | mid | 15 | 295 |
| 10 | 10 | med | 0 | mid | 3 | 35 |
| 11 | 11 | high | 1 | mid | 60 | 215 |
| 12 | 12 | high | 0 | mid | 10 | 30 |
| 13 | 13 | low | 1 | lar | 40 | 340 |
| 14 | 14 | low | 0 | lar | 5 | 9 |
| 15 | 15 | med | 1 | lar | 40 | 140 |
| 16 | 16 | med | 0 | lar | 0 | 2 |
| 17 | 17 | high | 1 | lar | 40 | 98 |
| 18 | 18 | high | 0 | lar | 4 | 8 |

Selected SAS output from running the model A is displayed next:

Model Fit Statistics

| Criterion | Intercept Only | Intercept and Covariates |
|-----------|----------------|--------------------------|
| AIC | 1602.415 | 1362.475 |
| SC | 1603.970 | 1371.807 |
| $-2$ Log L | 1600.415 | 1350.475 |

Testing Global Null Hypothesis: BETA=0

| Test | Chi-Square | DF | Pr > ChiSq |
|------|-----------|----|-----------| 
| Likelihood ratio | 249.9397 | 5 | <.0001 |
| Score | 261.2691 | 5 | <.0001 |
| Wald | 205.7533 | 5 | <.0001 |

Analysis of Maximum Likelihood Estimates

| Parameter | | DF | Estimate | Standard Error | Wald Chi-Square | Pr > ChiSq |
|---|---|---|---|---|---|---|
| Intercept | | 1 | −4.3348 | 0.2684 | 260.8223 | <.0001 |
| Toxin | high | 1 | 1.9898 | 0.1751 | 129.2018 | <.0001 |
| Toxin | med | 1 | 0.7288 | 0.1860 | 15.3544 | <.0001 |
| Feed | wet | 1 | 0.8637 | 0.2131 | 16.4325 | <.0001 |
| Hsize | large | 1 | 2.3400 | 0.2581 | 82.2175 | <.0001 |
| Hsize | medium | 1 | 1.0818 | 0.2487 | 18.9220 | <.0001 |

Odds Ratio Estimates

| Effect | Point Estimate | 95% Wald Confidence Limits | |
|---|---|---|---|
| Toxin high vs low | 7.314 | 5.190 | 10.308 |
| Toxin med vs low | 2.073 | 1.439 | 2.984 |
| Feed wet vs dry | 2.372 | 1.562 | 3.601 |
| Hsize large vs small | 10.381 | 6.260 | 17.215 |
| Hsize medium vs small | 2.950 | 1.812 | 4.803 |

Selected SAS output from running the model B is displayed next:

Model Fit Statistics

| Criterion | Intercept Only | Intercept and Covariates |
|---|---|---|
| AIC | 1602.415 | 1364.527 |
| SC | 1603.970 | 1376.970 |
| −2 Log L | 1600.415 | 1348.527 |

Analysis of Maximum Likelihood Estimates

| Parameter | | | DF | Estimate | Standard Error | Wald Chi-Square | Pr > ChiSq |
|---|---|---|---|---|---|---|---|
| Intercept | | | 1 | −4.3456 | 0.2711 | 256.9229 | <.0001 |
| Toxin | High | | 1 | 2.0114 | 0.1906 | 111.3244 | <.0001 |
| Toxin | Med | | 1 | 0.8249 | 0.1995 | 17.0977 | <.0001 |
| Feed | Wet | | 1 | 1.1221 | 0.3604 | 9.6937 | 0.0018 |
| Hsize | Large | | 1 | 2.3128 | 0.2578 | 80.4674 | <.0001 |
| Hsize | Medium | | 1 | 1.0529 | 0.2491 | 17.8723 | <.0001 |
| Toxin*feed | High | wet | 1 | −0.1618 | 0.4749 | 0.1160 | 0.7334 |
| Toxin*feed | Med | wet | 1 | −0.7283 | 0.5527 | 1.7365 | 0.1876 |

One should realize that the parameter estimates of the main effects of model B are not much different from those of model A. Now, to test for the significance of the interaction terms using the deviance, we use the likelihood ratio. In the model without interaction, −2log $L$ = 1350.475, and in the model with interaction, $-2\log L = 1348.527$. Therefore the value of the likelihood ratio test statistic is $1350.475 - 1348.527 = 1.948$. Asymptotically, the deviance has a chi-square distribution with degrees of freedom equal to $7 - 5 = 2$. Therefore, we do not have sufficient evidence to reject the null hypothesis of no interaction effect and conclude that the main effects model is adequate.

### General Comments

1. The main risk factor of interest in this study was the type of feed. Since the estimated coefficient of feed in model A is positive, then according to the way we coded that variable, dry feed seems to have sparing effect on the risk of listeriosis.
2. From model A, adjusting for herd size, the odds of a farm being a case with wet feed and high level of mycotoxin, relative to a farm with medium or low level of mycotoxin and dry feed is $\exp[0.864 + 1.990] = 17.35$.
3. From the model with interaction, the same odds ratio estimate is

$$\exp[1.122 + 2.0114 - 0.1618] = 19.52.$$

The difference, $19.52 - 17.35 = 2.17$, is the bias in the estimated odds ratio if the interaction effects were ignored. Note that epidemiologists are interested in measuring the magnitude of bias, in odds ratio estimate, due to omitted effects, even if these effects are not statistically significant.

In order to run these models in R, we need to enter the data first. The tabular data and the R code are given as

```
toxin <- gl(3,2, 18, label = c("Low","Medium","High"))
feed <- gl(2,1, 18, label = c("Dry","Wet"))
hsize <- gl(3,6, 18, label = c("Small","Medium","Large"))
disease <- c(2,1,5,4,2,8,15,5,15,3,60,10,40,5,40,0,40,4)
nodisease <- c(200,45,160,55,75,10,405,60,
280,32,155,20,300,4,100,2,58,4)
data.frame(toxin,feed,hsize,disease,nodisease)
outcome <- cbind(disease,nodisease)
# Model A with main effects only
Logist_a <-glm(outcome~toxin+feed+hsize,family=binomial)
summary(Logist_a)
confint(Logist_a)
exp(coef(Logist_a))
exp(cbind(OR=coef(Logist_a),confint(Logist_a)))
anova(Logist_a, test="Chisq")
# Model B with interations
Logist_b<-glm(outcome~toxin+feed+hsize+toxin*feed,
family=binomial)
> # Model A with main effects only
> Logist_a <-glm(outcome~toxin+feed+hsize,family=binomial)
> summary(Logist_a)
glm(formula = outcome ~ toxin + feed + hsize, family = binomial)
```

### Deviance Residuals:

```
    Min      1Q  Median      3Q     Max
-2.2190 -1.2675 -0.2793  0.2844  2.5191
```

**Coefficients:**

```
              Estimate  Std. Error z value Pr(>|z|)
(Intercept)    -4.3350     0.2684 -16.150   < 2e-16 ***
toxinMedium     0.7288     0.1860   3.918 8.91e-05 ***
toxinHigh       1.9898     0.1751  11.367   < 2e-16 ***
feedWet         0.8637     0.2131   4.054 5.04e-05 ***
hsizeMedium     1.0819     0.2487   4.350 1.36e-05 ***
hsizeLarge      2.3401     0.2581   9.068   < 2e-16 ***
(Dispersion parameter for binomial family taken to be 1)
```

**Null deviance: 284.66 on 17 degrees of freedom**
**Residual deviance: 34.72 on 12 degrees of freedom**
**AIC: 109.48**

> confint (Logist_a)

**Confidence interval on the estimated coefficients**

```
                  2.5 %     97.5 %

(Intercept) -4.8863625 -3.831539
toxinMedium  0.3640038  1.094440
toxinHigh    1.6509486  2.337990
feedWet      0.4382013  1.275354
hsizeMedium  0.6134811  1.592642
hsizeLarge   1.8540907  2.869391
```

**Confidence intervals on the odds ratios**

```
                    OR         2.5 %       97.5 %

(Intercept) 0.01310247 0.007548831  0.02167622
toxinMedium 2.07268339 1.439079669  2.98750895
toxinHigh   7.31432557 5.211921514 10.36038860
feedWet     2.37199521 1.549916813  3.57996784
hsizeMedium 2.95040995 1.846849331  4.91672406
hsizeLarge 10.38250307 6.385888610 17.62628381
```

**Analysis of Deviance using the ANOVA**

anova (Logist_a, test="Chisq")

**Analysis of Deviance Table**

```
        Df. Deviance Resid. Df  Resid.   Dev Pr(>Chi)

NULL                       17   284.66
toxin 2 124.237            15   160.42 <2e-16 ***
feed  1   2.385            14   158.04 0.1225
hsize 2 123.317            12    34.72 <2e-16 ***
```

Note that, the "glm" would use the first level of a factorial explanatory variable as a reference. In order for some other level to be the reference, the order of the levels should be changed. For example, if large herd size is desired as a reference, then the following statement should be run before fitting the model:

$$\text{hsize} < -\text{relevel}(\text{hsize}, \text{ref} = \} \text{Large})$$

The preceding sections demonstrated fitting the logistic regression to binary data with several explanatory variables. A crucial assumption for the likelihood inference to be valid is the independence of the responses. This assumption may be valid if the individuals are randomly sampled. However, there are situations where sampling individuals may not be feasible. The remainder of this chapter will be devoted to discussions concerned with fitting regression models for binary data obtained from clusters of subjects.

## 5.6 Modeling Correlated Binary Outcome Data

### 5.6.1 Introduction

Data from clustered samples arise frequently in many statistical and epidemiologic investigations. We have already discussed in Chapters 3 and 4 that clustering may be as a result of observations that are repeatedly collected on the experimental units as in cohort studies, or may be due to sampling blocks of experimental units such as families, herds, and litters. The data of interest consist of a binary outcome variable $y_{ij}$ where $i = 1, 2, \ldots, k$ indexes the clusters and $j = 1, 2, \ldots, n_i$ indexes units within clusters. A distinguishing feature of such clustered data is that they tend to exhibit intracluster correlation. To obtain valid inference, the analysis of clustered data must account for the effect of this correlation within a regression model. Ignoring the clustering effect will affect the validity of statistical inference on the regression parameters $\beta$. The following section illustrates the diversity of the problems in which the intracluster effect may be present through a number of examples.

Note that we shall investigate the problem of fitting regression models for clustered binary data when the units are naturally aggregated. For example, the cluster may be a herd, a flock, a family, or a litter of pups. However, there are special types of clusters of binary observations, such as longitudinal clustering that we deal with in Chapter 8. Correlated binary data from longitudinal studies where repeated measures of a binary outcome are gathered from independent samples of individuals need special modeling strategies, simply because the nature of the within-cluster (subject) correlation is different. Moreover, there usually are two types of covariates measured: The first set is measured at the baseline and do not change during the course of the study. The covariates of the second set are measured at each time point. For longitudinal studies we may be interested in how an individual's response changes over time, or more generally in the effect of change in covariates on the binary outcome.

The individual measurements within the cluster are called subunits of the cluster. Positive correlation among subunits is manifested by more homogeneous subunits and more variable cluster totals than would be expected with no correlation (i.e., simple Bernoulli sampling or binomial data). This effect is called extrabinomial variation or overdispersion. Ignoring this positive correlation in the analysis will result in statistical tests that overstate the significance of the differences seen among subunit responses. In Chapter 4 we focused on the analysis of correlated binary data from clinical and field trials. Here we focus primarily on regression-type methods, which can accommodate both continuous and categorical covariates.

Although modeling strategies for correlated binary data fall into several categories, in this chapter we shall focus on two models discussed briefly in Chapter 3:

1. Population average (PA) logistic regression models: Model the marginal probabilities in terms of covariates, treating the correlation among cluster members as nuisance parameters. For these models the generalized estimating equations is the method of choice to account for the correlation when covariates are measured at both the cluster and the unit level.

2. Cluster-specific logistic regression models: Allow the model for each cluster to differ by including cluster-specific parameters, which can describe the correlation structure within the cluster. Since the number of such parameters grows along with the number of clusters, a popular approach is to consider these cluster-specific parameters as a random sample from some underlying distribution.

Once a modeling strategy has been chosen, there is also the issue of which method or methods can be used to fit the model. Because of the complexity of specifying a complete joint distribution for the set of correlated responses and the associated computational burdens, maximum likelihood estimation is not always feasible. However, pseudo-likelihood and different types of approximations to the desired likelihoods have been used. We shall elaborate on this issue further in subsequent chapters.

### 5.6.2 Population Average Models: The Generalized Estimating Equation (GEE) Approach

The parameters of such models have a "population-averaged" interpretation in the sense that the effect of the covariates is averaged across clusters, which form the population. Models and methods have been formulated that deal specifically with only the mean and within-cluster correlation.

An estimation approach that places the emphasis on estimating marginal mean parameters while treating the association parameters as nuisance is called the generalized estimating equation (GEE) approach for which software is readily available in SAS PROC (GENMOD). The marginal means are modeled via any generalized linear model (GLM), which includes the familiar linear regression and logistic regression models. For binary data, the logit (i.e., logistic regression), probit, or complementary log–log links are commonly used to relate the marginal mean to the linear combination of the covariates (i.e., the linear predictor $x'_{ij}\beta$).

If the analyst incorrectly assumed that all observations, both within and between clusters, were independent, maximum likelihood estimation of the $\beta$ regression coefficients using standard software for generalized linear models would result in estimates that were consistent but not efficient. To obtain better efficiency, the association within clusters must be built into the estimation method. The GEE method provides a way to do this.

The introduction of GLM expanded the classical regression model by allowing the expected value of the response to be a nonlinear function of the linear predictor and the variance of the responses to depend on the expected value. The relationship between the variance and the expected value, however, is restricted to those found in exponential family distributions. The data distribution in GLM is completely specified, and thus maximum likelihood estimation is possible.

If the constraint that the marginal distributions have exponential family form is relaxed so that the variance can be an arbitrary function of the mean, then we obtain quasi-likelihood models (Wedderburn, 1974). For these models, which still make between- and within-cluster independence assumptions, a quasi-score equation is derived via a quasi-likelihood function.

The GEE approach extends quasi-likelihood models by including a within-cluster "working" correlation matrix in the quasi-score (estimating) equations. The analyst can specify a form of this within-cluster correlation matrix or allow it to be completely unspecified. For example, one could specify a common correlation for every pair of cluster members (which would be assumed the same in every cluster), called an "exchangeable" correlation structure, or an autoregressive structure, where cluster members closer in time or space would be assumed to be more highly correlated than those further apart. These correlation structures will be dealt with in Chapter 8. This method allows unequal cluster sizes, but any missing data are assumed missing completely at random (MCAR) in the sense of Little and Rubin (1987). The issues with missing data will be discussed in details with illustrative examples in Chapter 12.

The GEE method will produce consistent and asymptotically normal estimates of the $\beta$ parameters under some general conditions and the correct specification of the mean, even if the working correlation structure is incorrectly specified. The stronger the within-cluster correlation and the closer the working correlation is to the true underlying correlation, the higher the gain in efficiency. The resulting estimates of correlation parameters are biased and less efficient. If they are considered nuisance parameters of no scientific interest, this lack of useful estimates is of little concern.

### 5.6.3 Cluster-Specific Models (Random-Effects Models)

Cluster-specific models are differentiated from population average of marginal models by the inclusion of parameters, which are specific to a cluster. A cluster-specific model includes covariates, which are linearly related to the log odds of the marginal probability of a positive response. We might also expect the intercept and slope of the relationship to vary from cluster to cluster. A model with only one covariate would take a form similar to the mixed linear model discussed in Chapter 3. Recall that if the $\log[P_r(y_{ij} = 1 | x_{ij})]$ is a linear function of one covariate $x_{ij}$ such that

$$\text{logit}\left[\Pr\left(y_{ij} = 1 | x_{ij}\right)\right] = \beta_{i1} + \beta_{i2} x_{ij} \tag{5.13}$$

where $\beta_{i1}$ and $\beta_{i2}$ are the intercept and slope parameters for cluster $i$. Inference under this model is complicated by the fact that the number of parameters grows with the number of clusters. A popular approach to reducing the number of parameters in a cluster-specific model is to assume that the clusters are a random sample from some underlying population of clusters and that the parameter values for the clusters follow a distribution. A typical choice is the Gaussian distribution:

$$\begin{bmatrix} \beta_{i1} \\ \beta_{i2} \end{bmatrix} \sim N\left( \begin{bmatrix} \alpha_1 \\ \alpha_2 \end{bmatrix}, \Sigma \right)$$

where $\alpha_1$ and $\alpha_2$ are the mean intercept and slope values for the population of clusters, and $\Sigma$ is a $2 \times 2$ covariance matrix. This assumed distribution on the parameters makes this cluster-specific model a random-effect model. Models of this type are also commonly called mixed-effects, hierarchical, or random coefficients models. If we define $\beta_{i1} = \alpha_1 + u_{i1}$ and $\beta_{i2} = \alpha_2 + u_{i2}$, then $u_{i1}$ and $u_{i2}$ are the deviations from the mean intercept and slope term for the $i$th cluster. We can rewrite the model for the $i$th cluster's response in terms of the mean intercept and slope $\alpha_1$ and $\alpha_2$ (fixed effects) and the (unobserved) individual deviations as

$$\text{logit}\left[\Pr\left(y_{ij} = 1 | x_{ij}\right)\right]$$

$$= (\alpha_1 + u_{i1}) + (\alpha_2 + u_{i2})x_{ij}, \quad \text{where} \begin{bmatrix} u_{i1} \\ u_{i2} \end{bmatrix} \sim N(0, \Sigma)$$

In this formulation the fixed effects $\alpha_1, \alpha_2$ are interpreted as the typical parameter values for the population while the random effects modify the average parameters to be specific to that cluster. Fitting the aforementioned model is done quite easily using the GLIMMIX/SAS macro as we show in this chapter.

The mixed effects model can be written in a more compact form as

$$\text{logit}\left(p_{ij}\right) = X_{ij}\beta + Z_{ij}\gamma_i \tag{5.14}$$

where $\beta$ is a vector of fixed effects, $\gamma_i$ is a vector of random effects, $X_{ij}$ and $Z_{ij}$ are covariate vectors, corresponding to the fixed effects and the random effects, respectively. In most applications it is assumed that $\gamma_i$ has a multivariate normal distribution with mean 0 and covariance matrix $\sum$, that is, $\gamma_i \sim \text{MVN}\,(0, \sum)$.

To estimate the model parameters, the method of maximum likelihood can be used. The likelihood function is given by

$$
\begin{aligned}
L(\beta, b_i) &= \prod_{i=1}^{k}\prod_{j=1}^{n_i} p_{ij}^{y_{ij}}\left(1 - p_{ij}\right)^{1-y_{ij}} \\
&= \prod_{i=1}^{k}\prod_{j=1}^{n_i} \frac{\left[\exp\left(x_{ij}\beta + Z_{ij}\gamma_i\right)\right]^{y_{ij}}}{\exp\left(x_{ij}\beta + Z_{ij}\gamma_i\right)} \\
&= \prod_{i=1}^{k} l(l_i)
\end{aligned}
\tag{5.15}
$$

The standard approach to dealing with a likelihood function that contains random variables is to integrate the likelihood function with respect to the distribution of these variables. After integrating out $\gamma_i$, the resulting function is called a "marginal likelihood," which depends on the fixed effects parameters and the parameters of the covariance matrix $\sum$. The maximum likelihood estimates of these parameters are those values that maximize the marginal likelihood function given as

$$L(\beta, D) = \prod_{i=1}^{k} \int l_i \frac{\exp\left[-\frac{1}{2}b_i D^{-1} b_i^T\right]}{(2\pi)^{k/2}|D|^{1/2}}\, db_i \tag{5.16}$$

The two problems associated with the direct maximization of the marginal likelihood 5.16 are

a. Closed form expression for the integrals (5.16) is not available, so we cannot find exact maximum likelihood estimates.

b. The maximum likelihood estimator of the variance components does not take into account the loss in the degrees of freedom resulting from estimating fixed effects. This means that the ML estimates of the variance components are biased in small samples. However, the GLIMMIX uses the restricted maximum likelihood (REML), which corrects for bias in the estimates of the variance components.

Models with random effects are expected to be more efficient than population-average models, provided that we have correctly identified the correct distribution for the random component in the model. The random effects model can be extended to include more than one random component. For example, the health status (presence or absence of disease) of individuals in the same household, same counties, and in the same geographical area is often more alike than that of individuals from different households, counties, or regions. This may be due to common socioeconomic, environmental, and/or behavioral factors. The ability to quantify sources of unobserved heterogeneity in the health outcome of individuals is important for several reasons (see Katz et al., 1993). The within household, or counties or regional clustering of disease, alters the effective sample size needed to provide accurate estimates of disease prevalence. Estimates of disease prevalence at each level of an organization (household, county, region, etc.) can provide insight into the dynamics among the risk factors operating at each level. The ability to obtain separate estimates of the variance component for the random effect at each level of clustering may guide the policy makers as to which level of organization we should direct scarce health management dollars.

### 5.6.4 Interpretation of Regression Parameters

Since the two modeling approaches will often lead to different interpretations of the $\beta$ parameters, an understanding of their differences is crucial. The choice of an appropriate model will be guided by how well the interpretations address the research question of interest.

Comparing the interpretations of $\beta_k$ in the marginal model and as a fixed-effects parameter in a cluster-specific model can be illustrated with the Miall and Oldham data discussed in Chapter 3. The family constitutes a cluster and siblings are the subunits of the cluster. The mother's hypertensive status (Yes if SBP is over or equal 130 and No if SBP is under 130) was recorded at the beginning of the study. Thus the mother's hypertensive status would be a cluster level covariate, assumed not to change between subunits. Suppose that the siblings SBP levels are similarly dichotomized, so that the response variable is binary.

Consider first a marginal model with an intercept, mother's hypertensive status, and age of the child at the time of measurement. If there was independence among siblings' responses, the interpretation of the regression coefficient would be precisely that of a simple logistic regression model. In the marginal model, the dependence is recognized in the estimation methodology, but not in the model for the marginal mean. The parameter $\beta$ for hypertension status represents the difference in the log odds ratio for disease children with a hypertensive mother and those whose mother did not smoke at baseline. Mathematically, this is the difference in the log odds of the mean risk between these two groups, where the mean is taken over all children (subunits), weighted by the working dependency structure used in the estimation method.

In a random-effects model with fixed effects for the same covariates as listed earlier, plus a random effect for family (i.e., a random cluster effect), the interpretation is conditional on that random effect for the family. Within this family, the coefficient represents the magnitude of change in the log odds one would expect with sib's mother being diseased at baseline versus sib's mother healthy at baseline. Since the model specifies that this coefficient is the same for all families, it is estimated by combining information from different families, such as averaging over all families according to the distribution of that random effect. Because the effect we are trying to measure is not observable, its estimation is heavily model based.

It should be noted that if there are no covariates measured on the individuals within the cluster, rather that the covariates are measured only at the cluster level, we can summarize the binary outcomes for a cluster as a binomial proportion. In this case a positive correlation between $y_{ij}$ and $y_{il}$ is then manifest as overdispersion or extrabinomial variation.

Before we illustrate through several examples how to model correlated binary data using the GEE and the mixed effects models, we remind the reader that when a regression model if fitted using the weighted least squares, we used the reciprocal of variance as the weights. If we inflate these variances with an appropriate inflation factor that accounts for the within-cluster correlation, we may use the "weighted" logistic regression to fit clustered binary data. Therefore, both the GEE and the weighted logistic regression would be population-average models. Two methods of weighting will be used: the first utilizes Donner's adjustment and therefore is model based, and the second uses the inflation factor devised by Rao and Scott, as shown in Chapter 4.

### Example 5.4

In this example we analyze experimental data taken from Paul (1982) on the number of live fetuses affected by treatment (control, low dose, and medium dose). The high dose was removed for the purpose of this example (see Table 5.10).

To show how RS adjustment is used to correct for the effect of within-litter correlation in the framework of logistic regression, we first derive the inflation factor for each group. From Chapter 4, direct computations show that

$$\text{Control}: \ \hat{\rho}_1 = 0.135, \ v_1 = 0.00127, \ d_1 = 2.34$$

$$\text{Low}: \ \hat{\rho}_2 = 0.135, \ v_2 = 0.0017, \ d_2 = 1.94$$

$$\text{Medium}: \ \hat{\rho}_3 = 0.344, \ v_3 = 0.0036, \ d_3 = 2.41$$

Moreover, the ANOVA estimator of the intralitter correlation needed to construct Donner's inflation factor is $\hat{\rho} = 0.261$. The following is the SAS program that can be used to analyze the data:

```
data dose;
input group $ y n @@;
if group='con' then do; x=-1; d=2.34; end;
if group='low' then do; x= 0; d=1.94; end;
if group='med' then do; x= 1; d=2.41; end;
raoscott =1/d;
donner=1/(1+(n-1)*0.261);
litter=_n_;
cards;
con 1 12 con 1   7 con 4  6 con 0  6 con 0   7 con 0    8
con 0 10 con 0   7 con 1  8 con 0  6 con 2  11 con 0    7
con 5  8 con 2   9 con 1  2 con 2  7 con 0   9 con 0    7
con 1 11 con 0  10 con 0  4 con 0  8 con 0  10 con 3   12
con 2  8 con 4   7 con 0  8 low 0  5 low 1  11 low 1    7
low 0  9 low 2  12 low 0  8 low 1  6 low 0   7 low 1    6
low 0  4 low 0   6 low 3  9 low 0  6 low 0   7 low 1    5
low 5  9 low 0   1 low 0  6 low 3  9 med 2   4 med 3    4
med 2  9 med 1   8 med 2  9 med 3  7 med 0   8 med 4    9
med 0  6 med 0   4 med 4  6 med 0  7 med 0   3 med 6   13
med 6  6 med 5   8 med 4 11 med 1  7 med 0   6 med 3   10
med 6  6
;

* Model 1: Marginal Model: Logistic regression ignoring cluster-
ing;
proc logistic descending data=dose;
model y/n = x;

* Model 2: Marginal Model: Logistic regression accounting for
clustering using Rao-Scott's weights;
proc logistic descending data=dose;
model y/n = x;
weight raoscott;
run;

* Model 3: Marginal Model: Logistic regression accounting for
clustering using Donner's weights;
proc logistic descending data=dose;
model y/n =x;
weight donner;
run;

* Model 4: Cluster Specific GEE model with exchangeable correlation
structure;
proc genmod data=dose;
class litter;
model y/n = x / dist=bin link=logit;
repeated subject=litter /type=cs corrw;
run;
data new; set dose;
 noty=n-y;
 do i=1 to y;   dead= 1; output; end;
 do i=1 to noty; dead= 0; output; end;
run;

* Model 5: Cluster specific GLIMMIX model with exchangeable corre-
lation structure;
proc glimmix data=new;
```

**TABLE 5.10**

Data from Shell Toxicology Laboratory

| Control | | Low | | Medium | |
|---|---|---|---|---|---|
| $x^*$ | $n^{**}$ | $x$ | $n$ | X | $n$ |
| 1 | 12 | 0 | 5 | 2 | 4 |
| 1 | 7 | 1 | 11 | 3 | 4 |
| 4 | 6 | 1 | 7 | 2 | 9 |
| 0 | 6 | 0 | 9 | 1 | 8 |
| 0 | 7 | 2 | 12 | 2 | 9 |
| 0 | 8 | 0 | 8 | 3 | 7 |
| 0 | 10 | 1 | 6 | 0 | 8 |
| 0 | 7 | 0 | 7 | 4 | 9 |
| 1 | 8 | 1 | 6 | 0 | 6 |
| 0 | 6 | 0 | 4 | 0 | 4 |
| 2 | 11 | 0 | 6 | 4 | 6 |
| 0 | 7 | 3 | 9 | 0 | 7 |
| 5 | 8 | 0 | 6 | 0 | 3 |
| 2 | 9 | 0 | 7 | 6 | 13 |
| 1 | 2 | 1 | 5 | 6 | 6 |
| 2 | 7 | 5 | 9 | 5 | 8 |
| 0 | 9 | 0 | 1 | 4 | 11 |
| 0 | 7 | 0 | 6 | 1 | 7 |
| 1 | 11 | 3 | 9 | 0 | 6 |
| 0 | 10 | | | 3 | 10 |
| 0 | 4 | | | 6 | 6 |
| 0 | 8 | | | | |
| 0 | 10 | | | | |
| 3 | 12 | | | | |
| 2 | 8 | | | | |
| 4 | 7 | | | | |
| 0 | 8 | | | | |

\* Number of live fetuses affected by treatment.
\*\* Total number of live fetuses.

```
 class litter;
 model dead =x/ solution dist=bin link=logit;
 random intercept /subject=litter type=cs;
run;
```

Selected SAS output of fitting these models is given next.
**Model 1:** Marginal Model: Logistic Regression Ignoring Clustering

Analysis of Maximum Likelihood Estimates

| Parameter | DF | Estimate | Standard Error | Wald Chi-Square | Pr > ChiSq |
|-----------|-----|----------|----------------|-----------------|------------|
| Intercept | 1 | −1.4012 | 0.1165 | 144.6401 | <.0001 |
| X | 1 | 0.6334 | 0.1372 | 21.3253 | <.0001 |

Odds Ratio Estimates

| Effect | Point Estimate | 95% Wald Confidence Limits | |
|--------|----------------|----------------------------|---|
| X | 1.884 | 1.440 | 2.465 |

**Model 2:** Marginal Model: Logistic Regression Accounting for Clustering Using Rao-Scott's Weights

Analysis of Maximum Likelihood Estimates

| Parameter | DF | Estimate | Standard Error | Wald Chi-Square | Pr > ChiSq |
|-----------|-----|----------|----------------|-----------------|------------|
| Intercept | 1 | −1.4247 | 0.1752 | 66.1522 | <.0001 |
| X | 1 | 0.6354 | 0.2126 | 8.9351 | 0.0028 |

Odds Ratio Estimates

| Effect | Point Estimate | 95% Wald Confidence Limits | |
|--------|----------------|----------------------------|---|
| X | 1.888 | 1.245 | 2.864 |

**Model 3:** Marginal Model: Logistic Regression Accounting for Clustering Using Donner's Weights

Analysis of Maximum Likelihood Estimates

| Parameter | DF | Estimate | Standard Error | Wald Chi-Square | Pr > ChiSq |
|-----------|-----|----------|----------------|-----------------|------------|
| Intercept | 1 | −1.4058 | 0.1934 | 52.8505 | <.0001 |
| X | 1 | 0.6180 | 0.2287 | 7.3015 | 0.0069 |

Odds Ratio Estimates

| Effect | Point Estimate | 95% Wald Confidence Limits | |
|--------|----------------|----------------------------|---|
| X | 1.855 | 1.185 | 2.905 |

**Model 4:** Cluster-Specific GEE Model Using Exchangeable Correlation Structure

Criteria For Assessing Goodness Of Fit

| Criterion | DF | Value | Value/DF |
|---|---|---|---|
| Deviance | 65 | 168.1053 | 2.5862 |
| Scaled Deviance | 65 | 168.1053 | 2.5862 |
| Pearson's Chi-Square | 65 | 159.5761 | 2.4550 |
| Scaled Pearson X2 | 65 | 159.5761 | 2.4550 |
| Log Likelihood | | −237.4111 | |

Analysis Of GEE Parameter Estimates

| | | Empirical Standard Error Estimates | | | | |
|---|---|---|---|---|---|---|
| Parameter | Estimate | Standard Error | 95% Confidence Limits | | Z | Pr > \|Z\| |
| Intercept | −1.4012 | 0.1733 | −1.7409 | −1.0615 | −8.08 | <.0001 |
| X | 0.6334 | 0.2171 | 0.2078 | 1.0589 | 2.92 | 0.0035 |

**Model 5:** Cluster-Specific GLIMMIX Model Using Exchangeable Correlation Structure

| Fit Statistics | |
|---|---|
| -2 Res Log Pseudo-Likelihood | 2411.02 |
| Generalized Chi-Square | 355.66 |
| Gener. Chi-Square / DF | 0.72 |

Solutions for Fixed Effects

| Effect | Estimate | Standard Error | DF | t Value | Pr > \|t\| |
|---|---|---|---|---|---|
| Intercept | −1.5939 | 0.2003 | 65 | −7.96 | <.0001 |
| x | 0.6631 | 0.2343 | 432 | 2.83 | 0.0049 |

The R code to set up the data for analysis and run models 1 to 4 is shown in the Excel data file toxicology.xlsx, which may be imported to the R using the command:

```
Data=read.csv(file.choose()).
```

```
group <- c(rep("c",27),rep("l",19),rep("m",21))
litter <- 1:67
y <- c(1, 1, 4, 0, 0, 0, 0, 0, 1, 0, 2, 0, 5, 2, 1, 2, 0, 0, 1,
    0, 0, 0, 0, 3, 2, 4, 0, 0, 1, 1, 0, 2, 0, 1, 0, 1, 0, 0,
    3, 0, 0, 1, 5, 0, 0, 3, 2, 3, 2, 1, 2, 3, 0, 4, 0, 0, 4,
    0, 0, 6, 6, 5, 4, 1, 0, 3, 6)
n <- c(12, 7, 6, 6, 7, 8,10, 7, 8, 6,11, 7, 8, 9, 2, 7, 9, 7,11,
    10, 4, 8,10,12, 8, 7, 8, 5,11, 7, 9,12, 8, 6, 7, 6, 4, 6,
```

```
      9, 6, 7, 5, 9, 1, 6, 9, 4, 4, 9, 8, 9, 7, 8, 9, 6, 4, 6,
      7, 3,13, 6, 8,11, 7, 6,10, 6)
x    <- c(rep(-1,27),rep(0,19),rep(1,21))
d    <- c(rep(2.34,27),rep(1.94,19),rep(2.41,21))
raoscott=1/d
donner=1/(1+(n-1)*0.261)
data.frame(group,n,y,litter,x,d,raoscott,donner)
outcome <- cbind(y,n-y)

# Model 1
logistic1 <-glm(outcome~x,family=binomial("logit"))
summary(logistic1)

# Model 2
logistic2             <-glm(outcome~x,family=binomial("logit"),
weights=raoscott)
summary(logistic2)

# Model 3
logistic3             <-glm(outcome~x,family=binomial("logit"),
weights=donner)
summary(logistic3)

# Model 4
library(gee)
gee4      <-gee(outcome~x,id=litter,family=binomial("logit"),
corstr="exchangeable")
summary(gee4)
```

As can be seen from Table 5.11, the point estimates of the slope and intercept are almost the same across models 1 to 4, however, model 1 produces the smallest SE for the estimated slope because it ignores the within-litter correlation. The other four models provide an almost similar SE for the slope after adjusting for the within-litter correlation. The differences between population average model and the cluster specific or random effects models are documented in literature and the reader can find a lucid discussion in the papers by Zeger et al. (1988) and Neuhaus et al. (1991).

## 5.6.5 Multiple Levels of Clustering

As demonstrated in the previous section, clustered data may exhibit more than two levels of hierarchy. For example, radon is a naturally occurring radioactive gas known to cause lung cancer in high concentration. The distribution of radon varies considerably among homes with some homes having

**TABLE 5.11**

Results of Fitting the "Toxicology Data" Using R

| Model | Regression Coefficient | Standard Error | AIC | Residual Deviance |
|-------|------------------------|----------------|-------|-------------------|
| 1 | 0.633 | .137 | 250 | 168 |
| 2 | .635 | .213 | 113 | 74 |
| 3 | .618 | .229 | 100.2 | 68.14 |
| 4 | .633 | .217 | 100 | 68.14 |

considerably high levels of concentration. The data are organized in several levels: regions, counties, and households within counties. It is important to identify the level of hierarchy at which the intervention should be directed. The following example illustrates this situation within an experimental setting.

**Example 5.5**

The following data are taken from Schall (1991). Four hundred cells were placed on a dish and three dishes were irradiated at a time. After the cells were irradiated, the surviving cells were counted. Since cells would also die naturally, dishes with cells were put in the radiation chamber without being irradiated to establish the natural mortality. For the purpose of this example, only these zero-dose data will be analyzed. Twenty-seven dishes on nine time points, or three per time point, were available. The resulting 27 binomial observations are given in Table 5.12.

The GLIMMIX SAS code is

```
data dish;
input time dish y @@;
n=400; noty = 400-y;
cards;
1  1 178 1  2 193 1  3 217 2  4 109 2  5 112 2  6 115 3  7  66
3  8  75 3  9  80 4 10 118 4 11 125 4 12 137 5 13 123 5 14 146
5 15 170 6 16 115 6 17 130 6 18 133 7 19 200 7 20 189 7 21 173
8 22  88 8 23  76 8 24  90 9 29 121 9 30 124 9 31 136
;

data cell; set dish;
     do i=1 to y;   r=1;  output; end;
     do i=1 to noty; r = 0; output; end;
run;
```

**TABLE 5.12**

Cell Irradiation Data

| Occasion | Cells Surviving Out of 400 Placed | Occasion | Cells Surviving Out of 400 Placed |
|---|---|---|---|
| 1 | 178 | 6 | 115 |
| 1 | 193 | 6 | 130 |
| 1 | 217 | 6 | 133 |
| 2 | 109 | 7 | 200 |
| 2 | 112 | 7 | 189 |
| 2 | 115 | 7 | 173 |
| 3 | 66 | 8 | 88 |
| 3 | 75 | 8 | 76 |
| 3 | 80 | 8 | 90 |
| 4 | 118 | 9 | 121 |
| 4 | 125 | 9 | 124 |
| 4 | 137 | 9 | 136 |
| 5 | 123 | | |
| 5 | 146 | | |
| 5 | 170 | | |

```
* Model 1: One random effect;
proc glimmix data=cell;
 class time;
 model r = /solution;
 random intercept/ subject=time;
run;

* Model 2: Two random efects;
proc glimmix data=cell;
 class time dish;
 model r = /solution;
 random time dish/subject=time;
run;
```

The result of fitting the GLIMMIX is summarized in Tables 5.13 and 5.14.

**TABLE 5.13**

GLIMMIX Procedure Output with One Random
Effect

| Fit Statistics | |
| --- | --- |
| -2 Res Log Likelihood | 13892.10 |
| AIC (smaller is better) | 13896.10 |

Covariance Parameter Estimates

| Cov Parm | Subject | Estimate | Standard Error |
| --- | --- | --- | --- |
| Intercept | Time | 0.01036 | 0.005269 |
| Residual | | 0.2111 | 0.002874 |

Covariance Parameter Estimates

| Effect | Estimate | Stand Error | DF | t value | Pr > |t| |
| --- | --- | --- | --- | --- | --- |
| Intercept | 0.3277 | 0.03422 | 8 | 9.58 | <.0001 |

**TABLE 5.14**

GLIMMIX Procedure Output with Two Random
Effects

| Fit Statistics | |
| --- | --- |
| -2 Res Log Likelihood | 13886.93 |
| AIC (smaller is better) | 13892.93 |

| Covariance Parameter Estimates | | |
| --- | --- | --- |
| Cov Parm | Estimate | Standard Error |
| Time | 0.01019 | 0.005270 |
| Dish (time) | 0.000507 | 0.000345 |
| Residual | 0.2108 | 0.002872 |

| Covariance Parameter Estimates | | | | | |
| --- | --- | --- | --- | --- | --- |
| Effect | Estimate | Stand Error | DF | t value | Pr > |t| |
| Intercept | 0.3277 | 0.03422 | 8 | 9.58 | <.0001 |

The preceding two models do not have fixed effects except the intercept parameter. One way to select the model that provides the best fit is to compare the (-2 RES LIKELIHOOD) with the smaller value indicting a better fit. Since for the model with one variance component, $-2reslikelihood$ = 13892.1, and for the model with two variance components, $-2reslikelihood$ = 13886.93, the second model is a better model.

**Important Remark:** Note that we used the *residual likelihood* (or restricted maximum likelihood, REML) to compare the models in terms of their variance components. Once a model has been selected, the model building strategy with regard to the fixed effects cannot proceed using the model fitted with the REML. The model should be refitted using the option *method=ml* in the model statement of the GLIMMIX, that is, requesting that the model be fitted using the method of maximum likelihood. Therefore, only the difference between "−2 log-likelihood" of two nested models can be used to assess the goodness of it in the "generalized linear mixed model" with components of variance. It should also be noted that this remark holds true for the linear mixed model of Chapter 3.

Here is the R code:

```
# Must install MASS package

library(MASS)

time <- c(1,1,1,2,2,2,3,3,3,4,4,4,5,5,5,6,6,6,7,7,7,8,8,9,9,9)
dish <- 1:27
y <- c(178,193,217,109,112,115,66,75,80,118,125,137,123,146,
170,115,130, 133,200,189,173,88,76,90,121,124,136)
dishn <- data.frame(time,dish,y)

r1 <- data.frame(lapply(dishn, function(x) rep(x,dishn$y)),
r=1)[,c(1,2,4)]
r0 <- data.frame(lapply(dishn, function(x) rep(x,(400-
dishn$y))),r = 0)[,c(1,2,4)]
dish <- rbind(r1,r0)

glmm1 <-glmmPQL(r~1, random=~1|factor(time), family=binomial
(link = "logit"), data=dish)
summary(glmm1)
glmm2 <-glmmPQL(r~1, random=~time+dish| factor(time), family=
binomial(link = "logit"), data=dish)
summary(glmm2)
```

## 5.7 Logistic Regression for Case-Control Studies

### 5.7.1 Cohort versus Case-Control Models

Although initial applications of the logistic regression model were specific to cohort studies, this model can be applied to the analysis of data from case-control studies. The specification of the logistic model for case-control studies in which the presence or absence of exposure is taken to be the dependent variable was given by Prentice (1976). His approach assumes that we are interested in the effect of one factor. Suppose that the exposure factor, which is the focus of interest is dichotomous, say $x_1$, where $x_1 = 1$ (exposed) and $x_1 = 0$ (unexposed) and that $x_2$ is another potential risk factor or a confounder. Hence, the prospective logistic model corresponding to the retrospective study is such that

$$\text{logit}[\text{pr}(X_1 = 1|y, X_2)] = \beta_0 + \beta_1 y + \beta_2 X_2$$

The relative odds of exposure among diseased as compared to the nondiseased may be given as

$$\text{OR} = e^{\beta_0 + \beta_1 (1) + \beta_2 X_2} / e^{\beta_0 + \beta_1 (0) + \beta_2 X_2}$$

$$= e^{\beta_1}$$

The above OR is mathematically equivalent to the relative odds of disease among the exposed subjects as compared to the unexposed, as we have already shown.

The rationale behind this argument was provided by Mantel (1973) as follows: Let $f_1$ denote the sampling fraction of cases, that is, if $n_1$ cases were drawn from a population of size $N_1$, then $f_1 = n_1/N_1$. Similarly, we define $f_0$ as the sampling fraction of control. It is assumed that neither $f_1$ nor $f_0$ depend on the covariate vector $X$. Now consider the 2 × 2 Table 5.15.

In a case-control study we wish to model $\text{logit}[y = 1 \mid x, sampled]$ as a linear function of the covariate vector $x$.

Since

$$\text{pr}[y = 1|x, sampled] = \frac{\text{pr}[y = 1, s|x]}{\text{pr}(s)}$$

$$= \frac{f_1 p_x}{f_1 p_x + f_0 q_x}$$

and

$$\text{pr}[y = 0|x, sampled] = \frac{f_0 q_x}{f_1 p_x + f_0 q_x}$$

**TABLE 5.15**

Description of Sampling and Disease Occurrence in a 2 × 2 Layout

|                  | Case $Y = 1$     | Control $Y = 0$  | Total                         |
| ---------------- | ---------------- | ---------------- | ----------------------------- |
| Sampled ($s$)    | $f_1 p_x$        | $f_0 q_x$        | $f_1 p_x + f_0 q_x$           |
| Not Sampled ($s$)| $(1 - f_1)p_x$   | $(1 - f_0)q_x$   | $(1 - f_1)p_x + (1 - f_0)q_x$ |
| All              | $p_x$            | $q_x$            | 1                             |

*Note:* $p_x = \text{pr}[y = 1 \mid x]$.

Then

$$\text{logit}[\text{pr}(y = 1 | x, sampled)] = \log\left(\frac{f_1 p_x}{f_0 q_x}\right)$$

$$= \log\frac{f_1}{f_0} + \text{logit}[\text{pr}(y = 1 | x)]$$

$$= \log\frac{f_1}{f_0} + \beta_0 + \beta_1 x_1 + \dots \beta_x k_k$$

$$= \beta_0^* + \sum_{j=1}^{k}\beta_j x_j$$

where

$$\beta_0^* = \log\frac{f_1}{f_0} + \beta_0$$

As can be seen from the last equation, the logistic model for the case-control study has the same form as that of the logistic model for the cohort study. This means that the regression parameters, which measure the joint effects of groups of covariates on the risk of disease, can be estimated from the case-control study. The following remarks are emphasized:

1. If $\beta_0^*, \beta_1, \dots, \beta_k$ are estimated from a case control study, and since $\beta_0^*$ depends on the ratio $f_1/f_0$, the risk of disease $p_x$ (which depends on $\beta_0$) cannot be estimated unless $f_1/f_0$ is known. The situations in which $f_1/f_0$ is known are quite uncommon.
2. For a given $x$, Equation 5.15 represents the log-odds of disease in the sample of cases and controls, which is related to the log-odds of disease in the target population by the factor $(\log f_1/f_0)$. However, if we estimate the log-odds of disease in the sample of cases for a subject with covariate pattern $X^*$, relative to the sampled control whose covariate pattern is $\hat{X}$, then

$$\psi\left(X^*; \hat{X}\right) = \log\left[\frac{e^{\beta_0 + \sum_{j=1}^{k}\beta_j X_j^*}}{e^{\beta_0 + \sum_{j=1}^{k}\beta_j \hat{X}_j}}\right]$$

$$= \exp\sum_{j=1}^{k}\beta_j\left(X_j^* - \hat{X}_j\right)$$

This means that the estimate of $\beta_0^*$ is irrelevant to the estimation of the odds ratio.

### 5.7.2 Matched Analysis

We saw how data from a case-control study can be analyzed using the logistic regression model to measure the effect of a group of covariates on the risk of disease, after adjusting for potential confounders. We have also indicated (Chapter 4) that the primary objective of matching is the elimination of the biased comparison between cases and controls that results when confounding factors are not properly accounted for. A design that enables us to achieve this control over such factors is the matched case-control study. Unless the logistic regression model properly accounts for the matching used in the selection of cases and controls, the estimated odds ratios can be biased. Thus, matching is only the first step in controlling for confounding. To analyze matched case-control study data using logistic regression we will discuss two situations. The first is called 1:1 matching (which means that each case is matched with one control), and the other is 1:M matching (which means that each case is matched with M controls). Before we show how the data analysis is performed we describe the general setup of the likelihood function.

Suppose that controls are matched to cases on the basis of a set of variables $x_1, x_2, \ldots, x_k$. These variables may represent risk factors and those potential confounders that have not been used in the matching procedure. Moreover, the risk of disease for any subject may depend on the "matching variable" that defines a "matched set" to which an individual belongs. The values of these matching variables will generally differ between each of $n$ matched sets of individuals.

Let $p_j(x_{ij})$ denote the probability that the $i$th person in the $j$th matched set is a case (or diseased), $i = 0, 2, \ldots, M, j = 1, 2, \ldots, n$. The vector of explanatory variables for the case is $x_{0j}$, while the vector $x_{ij}$ denotes the explanatory variables for the $M$th control in the $j$th matched set. The disease risk $p_j(x_{ij})$ will be modeled as

$$p_j\left(x_{ij}\right) = \frac{e^{\alpha_j + \sum_{l=1}^{k} \beta_l X_{lij}}}{1 + e^{\alpha_j + \sum_{l=1}^{k} \beta_1 X_{lij}}} \tag{5.17}$$

Here $x_{lij}$ is the value of the $l$th explanatory variable $l = 1, 2, \ldots, k$ for the $i$th individual in the $j$th matched set. The term $\alpha_j$ represents the effects of a particular configuration of matching variables for the $j$th matched set on the risk of disease. It can be seen from Equation 5.17 that the relationship between each explanatory variable and the risk of disease is the same for all matched sets. From Equation 5.17, the odds of the disease are given by

$$\frac{p_j\left(x_{ij}\right)}{1 - p_j\left(x_{ij}\right)} = \exp\left[\alpha_j + \sum_{l=1}^{k} \beta_l x_{lij}\right]$$

In particular, for two individuals from the same matched set, the odds of disease for a subject with explanatory variable $x_{1j}$ relative to one with explanatory variable $x_{2j}$ is

$$\left\{ \frac{p_j(x_{1j})}{1-p_j(x_{1j})} \right\} \left\{ \frac{p_j(x_{2j})}{1-p_j(x_{2j})} \right\}$$

$$= \exp\left[ \beta_1(x_{11j} - x_{12j}) + \ldots + \beta_k(x_{k1j} - x_{k2j}) \right] \tag{5.18}$$

Clearly Equation 5.18 is independent of $\alpha_j$. This means that the odds of disease for two matched individuals with different explanatory variables do not depend on the actual values of the matching variables.

### 5.7.3 Fitting Matched Case-Control Study Data in SAS and R

The likelihood function based on matched case-control studies is known as *conditional likelihood*. For 1:M matched case control study, where the $j$th matched set contains M controls ($j = 1,2,\ldots,n$), we denote by $x_{0j}$ the vector of the explanatory variables for the case and $x_{1j},x_{2j},\ldots,x_{Mj}$ the vectors of explanatory variables for the M controls in the $j$th matched set. Breslow et al. (1978) derived the likelihood function under this setup and showed that it can be written as

$$L = \prod_{j=1}^{n} \left\{ 1 + \sum_{i=1}^{M} \exp\left[ \beta_1(x_{1ij} - x_{10j}) + \beta_2(x_{2ij} - x_{20j}) + \ldots + \beta(x_{kij} - x_{k0j}) \right] \right\}^{-1} \tag{5.19}$$

The main purpose of this section is to discuss, through an example, how the parameters of the conditional likelihood (Equation 5.19) are estimated. For the matched pair design, M = 1, the conditional likelihood (Equation 5.19) reduces to

$$L = \prod_{i=1}^{n} \left\{ 1 + \exp\left[ \sum_{r=1}^{k} \beta_r(x_{rij} - x_{r0j}) \right] \right\}^{-1}$$

$$= \prod_{i=1}^{n} \left\{ 1 + \exp\left[ -\sum_{r=1}^{k} \beta_r z_{rj} \right] \right\}^{-1} \tag{5.20}$$

where $z_{rj} = x_{r0j} - x_{r1j}$.

This likelihood (Equation 5.20) is identical to the likelihood function of a logistic regression for $n$ binary observations $y_i$, such that $y_i = 1$ for $i = 1, 2, \ldots, n$. Note that the explanatory variables become $z_{1j},z_{2j},\ldots,z_{kj}$, and there is no intercept. Therefore, using SAS, PROC LOGISTIC fits the 1:1 matched data by following these steps:

1. The number of matched sets $n$ is the number of observations.
2. The response variable $y_i = 1$ for all $i = 1, 2, \ldots, n$.
3. The explanatory variable $z_{rj} = x_{r0j} - x_{r1j}$ $(r = 1, 2, \ldots, k$ and $j = 1, 2, \ldots, n)$ is the difference between the value of the $r$th explanatory variable for a control and the $r$th explanatory variable for a case, within the same matched set. Note that if qualitative or factor variables are used, the explanatory variables in Equation 5.20 will correspond to dummy variables. Consequently, variables are the differences between the dummy variables of the case and control in the matched pair. Note also that the interaction terms can be included in the model by representing them as products of the corresponding main effects. The differences between these products for the case and control in a matched pair are included in the model.

**Example 5.6: "Hypothetical Data"**

In an investigation aimed at assessing the relationship between somatic cell counts (SCC) and the occurrence of mastitis, a 1:1 matched case-control study was postulated. A "case" cow was matched with a control cow from the same farm based on breed, number of lactations, and age as a possible confounder. The data summary is given in Table 5.16.

In this simple example we have one risk factor, namely, the SCC, which was dichotomized as high and low. Since it is believed that the incidence rate of the disease in younger cows is different from older cows, the matching variable age was divided into two distinct strata, the first for cows whose age is less than four years, and the second for those that are at least four years old.

As already mentioned, we cannot investigate the association between the disease and the age variable since age is a matching variable. However, we shall investigate the possible interaction between the SCC

**TABLE 5.16**

Hypothetical Mastitis Data

|  |  | Case | | |
|---|---|---|---|---|
|  |  | High | Low |  |
| Control | High | 5 | 110 | age < 4 |
|  | Low | 216 | 40 |  |
|  |  | Case | | |
|  |  | High | Low |  |
| Control | High | 5 | 212 | age ≥ 4 |
|  | Low | 308 | 21 |  |

and age. Since the risk factor and the confounder are factor variables and each factor has two levels, we define a single dummy variable for each factor. Let $X_1$ be the indicator variable for SCC, and $X_2$ for age, where

$X_1 = 1$ if the cow has high SCC or $= 0$ if the cow has low SCC
$X_2 = 1$ if the cow's age is <4 or $= 0$ if the cow's age is ≥4

We also define a third dummy variable $X_3$, obtained by multiplying $X_1$ and $X_2$ for each individual animal in the study. With this coding the data is structured as in Table 5.17.

The input variables $(Z_1, Z_2, Z_3)$ are created using variables as set up in Table 5.18, where

$$Z_1 = X_1(\text{Case}) - X_1(\text{Control})$$

$$Z_2 = X_2(\text{Case}) - X_2(\text{Control})$$

$$Z_3 = X_3(\text{Case}) - X_3(\text{Control})$$

Table 5.17 shows the result of coding the data.

**TABLE 5.17**

Coded Variable for the Data in Table 5.16

| $X_1$ (Case) | $X_1$ (Control) | $X_2$ (Case) | $X_2$ (Control) | $X_3$ (Case) | $X_3$ (Case) | Matched Pairs |
|---|---|---|---|---|---|---|
| 1 | 1 | 0 | 0 | 0 | 0 | 5 |
| 1 | 0 | 0 | 0 | 0 | 0 | 216 |
| 0 | 1 | 0 | 0 | 0 | 0 | 110 |
| 0 | 0 | 0 | 0 | 0 | 0 | 40 |
| 1 | 1 | 1 | 1 | 1 | 1 | 5 |
| 1 | 0 | 1 | 1 | 1 | 0 | 308 |
| 0 | 1 | 1 | 1 | 0 | 1 | 212 |
| 0 | 0 | 1 | 1 | 0 | 0 | 21 |

**TABLE 5.18**

Variables for the Logistic Regression Model

| $y$ | $Z_1$ | $Z_2$ | $Z_3$ | Count |
|---|---|---|---|---|
| 1 | 0 | 0 | 0 | 5 |
| 1 | 1 | 0 | 0 | 215 |
| 1 | −1 | 0 | 0 | 110 |
| 1 | 0 | 0 | 0 | 40 |
| 1 | 0 | 0 | 0 | 5 |
| 1 | 1 | 0 | 1 | 308 |
| 1 | −1 | 0 | −1 | 212 |
| 1 | 0 | 0 | 0 | 21 |

The following SAS statements describe how the logistic regression model can be fitted:

```
data match;
input y z1 z3 count @@;
cards;
1 0 0  5 1 1 0 216 1 -1 0 110 1 0 0 40
1 0 0  5 1 1 1 308 1 -1 -1 212 1 0 0 21
;

proc logistic data = match;
freq count;
model y = z1 z3 / noint covb;
run;
```

Note that the response variable $y_i = 1$ for all $i = 1, 2, ..., 8$. In the model statement of the SAS program does not include the matching variable $z_2$ since it is not possible to investigate its association with the disease status. Moreover the option "noint" is specified, so that the logistic regression is fitted without the intercept parameter.

Selected SAS output is shown next:

Analysis of Maximum Likelihood Estimates

| Parameter | DF | Estimate | Standard Error | Wald Chi-Square | Pr > ChiSq |
|-----------|----|----------|----------------|------------------|------------|
| z1 | 1 | 0.6747 | 0.1171 | 33.1793 | <.0001 |
| z3 | 1 | -0.3012 | 0.1473 | 4.1836 | 0.0408 |

Odds Ratio Estimates

| Effect | Point Estimate | 95% Wald Confidence Limits | |
|--------|----------------|----------------------------|---|
| z1 | 1.963 | 1.561 | 2.470 |
| z3 | 0.740 | 0.554 | 0.988 |

Estimated Covariance Matrix

| Parameter | Z1 | z3 |
|-----------|------|------|
| z1 | 0.01372 | -0.01372 |
| z3 | -0.01372 | 0.021684 |

The results indicate that there is a significant association between SCC ($\hat{\beta}_1$) and the disease. There is also significant interaction between the SCC and age ($\hat{\beta}_2$) on the risk of mastitis.

The following R code reads the data and runs the desired analysis:

```
y <- rep(1,8)
z1 <- c(0, 1, -1, 0, 0, 1, -1, 0)
z3 <- c(0, 0,  0, 0, 0, 1, -1, 0)
count <- c(5, 216, 110, 40, 5, 308, 212, 21)
```

```
match <-glm(y~-1+z1+z3,family=binomial("logit"),weights=count)
summary(match)$coef
# Computing the odds ratio
sum.coef<-summary(match)$coef

est<-exp(sum.coef[,1])
upper.ci<-exp(sum.coef[,1]+1.96*sum.coef[,2])
lower.ci<-exp(sum.coef[,1]-1.96*sum.coef[,2])

cbind(est,upper.ci,lower.ci)

# Estimated covariance matrix
summary(match)$cov.scaled
```

**Remark:** For N:M matching, where each case is matched with an arbitrary number of controls, the PROC PHREG in SAS can be used to fit the model. The data should be put in a special format. The following SAS code shows how to analyze the data of Example 5.8 using the proportional hazard Cox regression model. As should be expected there are some differences between the two models and this is attributed to differences in the methods of parameters estimation.

**Example 5.7: Analysis of Matched Data Using Proportional Hazard Regression**

```
data ph_match;
input age scc case count;
agescc=age*scc;
time=2-case;
cards;
0 1 1    5 0 1 0    5 0 1 1 216 0 0 0 216
0 0 1 110 0 1 0 110 0 0 1   40 0 0 0   40
1 1 1    5 1 1 0    5 1 1 1 308 1 0 0 308
1 0 1 212 1 1 0 212 1 0 1   21 1 0 0   21
;
proc phreg data=ph_match;
 model time*case(0)=scc agescc / ties=discrete;
 freq count;
 strata age;
run;
```

Selected SAS output is shown next:

Summary of the Number of Event and Censored Values

| Stratum | Age | Total | Event | Censored | Percent Censored |
|---------|-----|-------|-------|----------|------------------|
| 1 | 0 | 742 | 371 | 371 | 50.00 |
| 2 | 1 | 1092 | 546 | 546 | 50.00 |
| Total | | 1834 | 917 | 917 | 50.00 |

Model Fit Statistics

| Criterion | Without Covariates | With Covariates |
|-----------|--------------------|-----------------|
| –2 LOG L  | 2527.954           | 2432.091        |
| AIC       | 2527.954           | 2436.091        |
| SBC       | 2527.954           | 2445.733        |

Analysis of Maximum Likelihood Estimates

| Variable | DF | Parameter Estimate | Standard Error | Chi-Square | Pr > ChiSq | Hazard Ratio |
|----------|----|--------------------|----------------|------------|------------|--------------|
| Scc      | 1  | 1.18599            | 0.15413        | 59.2107    | <.0001     | 3.274        |
| agescc   | 1  | -0.47533           | 0.19717        | 5.8120     | 0.0159     | 0.622        |

Proportional hazard regression (PHREG) models will be discussed in detail in the survival analysis chapter in this book (Chapter 9), and the reader may delay reading the last example until then. However, it can be seen that the coefficients estimates are slightly different from Example 5.6. One explanation is that whereas the logistic regression is fitted using the method of maximum likelihood, the PHREG is fitted using the method of partial likelihood (see Cox, 1972).

## Example 5.8

This is another example on matched case control studies using PHREG in SAS. The data is a study on the effect of consanguinity on birth defects. Case mothers (women with at least one birth defect) were matched with control mothers who did not have any children with defect. The matching variables were mothers' age and parity. The first few lines of the data statement are given next:

```
weight      cluster    case    consang    folic    survt
1           1          1       1          0        1
1           1          0       0          0        1
1           2          1       1          0        1
1           2          0       0          0        1
1           3          1       1          0        1
1           3          0       0          0        1
22          4          1       1          1        1
22          4          0       0          0        1
14          5          1       1          0        1
```

The first column weight is the number of case = 1, control = 0 pairs that have the same covariate profile. Consanguinity is a categorical variable that = 1 when the mother is married to her first cousin and consanguinity = 0 otherwise. The "folic" column is whether the woman was taking the recommended dose of folic acid (folic = 1 or = 0 if not). The last column is a constant that takes the value 1, as required by SAS to be able to fit Cox regression. The SAS code is

```
Proc phreg data=folic;
model survt*case(0) = folic consang/ties = breslow;
strata cluster;
freq weight;
run;
```

Following is part of the SAS output. As can be seen both risk factors are significantly correlated with birth defect. Consanguinity increases the hazard of birth defect, whereas taking the recommended dose of folic acid reduces the risk.

### Model Fit Statistics

| Criterion | Without Covariates | With Covariates |
|-----------|--------------------|-----------------|
| -2 LOG L | 1487.175 | 1364.928 |
| AIC | 1487.175 | 1368.928 |
| SBC | 1487.175 | 1375.847 |

### Testing Global Null Hypothesis: BETA=0

| Test | Chi-Square | DF | Pr > ChiSq |
|------|-----------|-----|------------|
| Likelihood Ratio | 122.2472 | 2 | <.0001 |
| Score | 104.1115 | 2 | <.0001 |
| Wald | 71.0490 | 2 | <.0001 |

### Analysis of Maximum Likelihood Estimates

| Parameter | DF | Parameter Estimate | Standard Error | Chi-Square | Pr > ChiSq | Hazard Ratio |
|-----------|-----|-------------------|----------------|-----------|-----------|--------------|
| folic | 1 | −0.86646 | 0.24131 | 12.8932 | 0.0003 | 0.420 |
| consang | 1 | 2.28938 | 0.27234 | 70.66601 | <.000 | 9.869 |

Clearly both consanguinity and folic acid intake are risk factors for the risk of birth defects.

**R Code and the Corresponding Output**

**Here the data file is the "folic_matched_case_control"**

```
fit=coxph(Surv(y, case ==1)~consang + folic+strata(cluster),
data=data)
```

**R Output**

```
n= 471, number of events= 235

              coef    exp(coef)   se(coef)    Pr(>|z|)
consang     3.3741    29.1977     0.3037     < 2e-16 ***
folic      -1.3874     0.2497     0.2567     6.47e-08 ***

           exp(coef)   exp(-coef)   lower .95   upper .95
consang     29.1977     0.03425     16.100       52.951
```

```
folic         0.2497      4.00459   0.151         0.413
Concordance= 0.864      (se = 0.114 )
R-square= 0.396     (max possible= 0.945 )
Likelihood ratio test= 237.7 on 2 df,  p=0
Wald test              = 123.8 on 2 df,  p=0
Score (log-rank) test  = 221.8 on 2 df,  p=0
```

There some differences in the values of the regression coefficients in the R output when compared to the results from SAS. These differences are not substantial and may be attributed to the difference in the optimization algorithms.

### 5.7.4 Some Cautionary Remarks on the Matched Case-Control Designs

There has been extensive discussion regarding the advantages and disadvantages of pair matching in group or cluster-randomized experiments. Matching can increase power by reducing study population heterogeneity and can guarantee balance on selected confounders by matching on them. However, matching may lead to a loss of statistical power when matching is not effective. Diehr et al. (1995) showed that matching may lead to some loss of power when the number of pairs is between three and nine, and recommended breaking the matches in statistical analyses to gain power. Donner et al. (2004) cautioned that even though unmatched analysis for a pair-matched design is unbiased for the estimation of the intervention effect, it is biased for the estimation of the effects for other individual-level risk factors. They also pointed out that the pair-matching design makes the applications of generalized estimation equations or mixed-effects models no longer routine and further complicates statistical analyses. Moreover, with a pair-matching design, if one cluster is lost to follow-up, then the matched cluster is no longer useful and must be discarded. Due to these limitations, Donner et al. suggested that pair-matching designs should be used with caution.

## 5.8 Sample Size Calculations for Logistic Regression

We have shown in Chapters 3 and 4 that sample size determination is dominated by work concerned with calculating sample sizes required for comparing means or proportions in two groups. Statistical literature gives little guidance on how to use additional information which may often be available. Gail (1973) specifies calculations for trials with dichotomous exposure and outcome, which are stratified on a third variable. Whittemore (1981) gave approximate sample sizes for logistic regression with small probability when there are covariates. This approach is based on an approximation to the Fisher's information matrix for the estimated parameters in a multiple logistic regression. Wilson and Gordon (1986) showed that approximate sample size can be obtained from Fisher's information matrix without the restriction that the response probability is small.

It has been shown (see McCullagh and Nelder, 1989) that if the GLM is valid, the maximum likelihood estimator of $\hat{\beta}$ is asymptotically normal and variance–covariance matrix $V(\hat{\beta}) = (I(\beta))^{-1}$, where $I(\beta)$ is Fisher's information matrix, whose $(r,s)$ element is

$$I_{rs}(\beta) = \sum_{i=1}^{n} \left(\frac{\partial \mu_i}{\partial \eta_i}\right)^2 x_{ir} x_{is} (V(y_i))^{-1}$$

In discussing sample size required to detect for example, significant values of $\beta_1$, only the corresponding diagonal element $v_{11}$ of $V(\hat{\beta})$ is relevant. In this next section we discuss the contributions made by Wilson and Gordon (1986).

Suppose, without loss of generality, that we are interested in testing $H_0$: $\beta_1 = \beta_1^{(0)}$, with level of significance $\alpha$ and power $1 - \delta$ for specified alternative $H_1 : \beta_1 = \beta_1^{(1)}$. Let $\hat{\beta}_1^{(k)}$ be the value of the MLE of $\beta_1$ under $H_k; k = 0,1$. Further, let $\Phi(z_t) = 1 - t$, where $\Phi$ is the standard normal cumulative distribution function. Moreover, let $u_k = [nv_{11}(\hat{\beta})^{(k)}]^{1/2}$; $u_k$ does not involve $n$. By the standard power calculations we have

$$n \geq \frac{(z_\alpha u_0 + z_\delta u_1)}{\left(\beta_1^{(1)} - \beta_1^{(0)}\right)^2} \tag{5.21}$$

### Example 5.9

As an example we consider the case of a prospective study discussed by Schlesselman (1982). The main interest in the study was whether there is an increased risk of giving birth to a child with a congenital heart defect among mothers who have oral contraceptive exposure three months prior to or after conception. Both outcome ($y$) and exposure ($x$) are binary variables. Let $P_0$ represent the proportion in the control population of unexposed subjects who develop the disease and $P_1$ the corresponding proportion in the exposed population. We assume the logistic model:

$$\mu = E[y|x] = \exp(\beta_0 + \beta_1 x)/[1 + \exp(\beta_0 + \beta_1 x)]$$

Under the null hypothesis $\beta_0^{(0)} = 0$, and under the alternative hypothesis $\beta_0^{(1)} = \log(P_1 q_0 / P_0 q_1)$, and $\beta_1^{(1)} = \log(p_1 q_0 / p_0 q_1)$, where $q_j = 1 - p_j; j = 0,1$, $\beta_1$ is the log odds ratio. Therefore, the total number of subjects in each group is $n/2$ where

$$n \geq \frac{\left[z_\varepsilon (4/\bar{p}\bar{q})^{1/2} + z_\delta [(2/p_0 q_0) + (2/p_1 q_1)]^{1/2}\right]^2}{\left[\log(p_1 q_0 / p_0 q_1)\right]^2} \tag{5.22}$$

and $\bar{p} = (p_0 + p_1)/2$.

As an illustration, the incidence of all congenital heart diseases among controls is $p_0 = 0.008$. The relative risk of disease is $p_1/p_0$. If we assume that the meaningful relative risk to detect is 3, then $p_1 = 3(0.008) = 0.024$. Taking $\alpha = .025$ and power 0.90, we get $n \cong 2416$ for each group.

## 5.9 Sample Size for Matched Case Control Studies

This function gives you the minimum sample size necessary to detect a true odds ratio OR with power and a two-sided type I error probability $\alpha$. If we are using more than one control per case, then this function also provides the reduction in sample size relative to a paired study that you can obtain using your number of controls per case (Dupont, 1988).

Information required

- POWER—Probability of detecting a real effect
- ALPHA—Probability of detecting a false effect (two-sided; double this if you need one sided)
- r—Correlation coefficient for exposure between matched cases and controls
- $P_o$—Probability of exposure in the control group
- m—Number of control subjects matched to each case subject
- OR—Odds ratio

Practical Issues

- Usual values for POWER are 80%, 85%, and 90%; try several in order to explore/scope.
- 5% is the usual choice for ALPHA.
- r can be estimated from previous studies. Note that r is the phi (correlation) coefficient that is given for a 2 × 2 table if you enter it into the Stats-Direct "r by c chi-square" function. When r is not known from previous studies, some authors state that it is better to use a small arbitrary value for r, say 0.2, than it is to assume independence (a value of 0) (Dupont, 1988).
- $P_o$ can be estimated as the population prevalence of exposure. Note, however, that due to matching, the control sample is not a random sample from the population, therefore population prevalence of exposure can be a poor estimate of $P_o$ (especially if confounders are strongly associated with exposure) (Dupont, 1988).
- If possible, choose a range of odds ratios that you want to have the statistical power to detect.

Technical validation

The estimated sample size $n$ is calculated as

$$n = \frac{\left[(1/\sigma_\psi)Z_{\alpha/2} + Z_\beta\right]^2}{d^2}$$

$$\sigma_\psi = \sqrt{\sum_{k=1}^{m} \frac{kt_k \psi(m-k+1)}{(k\psi+m-k+1)^2}}$$

$$t_k = p_1(k-1)p_{0+}^{k-1}(1-p_{0+})^{m-k+1} + (1-p_1)kp_{0-}^k - (1-p_{0-})^{m-k}$$

$$p_{0+} = \frac{p_1 p_0 + r\sqrt{p_1(1-p_1)p_0(1-p_0)}}{p_1} \tag{5.23}$$

$$p_{0-} = \frac{p_0(1-p_1) - r\sqrt{p_1(1-p_1)p_0(1-p_0)}}{1-p_1}$$

$$d = \frac{\left[\sum_{k=1}^{m} \frac{kt_k\psi}{k\psi+m-k+1}\right]}{\sigma_\psi}$$

where $\alpha$ = alpha, $\beta$ = 1− power, $\psi$ = odds ratio, and $z_p$ is the standard normal deviate for probability $p$. The number $n$ is rounded up to the closest integer.

## Exercises

5.1. Let $y_i$ be binomially distributed random variable ($i = 1,2,\ldots,k$). Show $E(y_i) = n_i P_i$ and $var(y_i) = n_i P_i(1-P_i)$. Use the delta method to derive the variance of $z_i = \log\left(\frac{y_i}{n_i - y_i}\right)$.

5.2. The mycotoxin data has only group-level covariates. Use PROC Reg in SAS to fit linear regression, using the method of weighted least squares, using $w_i = [var(z_i)]^{-1}$ as weights. Compare the fitting of this linear regression model to the results of the fitting logistic regression model. Comment on your findings.

5.3. In a regression model, when two covariates have zero correlation, they are said to be orthogonal. For the logistic regression model the concept of orthogonality is absent. Show that, using any of the examples, that the order in which terms are included in a model is important (in some cases).

5.4. Formal testing for "extra-binomial variation": Consider the data reported in Elston (1977), where 12 strains of mice were treated with a carcinogen and the numbers with and without tumors were noted. The results are in the following table.

| Strain | | 1 | 2 | 3 | 4 | 5 | 6 | 7 | 8 | 9 | 10 | 11 | 12 |
|---|---|---|---|---|---|---|---|---|---|---|---|---|---|
| $y$; (# with tumor) | | 26 | 27 | 35 | 18 | 33 | 11 | 11 | 13 | 13 | 5 | 5 | 2 |
| $n - y$ (# without tumor) | | 1 | 3 | 14 | 9 | 20 | 11 | 11 | 15 | 22 | 19 | 30 | 24 |

Construct on 95% confidence interval on the tumor proportion.

5.5. Tarone (1979) constructed a one degree of freedom chi-square test on the null hypothesis that data are independently binomially distributed versus a correlated binomial alternative. His test statistic is given by

$$X_T^2 = \frac{\left[ \hat{p}\hat{q} \sum_{i=1}^{k} (y_i - n_i\hat{p})^2 - N \right]^2}{\left( 2\sum_{i=1}^{j} n_i^2 - N \right)}$$

where $\hat{p}_i = y_i/n_i$, $\hat{p} = \sum y_i / \sum n_i$, $N = \sum n_i$, and $\hat{q} = 1 - \hat{p}$. Show that

a. $\hat{p} = 0.526$

b. $X_T^2 = 315.11$ What do you conclude?

5.6. Use PROC GLM to fit a one-way random effects model. Show that

a. MSB (mean-square between strains) = 2.35

b. MSW (mean-square within strains) = 0.187

c. Intrastrain correlation = 0.27

5.7. In a study of the relationship between endometrial cancer and history of gall bladder, 70 matched case-control pairs yielded the data in the following table:

| | Cases | | |
|---|---|---|---|
| Control | Exposed | Unexposed | Total |
| Exposed | 5 | 6 | 11 |
| Unexposed | 15 | 44 | 59 |
| Total | 20 | 50 | 70 |

a. Find an estimate of odds ratio. Construct a95% confidence interval on the population odds ratio.

b. Use McNemar's test to verify the null hypothesis $H_0$ = odds ratio =0.

5.8. The data of Exercise 5.7 were further classified according to age according to the following table.

Cancer and History of Gall Bladder

|  |  | Cases | | |
| --- | --- | --- | --- | --- |
|  | Controls | Exposed | Unexposed | Total |
| Age ≥ 65 | Exposed | 3 | 3 | 6 |
|  | Unexposed | 7 | 27 | 34 |
|  | Total | 10 | 30 | 40 |
| Age < 65 | Exposed | 2 | 3 | 5 |
|  | Unexposed | 8 | 17 | 25 |
| Total |  | 10 | 20 | 30 |

  a. Calculate the odds ratio for the two groups (group 1 ≡ age ≥ 65, group 2 ≡ age < 65) and produce the *p*-values on the significance of each *p*-value.

  b. Test the equality of odds ratios in the two groups.

  c. Is age an effect modifier?

5.9. Fit the above data, first using logistic regression, and then using PHREG in SAS.

5.10. You are given the results of a logistic regression analysis with linear predictor $\hat{\eta} = -22.91 + 0.31X_1 + 0.52X_2 + 0.16X_3$. The SAS output produced the following variance–covariance matrix of the estimated regression parameters:

$$\begin{bmatrix} 11 & -1 & -4 & 3 \\ -1 & 9 & -3 & -2 \\ -4 & -3 & 4 & 0 \\ 3 & -2 & 0 & 5 \end{bmatrix}$$

Construct 95% on the odds ratio for someone whose covariate profile is $\hat{X} = (3, 2, 0)$ relative to someone whose covariate profile is $X^* = (0,0,1)$.

5.11. The following data resulted for a multicenter clinical trial where patients within each center are randomized to receive one of two treatments. In this exercise, each center will be treated as a cluster. We shall use PROC GENMOD to fit the data under three different correlation structures. The response of interest is denoted by "yes," otherwise it is denoted by "no". Therefore, the total number of subjects recruited in a center is n = y + n.

```
data clin;
 input clinic trt yes no;     /* GEE for multi-center clinical trial */
   n= yes+no;
 cards;
 1   1  111 125
 1   2  210 227
```

```
2   1  116 140
2   2  222 210
3   1  114 150
3   2  270  54
4   1   20 140
4   2  210 160
5   1   56 280
5   2  111  121
6   1   11  110
6   2   90  110
7   1   11  14
7   2  112  96
8   1   12  70
8   2  118  10
9   1   15  28
9   2  131  121
10   1   11  110
10   2  175  120
11   1   11  14
11   2  120 310
12   1   5  28
12   2   16  10
;
ODS RTF;
  data new;
    set clin;
   do i = 1 to yes;
    y= 1;
   output;
  end;
    do i = 1 to no;
     y = 0;
     output;
   end;

    proc genmod data=new;
      class clinic trt;
     model y = trt / dist= bin ;

    repeated subject= clinic / type=un corrw;


 run;
proc genmod data=new;
     class clinic trt;
    model y = trt / dist= bin ;

    repeated subject= clinic / type=exch ;

 run;
```

```
proc genmod data=new;
    class clinic trt;
    model y = trt / dist= bin ;

repeated subject= clinic / type=ind ;


run;
```

ODS RTF CLOSE;

5.12. The intraclass correlation estimator has an asymptotic variance given by

$$var(\hat{\rho}) = [(k-1)n_0N(N-k)]^2/\lambda^4 \times \left\{ 2k + \left( \frac{1}{\pi(1-\pi)} - 6 \right) \sum 1/n_i + \right.$$

$$+ \left[ \left( \frac{1}{\pi(1-\pi)} - 6 \right) \sum 1/n_i - 2N + 7k - \frac{8k^2}{N} - \frac{2k\left(1-\frac{k}{N}\right)}{\pi(1-\pi)} + \left( \frac{1}{\pi(1-\pi)} - 3 \right) \sum n_i^2 \right] \rho$$

$$+ \left[ \frac{N^2-k^2}{\pi(1-\pi)} - 2N - k + \frac{4k^2}{N} + \left( 7 - \frac{8k}{N} - \frac{2\left(1-\frac{k}{N}\right)}{\pi(1-\pi)} \right) \sum n_i^2 \right] \rho^2$$

$$+ \left[ \frac{1}{\pi(1-\pi)} - 4 \right) \left( \frac{N-k}{N} \right)^2 \left( \sum n_i^2 - N \right) ] \rho^3 \right\}$$

where

$$\lambda = (N-k)[N-1-n_0(k-1)]\rho + N(k-1)(n_0-1) \qquad (5.24)$$

(see Wu et al., 2012).

For the data in Table 4.19 (Chapter 4) construct a 95% confidence interval separately for the cases and the controls as $(1-\alpha)100\%$ *confidence interval on $\rho$ is given by* $\hat{\rho} \pm z_{\alpha/2}\sqrt{var(\hat{\rho})}$.

5.13. Construct a 95% confidence interval on the difference between the ICC of the cases and the controls.

5.14. Analyze the family data where the siblings' responses were dichotomized such that a person with SBP above 120 and DBP above 80 is coded as hypertensive (y = 1), else (y = 0). Take into account the within-family intracluster correlation.

# 6

## Analysis of Clustered Count Data

### 6.1 Introduction

In Chapters 4 and 5 we discussed the statistical analysis and modeling of binary and binomial data, and discussed some of the techniques designed to deal with clustered data. In this chapter we deal with situations when the response variable is an integer nonnegative random quantity representing counts. Such variables are of common occurrence, for example, in biological, medical, agricultural, and environmental fields. The book by Cameron and Trivedi (1998) is the most recent reference that describes in full detail statistical analysis and regression models for count data.

As an example, consider the data in Table 6.1 from Janardan et al. (1979) showing the distribution of sow bugs under boards. Here, the response variable is a nonnegative integer that takes the values 0, 1, 2, and 3. The frequencies of these counts are given in the second row.

As another example, the data displayed in Table 6.2 shows the relationship between virus concentration and pock counts, which is a convenient technique for the estimation of virus titers.

Roizman et al. (1960) estimated the relationship between virus concentration and pock counts using linear and parabolic regression, assuming that the pock counts follow a Poisson distribution.

The fundamental aim of this chapter is to model count data as functions of several covariates, particularly when the sampling units are clusters.

### 6.2 Poisson Regression

A random variable $Y$ has a Poisson distribution with mean $\mu$ if the probability that $Y$ equals $y$ is given by

$$\Pr(Y = y) = \frac{e^{-\mu}\mu^y}{y!} \tag{6.1}$$

for $y = 0, 1, \ldots$ and $\mu > 0$.

**TABLE 6.1**

Distribution of the Number of Sow Bugs

| Number of spiders per board | 0 | 1 | 2 | 3 |
|---|---|---|---|---|
| Frequency | 159 | 64 | 13 | 4 |

**TABLE 6.2**

Pock Counts Data Corresponding to Each
Concentration Level

| Concentration Factor | Pock Count |
|---|---|
| 1 | 5, 6, 9, 10, 11, 11, 12, 13 |
| 2 | 18, 23, 26, 31, 33, 34, 47, 53 |
| 4 | 32, 32, 37, 42, 48, 53, 54, 58 |
| 8 | 79, 88, 103, 107 |

The Poisson distribution enjoys a variety of desirable properties that account for its popularity. The most important feature is the widespread situations where count data arise. Bishop et al. (1975) have focused their investigations almost exclusively on the analysis of data obtained from sampling models based on the Poisson, multinomial, and product multinomial sampling schemes.

To illustrate the application of the Poisson model and to describe the impact of explanatory variables on the mean count, we use the data in Table 6.2 to define a Poisson regression model, assuming that (a) the number of Pocks ($Y$) is a random variable distributed as Poisson with mean $\mu$; and (b) $\mu$ is some function of the virus concentration $X$. Plotting logarithms of counts against the virus concentration (see Figure 6.1) suggests a relationship of the form

$$\log \mu_i = \alpha + \beta x_i \tag{6.2}$$

For this log-linear model, the mean satisfies the relationship

$$\mu_i = e^{\alpha + \beta x_i} = e^{\alpha} \left( e^{\beta} \right)^{x_i} \tag{6.3}$$

Equation 6.3 means that a unit increase in $X$ has a multiplicative effect of $e^{\beta}$ on $\mu$, that is, the mean of $Y$ at $X + 1$ equals the mean of $Y$ at $X$ multiplied by $e^{\beta}$.

The case of multiple explanatory covariates is treated within the framework of the generalized linear model (GLM) as discussed in Chapters 3 and 5. For the Poisson model, the log-link is

$$\log \mu_i = \sum_{j=1}^{p} x_{ij} \beta_j \tag{6.4}$$

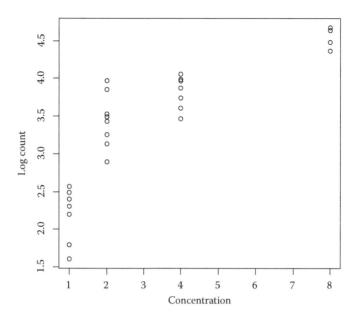

**FIGURE 6.1**
Scatter plot of log count against concentration.

When employing a specific parametric model, it is natural to estimate the parameters of the model by maximizing the likelihood function. The likelihood function contains all the relevant information about the mechanism that generated the data. For the Poisson regression model, the maximum likelihood estimation is the most often used method.

The log-likelihood function of the Poisson regression mode is given by

$$l(\beta) = \sum_{i=1}^{n} \{y_i \log[\eta_i(\beta)] - \eta_i(\beta)\} \tag{6.5}$$

where

$$\mu_i = \eta_i(\beta) = \exp\left[\sum_{j=1}^{p} x_{ij}\beta_j\right] \tag{6.6}$$

The estimating equations are obtained by differentiating Equation 6.5 with respect to the model parameters:

$$U_j(\beta) = \frac{\partial l(\beta)}{\partial \beta_j} = \sum_{i=1}^{n} \frac{\partial \eta_i(\beta)}{\partial \beta_j} \left(\frac{y_i - \eta_i(\beta)}{\eta_i(\beta)}\right) = 0$$

The preceding equations are solved to obtain the estimates of the regression parameters. When the log-link function 6.6 is used, $U_j(\beta)$ reduces to

$$U_j(\beta) = \sum_{i=1}^{n} \left[ y_i - \exp\left( \sum_{j=1}^{p} x_{ij}\beta_j \right) \right] x_{ij} = 0$$

The GENMOD procedure in SAS can be used to estimate the parameters of a Poisson model with logarithmic link transformation.

**Example 6.1: "pockcount"**

Consider the pock count data shown in Table 6.2. The following SAS code reads the data and runs the Poisson regression models:

```
data pock;
input concent count @@;
logcount=log(count);
cards;
1 5 1 6 1 9 1 10 1 11 1 11 1 12
1 13 2 18 2 23 2 26 2 31 2 33 2 34
2 47 2 53 4 32 4 32 4 37 4 42 4 48
4 53 4 54 4 58 8 79 8 88 8 103 8 107
;
ODS RTF;
* Plotting log count vs. concent;
proc gplot;
 plot logcount*concent/vaxis=axis1 haxis=axis2 hminor=0 frame;
   axis1 label=(angle=-90 rotate=90 'Log count')
      order = (1 to 5 by 1) minor = none
      offset = (0.5)pct;
   axis2 label= ('Concent') order=(1 to 8 by 1);
   run;

* Linear regression;
proc reg data=pock;
 model logcount =concent;
run;

* Poisson model with log link;
proc genmod data=pock;
  model count=concent / dist=poisson link=log scale=deviance
type3;
 output out=virus pred=predicted resdev=resid;
run;

ODS RTF CLOSE;
```

The following is the partial output of the SAS program for fitting simple linear regression:

**Parameter Estimates**

| Variable | DF | Parameter Estimate | Standard Error | t-Value | Pr > \|t\| |
|----------|-----|--------------------|----------------|---------|-----------|
| Intercept | 1 | 2.39515 | 0.15532 | 15.42 | <.0001 |
| Concent | 1 | 0.30192 | 0.03991 | 7.56 | <.0001 |

The following is the SAS output running GENMOD when the dispersion parameter is estimated as the square root of deviance divided by the number of degrees of freedom:

The Genmod Procedure

**Criteria for Assessing Goodness of Fit**

| Criterion | DF | Value | Value/DF |
|---|---|---|---|
| Deviance | 26 | 134.2770 | 5.1645 |
| Scaled Deviance | 26 | 26.0000 | 1.0000 |
| Pearson Chi-Square | 26 | 135.6029 | 5.2155 |
| Scaled Pearson X2 | 26 | 26.2567 | 1.0099 |
| Log Likelihood | | 591.2394 | |

**Analysis of Parameter Estimates**

| Parameter | DF | Estimate | Standard Error | Wald 95% Confidence Limits | | Chi-Square | Pr > ChiSq |
|---|---|---|---|---|---|---|---|
| Intercept | 1 | 2.7418 | 0.1371 | 2.4732 | 3.0105 | 400.14 | <.0001 |
| Concent | 1 | 0.2340 | 0.0252 | 0.1846 | 0.2834 | 86.32 | <.0001 |
| Scale | 0 | 2.2726 | 0.0000 | 2.2726 | 2.2726 | | |

*Note:* The scale parameter was estimated by the square root of DEVIANCE/DF.

In the first model, we fitted linear regression using the log-count as the dependent variable and the second model fits the Poisson model with logarithmic link transformation. If the scale = Pearson option is specified, another Poisson model may be fitted where the dispersion parameter is estimated by Pearson's chi-square statistic divided by its degrees of freedom, and the parameter estimates are adjusted accordingly. When we compare the outputs of the three models, we find no real differences among them. The slight difference between the two Poisson models is because of the way the scale parameter is estimated. In one, the scale parameter is estimated as $\sqrt{5.1645}$, where as in the other it is estimated as $\sqrt{5.2115}$.

There is no real quantitative measure according to which a specific model is preferred to the other. However, the most common approach to estimating the scale parameter is the square root of DEVIANCE/DF.

The following R code reads the data and the runs the linear regression and the Poisson model:

```
# Entering the data
concent <- c(rep(1,8),rep(2,8),rep(4,8),rep(8,4))
count <- c( 5, 6, 9,10,11,11,12,13,18,23,26,31, 33, 34,
     47,53,32,32,37,42,48,53,54,58,79,88,103,107)
logcount=log(count)
pock <- data.frame(concent,count,logcount)

# Plotting log count vs. concentration;
plot(logcount~concent,xlab="Concentration",ylab="Log Count",
data=pock)

# Linear regression
summary(lm(logcount ~ concent, data=pock))
```

```
# Poisson model with log link
pock.glm <- glm(count ~ concent,data=pock,family=poisson(link =
"log"))
summary(pock.glm)
```

**R: OUPUT**

```
glm(formula = count ~ concent, family = poisson(link = "log"), data
= pock)
Deviance Residuals:
```

| Min | 1Q | Median | 3Q | Max |
|-----|-----|--------|-----|-----|
| −3.9430 | −1.9190 | −0.3862 | 1.3664 | 4.9153 |

Coefficients:

| | Estimate | Std. Error | z Value | Pr(>\|z\|) |
|---|----------|-----------|---------|---------|
| (Intercept) | 2.74180 | 0.06031 | 45.46 | <2e-16 *** |
| concent | 0.23400 | 0.01108 | 21.11 | <2e-16 *** |

Null deviance: 548.27 on 27 degrees of freedom
Residual deviance: 134.28 on 26 degrees of freedom
AIC: 283.61

## 6.2.1 Model Inference and Goodness of Fit

Similar to the logistic regression, one may compare models using $-2l$. This kind of analysis is done in SAS using the "type 3" option in the model statement (see the earlier output of the GEMOD procedure). The degrees of freedom for these "type 3" tests are simply the difference in the number of parameters estimated in the two competing nested models. In these models and the remainder of this chapter, hypothesis testing on regression coefficients and dispersion parameters in count data involves application of the large sample theory of the likelihood inference. A general approach is the Wald test, calculated from the z-score of the estimated parameter, that is, $z =$ estimate/SE(estimate). Large sample confidence intervals are computed as estimate $\pm z_{\alpha/2}$SE(estimate), where $z_{\alpha/2}$ is the $(1 - \alpha/2)100\%$ cut-off point of the standard normal distribution. A statistic capable of measuring the amount of support given by the data to a particular value of the parameter compared to its maximum likelihood estimate is the "deviance" statistic, defined as minus two times the logarithm of the normed-likelihood, that is,

$$D_y(\beta) = -2\left[l_y(\beta) - l_y\left(\hat{\beta}\right)\right].$$

$$= -2\sum_{i=1}^{n}\left[y_i \log\left[\frac{\eta_i(\beta)}{\eta_i\left(\hat{\beta}\right)}\right] - \left[\eta_i(\beta) - \eta_i\left(\hat{\beta}\right)\right]\right] \tag{6.7}$$

Assuming $\beta$ to be the true parameter, the deviance has an asymptotic $\chi^2$ distribution with $p$ degrees of freedom, where $p$ is the dimension of $\beta$ vector.

To obtain a measure of model "parsimony," the likelihood for the maximal model that perfectly fits the data may be compared to the likelihood of the fitted model. The measure is written as

$$Dev_y = 2 \sum_{i=1}^{n} \left[ y_i \log \left( \frac{y_i}{\eta_i(\hat{\beta})} \right) - \left( y_i - \eta_i(\hat{\beta}) \right) \right] \tag{6.8}$$

This index is known as the "deviance" as well. It can be used to compare nested models, as has been described for the logistic regression model in Chapter 3.

## 6.2.2 Overdispersion in Count Data

The equality of mean and variance of the Poisson distribution places restriction on the applicability of this model to real-world data. Unfortunately, the popularity of the Poisson models and the relative ease of fitting them make it tempting to adopt these models without adequate attention to their applicability. An issue of importance is when empirical variance in the data exceeds the nominal variance under the presumed model. Support for the presence of overdispersion is obtained when Pearson deviance is large. Therefore, an alternative to the Poisson distribution to model count data must be sought. It can be argued, however, if the lack of fit is negligible, and the efforts involved in fitting a more appropriate model are costly, then approximate inference under the Poisson model may still be valid.

The negative binomial distribution has its variance larger than the mean and can be easily fitted to over dispersed count data using SAS.

Definition: The random variable $Y$ is said to follow a negative binomial distribution (NBD) with mean $\mu$ and dispersion parameter $\lambda$, denoted by $Y \sim NBD(\mu, \lambda)$, if

$$\Pr[Y = y] = \frac{\Gamma(y + \lambda^{-1})}{y! \, \Gamma(\lambda^{-1})} \left( \frac{\lambda \mu}{1 + \lambda \mu} \right)^y \left( \frac{1}{1 + \lambda \mu} \right)^{\lambda^{-1}} \tag{6.9}$$

for $y = 0, 1, 2, \ldots, \quad 0 < \mu < \infty$, and $0 \le \lambda < \infty$.

This form of the NBD reduces to the Poisson distribution when $\lambda = 0$. For the NBD $E(Y) = \mu$ and var $(y) = \mu(1 + \lambda \mu)$. It is understood, by convention that $\lambda = 0$ means that as $\lambda \to 0$, the distribution in Equation 6.9 approaches Poisson with mean $\mu$.

The problem of detecting departure from the Poisson distribution in the direction of a NBD is known as the problem of homogeneity testing.

When $k$ groups, each of size $n_i$, are available, several statistical tests have been developed to test $H_0 : \lambda = 0$ against $H_a : \lambda > 0$. For example, Collings and Margolin (1985) showed that $H_0$ is rejected for large values of

$$T = \sum_{i=1}^{k} \sum_{j=1}^{n_i} \left\{ \left( y_{ij} - \bar{y}_{i.} \right)^2 - y_{ij} \right\} \tag{6.10}$$

Since under $H_0 : \lambda = 0$, the Poisson distribution is a member of the regular exponential family so there exists a complete sufficient statistic (see Collings and Margolin, 1985) for $(\mu_1, \mu_2, ... \mu_k)$, namely, $(\bar{y}_{1.}, \bar{y}_{2.}, ... \bar{y}_{k.})$. Upon conditioning on the conditional statistics, one can find that large values of Equation 6.10 are equivalent to large values of

$$T_c = \sum_{i=1}^{k} \sum_{j=1}^{n_i} (y_{ij} - \bar{y}_{i.})^2 / y_{..} \tag{6.11}$$

where $\bar{y}_{i.} = \sum_{j=1}^{n_i} y_{ij} / n_i$ and $y_{..} = \sum_{i=1}^{k} \sum_{j=1}^{n_i} y_{ij}$.

This test statistic is valid only when all the observation are independent, and a new test statistic should be developed when observations within groups form clusters. This test statistic can be developed using random effect models for discrete data similar to the normal random effects model.

---

### 6.2.3 Count Data Random Effects Models

Let $y_{ij}$ denote independent Poisson random variables such that

$$\Pr\left[ y_{ij} | a_i \right] = e^{-a_i} a_i^{y_{ij}} / y_{ij}$$

where $i = 1, 2, ..., k$ and $j = 1, 2, ..., n_i$.

Suppose that for some $\lambda \geq 0$,

$$a_i \underset{iid}{\sim} \text{Gamma}\left( \lambda^{-1}, \mu \right)$$

where Gamma $(\alpha, \beta)$ represents a gamma distribution with mean $\alpha\beta$ and variance $\alpha\beta^2$.

The unconditional distribution of $y_{ij}$ is $NBD(\mu,\lambda)$ with

$$E(y_{ij}) = \mu$$

$$\text{var}\left(Y_{ij}\right) = \mu(1 + \lambda\mu) \tag{6.12}$$

$$\text{cov}\left(y_{ij}, y_{il}\right) = \lambda\mu^2 \qquad j \neq l$$

Recall that in the one-way random effects model for normally distributed data, the hypothesis of no group effect is equivalent to $H_0 : V(a_i) = 0$. In the discrete data this is equivalent to testing $H_0 : \lambda\mu^2 = 0$, which is equivalent to testing $H_0 : \lambda = 0$. The test statistic in this case was developed by Collings and Margolin (1985) and is given by

$$T_r = \sum_{i=1}^{k}(y_{i.} - n_i\tilde{\mu})^2/y_{.} \tag{6.13}$$

where

$$\tilde{\mu} = \sum_{i=1}^{k}\sum_{j=1}^{n_i}y_{ij} \Bigg/ \sum_{i=1}^{k}n_{i.}$$

Another way to detect overdispersion is to perform a least squares regression (without intercept)

$$\gamma_{ij} = \frac{\left(y_{ij} - \hat{\mu}_i\right)^2 - y_{ij}}{\hat{\mu}_i} = \lambda\hat{\mu}_i + u_i \tag{6.14}$$

where $\hat{\mu}_i = \frac{1}{n_i}\sum_{j=1}^{n_i}y_{ij}$ is the estimated group mean and $u_i$ is an error term (see Cameron and Trivedi, 1998). Under the hypothesis of no overdispersion, the statistic

$$t = \hat{\lambda}_{\text{ols}}/SE(\hat{\lambda}_{\text{ols}}) \tag{6.15}$$

is asymptotically normal. In the case of Poisson regression, the fitted values $\hat{\mu}_i = \frac{1}{n_i}\sum_{j=1}^{n_i}\exp(\hat{\beta}_0 + x_{ij1}\hat{\beta}_1 + \ldots + x_{ijp}\hat{\beta}_p)$ are used in Equation 6.6.

In summary, we first fit a Poisson regression model using PROC GENMOD. Second, the fitted (predicted) values $\hat{\mu}_{ij}$ are saved, and with few programming statements in SAS we fit the regression Equation 6.6. If $t$ is sufficiently large (above 3, say), then we conclude that overdispersion is present, and an alternative to the Poisson regression model should be investigated.

Count data may be modeled under one-way layout. With the log-link, log $(\mu_i) = \beta_o + \beta_i$. The mean is estimated by

$$\hat{\mu}_i = \bar{y}_i = \frac{1}{n_i} \sum_{j=1}^{n_i} y_{ij}.$$

and $\hat{\beta}_i = \log \bar{y}_i$. Using the delta method we can show that

$$\mathrm{Var}\left(\hat{\beta}_i\right) = \frac{1}{n_i \mu_i}$$

Therefore, a 95% CI on $\beta_i$ is then given as

$$\exp\left[\hat{\beta}_i \pm 1.96 \sqrt{\left(n_i \bar{y}_i\right)^{-1}}\right]$$

Modeling incidence rate: As we showed in Chapter 4, the incidence rate is estimated by $y_i/t_i$; here $y_i$ is the count of events of interest and $t_i$ is the person-time. The log-link is

$$\log\left(\frac{\mu_i}{t_i}\right) = \sum_{j=1}^{p} \beta_i x_{ij}$$

Since $\log\left(\dfrac{\mu_i}{t_i}\right) = \log \mu_i - \log t_i$, the term $(-\log t_i)$ is called "offset."

Therefore,

$$\mu_i = t_i \exp\left(\sum_{j=1}^{p} \beta_j x_{ij}\right)$$

**Example 6.2: Fat Intake and Prostate Cancer Mortality**

Prostate cancer mortality is believed to be related to fat intake. The data on cancer mortality from five regions, with information regarding the average fat intake, the Karnofsky scale at the time of diagnosis, and the person-time contributed by each person are in the study. It is clear that this is an ecological study and the results should not be extended to draw inference on the relationship between fat intake and prostate cancer mortality, at the individual level. The response variable is the incidence or "count." The R code to fit a Poisson regression model with region is taken as the cluster. Note also, fat intake <60 g is coded as 1, fat intake 60 to 90 is coded as 2, and fat intake above 90 is coded as 3. Moreover, for the Karnofsky scale, <30 is coded as 1, scale 30 to 60 is coded as 2, and is coded as 3 if the scale is 60 to 100.

**R Code**
**Library (gee)**

```
    fit<-gee(count~fat+ks+offset(log(persontime)),id=region,
family=poisson,data=fat,corstr="exchangeable")
```

It is left as an exercise (Exercise 6.10) to run the above code in R.

## 6.2.4 Introducing the Generalized Linear Mixed Model (GLMM)

In the previous sections we accounted for the extra variation in counts data by mixing the Poisson distribution with a gamma random variable to obtain a specific form for the NBD. Since the NBD has a closed form, the likelihood inference is possible. Shoukri et al. (2004) used the inverse Gaussian distribution as an alternative to the gamma density to obtain what was termed by them the *Poisson inverse Gaussian model*. They used the log-link function to model clustered count data as a function of a set of measured covariates. Such an outcome is somewhat artificially generated. In many, otherwise appealing, models, a closed-form marginal density may not be generated. In this case approximate or pseudo-likelihood methods of estimation may be used. The generalized linear mixed effects models (GLMMs) for correlated counts assume a Poisson model with normally distributed multiplicative heterogeneity terms.

Assuming that $\log \mu_i = X_i^T \beta + \sigma u_i$, with $u_i \sim N(0, 1)$, then conditional on $u_i$, $y_i$ follows a Poisson distribution with mean $\mu_i = \exp(x_i^T \beta + \sigma u_i)$.

The marginal distribution is

$$\Pr[Y_i = y_i] = \frac{1}{\sqrt{2\pi}} \int_{\infty}^{\infty} (y_i!)^{-1} \exp\left[-\exp(x_i^T \beta + \sigma u_i)\right]$$

$$\times \exp\left[(x_i^T \beta + \sigma u_i)^{y_i}\right] \bar{e}^{u^2/2} du_i \qquad (6.16)$$

There is no closed form solution for $\Pr[Y_i = y_i]$, unlike for the gamma heterogeneity term considered in previous sections. Two commonly used estimation procedures are (1) to approximate the objective function and (2) to approximate the model. Algorithms in the second approach can be expressed in terms of Taylor expansion (linearization). In such an approach, we employ expansions to approximate the model by one based on pseudo-data with fewer nonlinear components. The advantages of linearization-based methods include a relatively simple form of the linearized model that typically can be fit based on only the mean and variance in the linearized model. Models for which the marginal distribution in impossible to obtain can be fit with the linearization-based approach. Models with correlated errors (discussed in Chapter 7), a large number of random effects, crossed random effects, and multiple types of clusters are models that linearization is applied to.

The SAS fitting procedure that we use to fit models for clustered count data known as "GLIMMIX" fits the generalized linear mixed model based on linearization. The default estimation method in GLIMMIX for models containing random effects is a technique known as restricted pseudo-likelihood (RPL) estimation (see Wolfinger and O'Connell, 1993) with an expansion around the current estimate of the best linear unbiased predictors of the random effects.

## 6.2.5 Fitting GLMM Using SAS GLIMMIX

When nonnormally distributed data with random effects are analyzed by the GLIMMIX procedure in SAS, certain assumptions are made:

1. We assume that conditional on the random effect, the data have a known distribution. In models without random effects, the marginal distribution is assumed to be known so that maximum likelihood estimation is used. If quasi-likelihood methods are used, we assume that the mean and variance are known.

2. The link function is known as well.

3. Once the parameters are estimated, we assume that the large sample theory of likelihood inference applies. Test of hypotheses for the fixed effects are based on Wald-type tests and the estimated variance-covariance matrices.

4. The link function $\eta$ has the form $\eta = x\beta + z\gamma$.

The variance of the observations, conditional on the random effects, is

$$V(Y/\gamma) = A^{1/2}RA^{1/2} \tag{6.17}$$

The matrix $A$ is a diagonal matrix and its elements are the variance functions of the model, which are determined from the specification of the distribution of the response variable $Y$. For example, if $Y$ is Poisson, then the diagonal elements of $A$ are $\mu$ (since the mean equals the variance in this case). The matrix $R$ is a variance matrix specified by the RANDOM statement. The SAS GLIMMIX procedure distinguishes two types of random effects, the *G-side* and the *R-side*. The R-side effects are called residual effects. In other words, if a random effect is an element of $\gamma$, it is a G-side effect, otherwise it is an R-side effect. Models without G-side effects are actually the population-averaged, or PA, models. These models have been described in detail in Chapter 3. The columns of the design matrix $X$ are the fixed effects (covariates) specified in the right side of the MODEL statement. Columns of $Z$ are the variance matrices $G$ and $R$. These are determined from the RANDOM statement.

The $R$ matrix is by default the scaled identity matrix, $R = \phi I$. The scale parameter $\phi$ is set to 1 if the distribution does not have a scale parameter, for example, in the case of the binary response, binomial, Poisson, and the geometric distribution. To specify a different $R$ matrix, we use the RANDOM statement with the RESIDUAL keyword or the RESIDUAL option.

As a final remark, we note that the GLM (McCullagh and Nelder, 1989) is a special case of the GLMM. If the random effect $\gamma = 0$, and $R = \phi I$, with GLMM reduces to the GLM or a GLM with over dispersion. For example, if $A$ is a diagonal matrix with $E(Y) = V(y) = \mu$ on the diagonal, then the model is Poisson with $\phi = 1$, and we can model Poisson data with overdispersion by adding the statement "Random _ residual_;".

Similar to "Proc Mixed", the GLIMMIX constructs a linear mixed model according to the specifications in the CLASS, MODEL, and RANDOM statements. Each effect in the MODEL statement generates one or more columns in the design matrix $X$ of the fixed effects, and each G-side effect in the RANDOM statement generates one or more columns in the Z matrix. R-side effects in the RANDOM statement do not generate model matrices; they serve only to index observations within clusters. By default, all models automatically include a column of 1's in $X$ to estimate a fixed effect intercept parameter. Similar to "Proc mixed" we can use the NOINT option in the Model statement. The usefulness of the NOINT option has been explained in Chapter 3.

### Example 6.3: ch6_mastitis

Clinical mastitis (CM) is one of the endemic diseases and conditions of dairy cattle in many countries. The disease causes significant losses to the dairy industry both in terms of the reduction in output levels and wastage of resources incurred, in addition to the costs of disease prevention and treatment. Farm management practices play an important role in controlling and reducing the incidence of the disease. Hygiene is identified as an important factor. The use of organic (straw and sawdust) versus inorganic (sand) is associated with an increase in intramammary infections. Tail ducking is practiced to prevent the tail from hitting the udder and spread of the disease-causing pathogen. The milking technique is also identified as a route to spread mastitis from cow to cow. Some farmers milk the teats by hand (postmilking) after the machine milking to reduce residual milk in the quarters, and such practice might serve as an ideal growth flora for bacteria. Table 6.3 provides a summary of the data on total mastitis cases available from 57 Ontario dairy farms, each farm visited three to six times.

Let $y_{ij}$ denote the number of mastitis cases on $i$th farm at $j$th visit. Here, the farm is a cluster ($i = 1, 2,...,57$), and the number of units within the cluster is the number of visits $n_i$. Note that in this example we shall assume that the within-cluster correlation is exchangeable, and we shall discuss the issue of other correlation structures in Chapter 7.

**TABLE 6.3**

Summary of the Mastitis Data

| Visit | Number of Farms | Cultured Cows | Diseased Cows |
|-------|-----------------|---------------|---------------|
| 1 | 57 | 2722 | 268 |
| 2 | 57 | 2916 | 366 |
| 3 | 57 | 2837 | 340 |
| 4 | 55 | 2583 | 262 |
| 5 | 48 | 2313 | 306 |
| 6 | 3 | 141 | 29 |

Here we model $y_{ij}$ as a count variable using several modeling strategies: We first specify the fixed effects component of the model as $log(\mu_{ij}) =$
$\beta_0 + \beta_1 m_{ij} + \beta_0 bed_{ij} + \beta_3 pm_{ij} + \beta_4 tail_{ij}$, where $\mu_{ij} = E(y_{ij})$; $\beta_0 =$ intercept;
$m_{ij}$ = number of cows cultured on $i$th farm at $j$th visit; $bed_{ij} = 1$ if bedding
is organic, and is 0 otherwise; $pm_{ij} = 1$ if postmilking was not practiced, 0
otherwise; and tail = 1 if tail docking is practiced, 0 otherwise.
The four models are

> Model 1: Poisson without overdispersion
> Model 2: Poisson with overdispersion
> Model 3: GLMMIX Poisson
> Model 4: Negative binomial

The first two models are marginal or population average, model 3 is the
cluster-specific model, and model 4 is a random effects model, with a
gamma distribution for the random effect.
Following Cameron and Trivedi (1998, p. 78), a test of overdispersion
can be carried out empirically by using the method of least squares to fit
the regression equation

$$D_{ij} = \frac{(y_{ij} - \bar{y}_{i.})^2 - y_{ij}}{\bar{y}_{i.}} = \lambda \bar{y}_{i.} + \varepsilon_{ij}$$

where $\bar{y}_{i.} = \sum_{j=1}^{n_i} y_{ij}/n_i$ and $\varepsilon_{ij}$ is an error term. The reported $t$-statistic for $\tilde{\lambda}$ is
asymptotically normal under the null hypothesis $H_0: \lambda = 0$ versus $H_1: \lambda >$
0. For the mastitis data, it was found that $\tilde{\lambda} = 0.1169$ and the standard
error $SE(\tilde{\lambda}) = 0.0232$, from which $t = 0.1169/0.0232 = 5.03$ ($p$-value $<$
0.001). Therefore, based on the evidence in the data, we conclude that
overdispersion should not be ignored.
A short version of the SAS data step to read in the mastitis data is
shown next; however, the complete data set is available in the data
sources in the book site.

```
data mastitis;
input herd visit numcult y bedorg postmilk tailed;
cards;
10247 1 54  3 1 1 1
10247 2 64  6 1 1 1
10247 3 28  3 1 1 1
10247 4 3   2 1 1 1
  .    .  .   . . . .
  .    .  .   . . . .
30791 3 34 3 1 1 1
30791 5 32 2 1 1 1
;
```

Model 1: SAS code and selected output of fitting Poisson model
ignoring the clustering effect and accounting for overdispersion.

```
proc genmod data=mastitis;
class herd;
model y = visit numcult bedorg postmilk tailed/   dist=poisson
link=log dscale;
run;
```

| Criteria For Assessing Goodness of Fit | | | |
|---|---|---|---|
| Criterion | DF | Value | Value/DF |
| Deviance | 271 | 724.4622 | 2.6733 |
| Scaled Deviance | 271 | 271.0000 | 1.0000 |
| Pearson Chi-Square | 271 | 781.0768 | 2.8822 |
| Scaled Pearson X2 | 271 | 292.1779 | 1.0781 |
| Log Likelihood | | 520.1398 | |

| Analysis of Parameter Estimates | | | | | | | |
|---|---|---|---|---|---|---|---|
| | | | Standard | Wald 95% Confidence | | | |
| Parameter | DF | Estimate | Error | Limits | | Chi-Square | Pr > ChiSq |
| Intercept | 1 | 0.5855 | 0.4748 | −0.3451 | 1.5162 | 1.52 | 0.2175 |
| visit | 1 | 0.0415 | 0.0290 | −0.0155 | 0.0984 | 2.04 | 0.1536 |
| numcult | 1 | 0.0161 | 0.0014 | 0.0133 | 0.0190 | 123.57 | <.0001 |
| bedorg | 1 | 1.3796 | 0.3749 | 0.6449 | 2.1143 | 13.55 | 0.0002 |
| postmilk | 1 | −0.9117 | 0.1477 | −1.2013 | −0.6222 | 38.08 | <.0001 |
| tailed | 1 | −0.4044 | 0.1098 | −0.6197 | −0.1892 | 13.56 | 0.0002 |
| Scale | 0 | 1.6350 | 0.0000 | 1.6350 | 1.6350 | | |

*Note*: The scale parameter was estimated by the square root of DEVIANCE/DOF.

Model 2: SAS code and selected output of fitting Poisson regression model accounting for the clustering effect.

```
proc genmod data=mastitis;
class herd;
model y = visit numcult bedorg postmilk tailed/ dist=poisson
link=log dscale ;
repeated subject=herd/ type=cs corrw;
run;
```

| Exchangeable Working Correlation | |
|---|---|
| Correlation | 0.4163 |

| Analysis of GEE Parameter Estimates Empirical Standard Error Estimates | | | | | | |
|---|---|---|---|---|---|---|
| Parameter | Estimate | Standard Error | 95% Confidence Limits | | Z | Pr > |Z| |
| Intercept | 0.4237 | 0.5547 | −0.6635 | 1.5109 | 0.76 | 0.4450 |
| Visit | 0.0400 | 0.0347 | −0.0281 | 0.1081 | 1.15 | 0.2498 |
| numcult | 0.0168 | 0.0019 | 0.0131 | 0.0204 | 8.91 | <.0001 |
| bedorg | 1.4803 | 0.1960 | 1.0962 | 1.8644 | 7.55 | <.0001 |
| postmilk | −0.8572 | 0.2589 | −1.3646 | −0.3498 | −3.31 | 0.0009 |
| tailed | −0.4241 | 0.1504 | −0.7189 | −0.1292 | −2.82 | 0.0048 |

Model 3: SAS code and selected output of fitting Poisson regression model using the generalized linear mixed model.

```
proc glimmix data=mastitis;
class herd;
model y = visit numcult bedorg postmilk tailed/ s dist=poisson
link=log;
random intercept /subject=herd type=cs residual; run;
```

### Fit Statistics

| | |
|---|---|
| -2 Res Log Pseudo-Likelihood | 585.49 |
| Generalized Chi-Square | 452.33 |
| Gener. Chi-Square/DF | 1.67 |

### Covariance Parameter Estimates

| Cov Parm | Subject | Estimate | Standard Error |
|---|---|---|---|
| CS | herd | 1.2628 | 0.3141 |
| Residual | | 1.6691 | 0.1595 |

### Solutions for Fixed Effects

| Effect | Estimate | Standard Error | DF | t Value | Pr > \|t\| |
|---|---|---|---|---|---|
| Intercept | 0.4157 | 0.7755 | 53 | 0.54 | 0.5942 |
| visit | 0.03998 | 0.02319 | 218 | 1.72 | 0.0861 |
| numcult | 0.01679 | 0.001974 | 218 | 8.50 | <.0001 |
| bedorg | 1.4856 | 0.6425 | 53 | 2.31 | 0.0247 |
| postmilk | -0.8540 | 0.2553 | 53 | -3.34 | 0.0015 |
| tailed | -0.4258 | 0.1820 | 53 | -2.34 | 0.0231 |

Model 4: SAS code and selected output for fitting the count data using negative binomial regression model.

```
proc glimmix data=mastitis;
class herd;
model y = visit numcult bedorg postmilk tailed/ s dist=NB link=log;
random intercept /subject=herd;
run;
```

### Fit Statistics

| | |
|---|---|
| -2 Res Log Pseudo-Likelihood | 550.09 |
| Generalized Chi-Square | 248.17 |
| Gener. Chi-Square/DF | 0.92 |

### Covariance Parameter Estimates

| Cov Parm | Subject | Estimate | Standard Error |
|---|---|---|---|
| Variance | herd | 0.1962 | 0.04878 |
| CS | herd | -0.01132 | . |
| Scale | | 0.1017 | 0.02459 |

| Solutions for Fixed Effects | | | | | |
|---|---|---|---|---|---|
| Effect | Estimate | Standard Error | DF | t Value | Pr > \|t\| |
| Intercept | −0.09609 | 0.7645 | 53 | −0.13 | 0.9005 |
| Visit | 0.03739 | 0.02391 | 218 | 1.56 | 0.1193 |
| numcult | 0.01980 | 0.002491 | 218 | 7.95 | <.0001 |
| bedorg | 1.5773 | 0.5459 | 218 | 2.89 | 0.0043 |
| postmilk | −0.7035 | 0.3498 | 218 | −2.01 | 0.0455 |
| tailed | −0.3762 | 0.2146 | 218 | −1.75 | 0.0810 |

Remarks on the SAS output:

1. We note that all the covariates are measured at the cluster level. Except for the visit, all covariates are significantly correlated with the response variable.
2. When the Poisson model is fitted using the GEE, the moment estimator of the intracluster correlation (0.42) is almost identical to its restricted maximum likelihood estimate of the GLIMMIX procedure. This is obtained from the variance components estimators: 1.2628/(1.2628 + 1.6691).
3. While "tailed" was significant for the Poisson model, it was nonsignificant under the negative binomial (NBM) model.
4. The dispersion parameter estimate of the NBM is $\lambda = 0.1017$ and its standard error is 0.02459. Using the Wald test on the Poisson hypothesis $H_0 : \lambda = 0$ against the NBM alternative we have $z = \dfrac{0.1017}{.02459} = 4.14$ with $p$-value <.001, a result similar to the regression test.

Models 1 to 3 may be fitted in R as follows:

```
mastitis<-read.csv(file.choose())
The following packages are needed:
library(gee)
library(car)
library(lme4)
library(MASS)
library(arm)
library(nlme)
# Model 1: Poisson Model ignoring clustering effect
glm_mast<-glm(y~visit+numcult+bedorg+postmilk+tailed,
family=poisson(link="log"),data=mastitis)
summary(glm_mast)

# Model 2: Poisson Model accounting for clustering effect
gee_mast<-gee(y~visit+numcult+bedorg+postmilk+tailed,
id=herd, family=poisson(link="log"),data=mastitis,corstr=
"exchangeable")
summary(gee_mast)

Model 3: Fitting Poisson Model using generalized mixed modeling
cs <- corCompSymm(0.5, form = ~ 1|herd)
glmm_mast<-glmmPQL(y~visit+numcult+bedorg+postmilk+tailed,
random=~1|herd, family=poisson(link="log"),data=mastitis,cs)
summary(glmm_mast)
```

The results of fitting model 3 in R are slightly different from those in SAS because of difference in estimation methods.

Fitting the negative binomial model in R is node in two steps. First we need to upload the library (MASS), and then use the function (glm.nb).

### R output for the mastitis data:
### Model 1; Poisson

```
Call:  glm(formula = y ~ visit + numcult + bedorg + postmilk +
tailed,
    family = poisson(link = "log"), data = mastitis)
```

### Coefficients:

| (Intercept) | visit | numcult | bedorg | postmilk | tailed |
|---|---|---|---|---|---|
| 0.58554 | 0.04145 | 0.01611 | 1.37964 | -0.91174 | -0.40444 |

Degrees of Freedom: 276 Total (i.e. Null); 271 Residual
Null Deviance: 1195
Residual Deviance: 724.5 AIC: 1620

### Model 2: GEE:

```
Correlation Structure:    Exchangeable
Call:
gee(formula = y ~ visit + numcult + bedorg + postmilk + tailed,
    id = id, data = mastitis, family = poisson(link = "log"),
    corstr = "exchangeable")
```

### Coefficients:

|  | Estimate | Naive S.E. | Naive z | Robust S.E. | Robust z |
|---|---|---|---|---|---|
| (Intercept) | 0.423 | 0.761 | 0.556 | 0.554 | 0.764 |
| visit | 0.04 | 0.023 | 1.719 | 0.035 | 1.151 |
| numcult | 0.017 | 0.002 | 8.572 | 0.002 | 8.907 |
| bedorg | 1.480 | 0.630 | 2.35 | 0.196 | 7.553 |
| postmilk | -0.857 | 0.250 | -3.425 | 0.259 | -3.311 |
| tailed | -0.4240 | 0.179 | -2.372 | 0.150 | -2.819 |

Estimated Scale Parameter: 2.878
Working common Correlation = 0.42

### Model 3: Linear mixed-effects model fit by maximum likelihood

```
Random effects:
 Formula: ~1 | id
     (Intercept) Residual
StdDev:  0.462
Correlation Structure: Compound symmetry
Parameter estimate(s):
    Rho : -0.2
Variance function:
 Structure: fixed weights
 Formula: ~invwt
Fixed effects: y ~ visit + numcult + bedorg + postmilk + tailed
```

|  | Value | Std. Error | DF | t-Value | *p*-value |
|---|---|---|---|---|---|
| (Intercept) | 0.007 | 0.737 | 218 | 0.010 | 0.992 |
| visit | 0.036 | 0.022 | 218 | 1.602 | 0.111 |
| numcult | 0.019 | 0.002 | 218 | 8.008 | 0.000 |
| bedorg | 1.500 | 0.522 | 53 | 2.872 | 0.006 |
| postmilk | −0.694 | 0.344 | 53 | −2.019 | 0.048 |
| tailed | −0.405 | 0.211 | 53 | −1.917 | 0.061 |

**Model 4: Negative binomial model**
Call:

```
glm.nb(formula = y ~ visit + numcult + bedorg + postmilk + tailed,
data = mastitis, init.theta = 3.60, link = log)
```

Deviance Residuals:

| Min | 1Q | Median | 3Q | Max |
|---|---|---|---|---|
| −3.041 | −0.856 | −0.257 | 0.403 | 2.661 |

Coefficients:

|  | Estimate | Std. Error | z Value | Pr(>\|z\|) |
|---|---|---|---|---|
| (Intercept) | −0.108 | 0.495 | −0.218 | 0.827 |
| visit | 0.034 | 0.029 | 1.167 | 0.243 |
| numcult | 0.020 | 0.002 | 11.289 | < 2e-16 |
| bedorg | 1.657 | 0.352 | 4.700 | 2.6e-06 |
| postmilk | −0.733 | 0.200 | −3.658 | 0.000 |
| tailed | −0.32161 | 0.128 | −2.516 | 0.012 |

(Dispersion parameter for Negative Binomial (3.6002) family taken to be 1)
Null deviance: 457.70 on 276 degrees of freedom.
Residual deviance: 282.21 on 271 degrees of freedom. AIC: 1412.4
Theta: 3.600, Std. Err.: 0.510 ....... 2 × Log-likelihood = −1398.442
Note that there are very slight differences between the SAS and the R
outputs.

## 6.3 Other Models: Poisson Inverse Gaussian and Zero Inflated Poisson with Random Effects

There are many examples in the statistical literature where a mixture of count models is applied. Although the NBD is one of the most popular distributions, other distributions have been generated. For example, mixing Poisson with the inverse Gaussian distribution (Wilmot, 1987; Dean et al., 1989) and more recently the Poisson inverse Gaussian distribution (Shoukri et al., 2004) were discussed.

Quite often, particularly when the sample size is small, it may be difficult to distinguish between alternative mixing distributions, and the choice may be based on the ease of computation. This issue is illustrated by comparing the Poisson gamma (producing NBD) and Poisson inverse Gaussian (PIG) mixtures. Although the PIG does not have a closed form, the first two moments of the PIG mixture are the same as those of the NBD. Hence, the mixing distributions can be distinguished when other information such as higher moments become available. It was conjectured by Cameron and Trivedi (1998) that in small samples such information may lead to inconclusive results. As an example, the mastitis data, analyzed in Example 6.2. In that example a comparison between the PIG and the NBD regression models was made through the Q-Q plot of the residuals quantiles against the normal quantiles (see Figure 6.2, taken from Shoukri et al., 2004). Both plots are similar and one can hardly distinguish between the two models. However, because the NBD has closed form and programmed in commercial software, it is preferable to the PIG model.

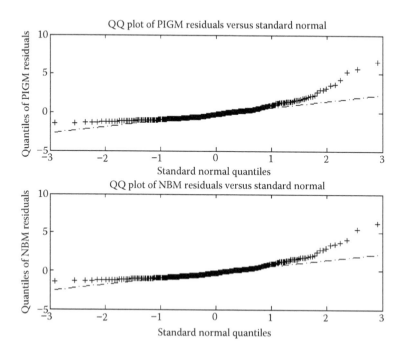

**FIGURE 6.2**
Q-Q plot of residuals for the Poison inverse Gaussian and the negative binomial regression models.

Another issue is the comparison among nonnested models. Two models are nonnested if neither model can be represented as a special case of the other model. Note that models that have some covariates in common and some covariates not in common are overlapping models. The usual method of discriminating among models by hypothesis test of the parameter restrictions that specializes one model to the other, for example, whether the dispersion parameter is zero in the transition from the negative binomial to the Poisson, is not strictly valid. Instead, we use the Akaike information criterion (AIC) to discriminate among nonnested models. Because the likelihood is expected to increase as parameters are added to the model, this criterion penalizes the likelihood with large number of parameters ($p$). Akaike (1973) proposed the information criterion AIC = $-2logl + p$.

The model with the lowest AIC is preferred. For example, the AIC for the Poisson GLMM is about 585.5, whereas the AIC for the negative binomial model is about 550, indicating that the negative binomial model provides a better fit within the class of cluster-specific models.

Other important application of count data with overdispersion is when there are excess zeros. The general term given to these models is *zero-inflated count models*. The zero-inflated Poisson (ZIP) distribution is given by

$$P_r[y_i = 0] = \omega + (1 - \omega)e^{-\mu_i}$$

$$P_r[y_i = r] = (1 - \omega)e^{-\mu_i}\mu_i/r!, \quad r = 1, 2, \ldots$$

Here, $\omega$, the proportion of zeros is added. Because $E(\mu_i) = \mu_i(1 - \omega)$ is smaller than $V(\mu_i) = (1 - \omega)(\mu_i + \omega\mu_i^2)$, excess zeros imply overdispersion.

Lambert (1992) introduced the ZIP model in which $\mu_i = \mu(x_i, \beta)$ and the probability $\omega$ is parameterized as a logistic function of the observable vector of covariates $z_i$, thereby ensuring nonnegativity of $\omega$, that is, $y_i = 0$ with probability $\omega$ and $y_i = e^{-\mu_i}\mu_i^{y_i}/y_i!$ with probability $1 - \omega$, and $\omega = \dfrac{\exp(z_i'\gamma)}{1 + \exp(z_i'\gamma)}$.

This model has been fitted to count data with zero inflated class by several packages.

If the data have hierarchical structure or the counts are from individuals belonging to the same cluster, then the ZIP must be extended to account for the within cluster correlation. That is, we need to construct a mixed-effect ZIP model. This is achieved by introducing two random effects ($U_1$ and $U_2$) into the linear predictors of the link functions, with additional variance components parameters, so that, for the logistic component

$$\omega = \frac{\exp(z_i'\gamma + u_1)}{1 + \exp(z_i'\gamma + u_1)}$$

and $u_2$, or the Poisson component. Here we assume that the random components are drawn from a bivariate normal distributed with $E(U_1) = E(U_2) = 0$, $var(U_1) = \sigma_1^2$, and $cov(U_1, U_2) \neq 0$. This will complete the model specifications of the mixed-effect ZIP model. We shall illustrate the mixed-effect ZIP model on insurance claims data.

### Example 6.4

Here we analyze the motor vehicle insurance claims data from Sweden for the year 1977. The data has seven variables:

1. Kilometers traveled per year—1: <1000, 2: 1000–15000, 3: 15000–20000, 4: 20000–25000, and 5: >25000.
2. Geographical zone—There are seven geographical zones.
3. Bonus—No claims bonus. Equal to the number of years, plus one, since last claim.
4. Make—1 to 8 represent eight different common car models. All other models are combined in class 9.
5. Insured—Number of insured in policy years.
6. Claims—Number of claims.
7. Payment—Total value of payments in SKR (Swedish krona).

The "Claims" is a count variable and its histogram is given in Figure 6.3. As can be seen from the histogram of the distribution of the number of claims (y), it is zero-inflated and is quite skewed to the right.

Again, we shall analyze the data using a series of models. From the simple descriptive statistics, we find that the standard deviation is about four times larger than the mean. We can guess that the possible sources of this overdispersion are (1) clustering, (2) inflation in the zero class,

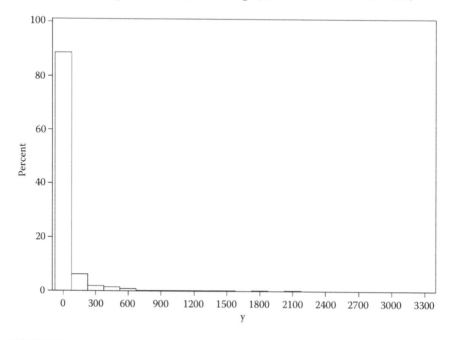

**FIGURE 6.3**

Histogram of the number of claims from the insurance data. (From Andrews, D. F., and Herzberg, A. M., 1985, *Data: A collection of problems from many fields for the student and research worker*, Springer, New York.)

(3) possibility of unmeasured covariates, or at least one of the above reasons. We fit the data with only two covariates: "kilo" modeled as a quantitative variable, and "zone" modeled as a categorical variable.

- SAS code to compute basic descriptive statistics and fit a Poisson regression model using the GEE:

```
* Basic descriptive statistics;
proc means data=insur mean median std maxdec=2; var
claims; run;


* Poisson model fitting using GEE;
proc genmod data=insur;
class make zone;
model y = kilo zone / dist=poisson link=log dscale;
repeated subject=make /type=cs ;
run;
```

Partial SAS output:

| Analysis Variable: Claims | | |
|---|---|---|
| Mean | Median | Std Dev |
| 51.86 | 5.00 | 201.71 |

| Criteria For Assessing Goodness of Fit | | | |
|---|---|---|---|
| Criterion | DF | Value | Value/DF |
| Deviance | 2174 | 352784.0607 | 162.2742 |
| Scaled Deviance | 2174 | 2174.0000 | 1.0000 |
| Pearson Chi-Square | 2174 | 984061.7447 | 452.6503 |
| Scaled Pearson $X^2$ | 2174 | 6064.1919 | 2.7894 |
| Log Likelihood | | 2311.2924 | |

| Exchangeable Working Correlation | |
|---|---|
| Correlation | 0.3537 |

| Analysis of GEE Parameter Estimates | | | | | | | |
|---|---|---|---|---|---|---|---|
| Empirical Standard Error Estimates | | | | | | | |
| | | | | 95% Confidence | | | |
| Parameter | | Estimate | Standard Error | Limits | | Z | Pr > \|Z\| |
| Intercept | | 3.0142 | 0.6415 | 1.7570 | 4.2715 | 4.70 | <.0001 |
| Kilo | | −0.3163 | 0.0186 | −0.3529 | −0.2798 | −16.97 | <.0001 |
| Zone | 1 | 2.4286 | 0.0741 | 2.2833 | 2.5739 | 32.76 | <.0001 |
| Zone | 2 | 2.3569 | 0.0579 | 2.2435 | 2.4703 | 40.73 | <.0001 |
| Zone | 3 | 2.3009 | 0.0565 | 2.1903 | 2.4116 | 40.76 | <.0001 |
| Zone | 4 | 2.7054 | 0.0439 | 2.6193 | 2.7915 | 61.58 | <.0001 |
| Zone | 5 | 1.3474 | 0.0366 | 1.2756 | 1.4192 | 36.80 | <.0001 |
| Zone | 6 | 1.7597 | 0.0231 | 1.7145 | 1.8049 | 76.31 | <.0001 |
| Zone | 7 | 0.0000 | 0.0000 | 0.0000 | 0.0000 | . | . |

- SAS code to fit a negative binomial regression model using GEE:

```
proc genmod data=insur;
class make zone;
model y = kilo zone / dist=nb link=log dscale;
repeated subject=make /type=cs;
run;
```

Partial SAS output:

| Criteria For Assessing Goodness of Fit | | | |
|---|---|---|---|
| Criterion | DF | Value | Value/DF |
| Deviance | 2174 | 2572.0563 | 1.1831 |
| Scaled Deviance | 2174 | 2174.0000 | 1.0000 |
| Pearson Chi-Square | 2174 | 6576.9273 | 3.0253 |
| Scaled Pearson $X^2$ | 2174 | 5559.0697 | 2.5571 |
| Log Likelihood | | 462042.7705 | |

| Exchangeable Working Correlation | |
|---|---|
| Correlation | 0.4243 |

| Analysis of GEE Parameter Estimates | | | | | | | |
|---|---|---|---|---|---|---|---|
| Empirical Standard Error Estimates | | | | | | | |
| Parameter | | Estimate | Standard Error | 95% Confidence Limits | | Z | Pr > \|Z\| |
| Intercept | | 2.0253 | 0.7147 | 0.6246 | 3.4261 | 2.83 | 0.0046 |
| Kilo | | −0.4721 | 0.0153 | −0.5021 | −0.4421 | −30.82 | <.0001 |
| Zone | 1 | 3.6308 | 0.0233 | 3.5850 | 3.6765 | 155.56 | <.0001 |
| Zone | 2 | 3.5694 | 0.0219 | 3.5264 | 3.6124 | 162.63 | <.0001 |
| Zone | 3 | 3.4999 | 0.0271 | 3.4469 | 3.5530 | 129.25 | <.0001 |
| Zone | 4 | 4.0206 | 0.0420 | 3.9382 | 4.1030 | 95.66 | <.0001 |
| Zone | 5 | 2.2214 | 0.0320 | 2.1588 | 2.2840 | 69.52 | <.0001 |
| Zone | 6 | 2.8592 | 0.0587 | 2.7442 | 2.9742 | 48.73 | <.0001 |
| Zone | 7 | 0.0000 | 0.0000 | 0.0000 | 0.0000 | . | . |

- SAS code to fit the random effects Poisson model using SAS GLIMMIX. Here "make" is used as the clustering factor.

```
* GLIMMIX fitting;
proc glimmix data=insur;
class make zone;
model y = kilo zone / s dist=poisson link=log ;
random intercept/ subject=make;
run;
```

Partial SAS output:

| Fit Statistics | |
| --- | --- |
| -2 Res Log Pseudo-Likelihood | 121748.1 |
| Generalized Chi-Square | 122566.0 |
| Gener. Chi-Square / DF | 56.38 |

| Covariance Parameter Estimates | | | |
| --- | --- | --- | --- |
| Cov Parm | Subject | Estimate | Standard Error |
| Intercept | make | 1.7092 | 0.8547 |

| Solutions for Fixed Effects | | | | | | |
| --- | --- | --- | --- | --- | --- | --- |
| Effect | Zone | Estimate | Standard Error | DF | t Value | Pr > \|t\| |
| Intercept | | 0.5742 | 0.4377 | 8 | 1.31 | 0.2260 |
| Kilo | | −0.3813 | 0.002300 | 2166 | −165.77 | <.0001 |
| Zone | 1 | 3.6154 | 0.04069 | 2166 | 88.84 | <.0001 |
| Zone | 2 | 3.5312 | 0.04074 | 2166 | 86.67 | <.0001 |
| Zone | 3 | 3.4650 | 0.04078 | 2166 | 84.97 | <.0001 |
| Zone | 4 | 3.9354 | 0.04055 | 2166 | 97.05 | <.0001 |
| Zone | 5 | 2.2582 | 0.04220 | 2166 | 53.52 | <.0001 |
| Zone | 6 | 2.8008 | 0.04136 | 2166 | 67.72 | <.0001 |
| Zone | 7 | 0 | . | . | . | . |

| Type III Tests of Fixed Effects | | | | |
| --- | --- | --- | --- | --- |
| Effect | Num DF | Den DF | F Value | Pr > F |
| Kilo | 1 | 2166 | 27480.2 | <.0001 |
| Zone | 6 | 2166 | 4701.37 | <.0001 |

- SAS code to fit the random effects negative binomial model using SAS GLIMMIX. Here "make" is used as the clustering factor.

```
proc glimmix data=insur;
class make zone;
model y = kilo zone / s dist=nb link=log ;
random make;
run;
```

Partial SAS output:

| Fit Statistics | |
| --- | --- |
| -2 Res Log Pseudo-Likelihood | 7897.09 |
| Generalized Chi-Square | 2223.39 |
| Gener. Chi-Square / DF | 1.02 |

| Covariance Parameter Estimates | | |
|---|---|---|
| Cov Parm | Estimate | Standard Error |
| make | 1.6038 | 0.8058 |
| Scale | 1.7335 | 0.06058 |

| Solutions for Fixed Effects | | | | | | |
|---|---|---|---|---|---|---|
| Effect | Zone | Estimate | Standard Error | DF | t Value | Pr > \|t\| |
| Intercept | | 0.6223 | 0.4402 | 8 | 1.41 | 0.1952 |
| Kilo | | −0.4174 | 0.02151 | 2166 | −19.40 | <.0001 |
| Zone | 1 | 3.5942 | 0.1353 | 2166 | 26.56 | <.0001 |
| Zone | 2 | 3.5682 | 0.1353 | 2166 | 26.36 | <.0001 |
| Zone | 3 | 3.5551 | 0.1354 | 2166 | 26.26 | <.0001 |
| Zone | 4 | 4.0803 | 0.1350 | 2166 | 30.23 | <.0001 |
| Zone | 5 | 2.2952 | 0.1376 | 2166 | 16.68 | <.0001 |
| Zone | 6 | 2.9745 | 0.1361 | 2166 | 21.86 | <.0001 |
| Zone | 7 | 0 | . | . | . | . |

| Type III Tests of Fixed Effects | | | | |
|---|---|---|---|---|
| Effect | Num DF | Den DF | F Value | Pr > F |
| Kilo | 1 | 2166 | 376.49 | <.0001 |
| Zone | 6 | 2166 | 191.76 | <.0001 |

We shall provide the R codes, but not the results for the following reasons:

1. Negative binomial models did not converge, but the Poisson model converged quite rapidly:

```
glm(formula = claims ~ kilo + zone + bonus + make +
insured +  payment + offset(log(insured)), data =
insurance)
```

2. GEE with exchangeable correlation structure did not converge.
3. GEE with independence correlation structure converged.

```
gee(formula = claims ~ kilo + bonus + make + insured +
payment, id = zone, data = insurance, family = poisson
(link = "log"), corstr = "independence")
```

Comments on the results:

1. It is clear from all the models that the number of claims significantly depends on the number of kilometers driven. The more the car has been driven, the less the number of claims, and there are wide variations among zones. Note that a large value for the intracluster correlation (0.353 for the Poisson and 0.424 for the negative binomial model) indicates that a significant percentage in the variability in the number of claims may be attributed to the "make" of the car.

2. If we take the deviance as a measure of goodness of fit in the GENMOD, the negative binomial provides a better fit. In fact the deviance Poisson is 352784, and the deviance negative binomial is 2572. The same remark holds when the models are fitted by the GLIMMIX. For the Poisson with random effect, the generalized chi-square/DF is 56.38, and the same index in the case of the negative binomial is 1.02.

3. The aforementioned fitted models do not take into account the two-part nature of the distribution of the number of claims caused by the zero inflation. Rather they deal with the issue of extra variability in the count variable. Following the Lambert (1992) approach and fitting the two-part negative binomial model using STATA, we get

| Parameter | Kilo | zone 1 | zone 2 | zone 3 | zone 4 | zone 5 | zone 6 |
|-----------|------|--------|--------|--------|--------|--------|--------|
| Estimate  | −0.46 | 3.25 | 3.19 | 3.12 | 3.64 | 1.85 | 2.48 |
| SE        | 0.03 | 0.18 | 0.18 | 0.18 | 0.18 | 0.18 | 0.18 |

All parameter estimates are significantly different from zero. Moreover, all parameter estimates of $\gamma$ in $\omega = \dfrac{\exp(z_i'\gamma)}{1 + \exp(z_i'\gamma)}$ are not significant. It is interesting to note as well that the zero-inflated negative binomial provides information that are similar to the information obtained from the GLIMMIX negative binomial.

## Example 6.5

As can be seen from the histogram of the number of claims, there seems to be an inflated-zero class. In what follows, we fit the insurance claims data under the Non-Linear Mixed Model (NLMIXED) procedure in SAS.

**SAS code: "ZERO INFLATED POISSON WITH RANDOM EFFECTS"**

```
/* Our attempt to fit Zero Inflated Poisson Distribution first we get
initial values
     for the prameter by fitting separate poisson and logistic
modles*/
   data zip;
   set insur;
   if y=0 then h=0;
   else if y> 0 then h=1;
   run;
   proc genmod data=zip descending;
   class make zone;
   model h = kilo zone / dist=bin link=logit;
   repeated subject=make /type =cs;
   run;

   /*NLMIXED FOR ZIP */

   /* fitting ZIP to the Insurance data*/

/* converting categorical variable into dummy variables
 Since the NLMIXED does not handle "CLASS" statement */
data new;
set insur;
if zone =1 then x1 =1;
```

```
    else x1=0;
    if zone =2 then x2 =1;
    else x2=0;
    if zone =3 then x3=1;
    else x3=0;
    if zone =4 then x4 =1;
    else x4=0;
    if zone =5 then x5 =1;
    else x5=0;
    if zone =6 then x6 =1;
    else x6=0;
    run;

    PROC NLMIXED data=new NOSORTSUB;
    PARMS

    b0= .5 bkilo=-.4 /* The only covariate that we shall include in the
    Logistic regression part */

    a0= .7 akilo=-.4 azone1=1 azone2=2 azone3=2 azone4=1 azone5=2.3
    azone6=4

    V_u1=.5 Cov_u12=0.0 V_u2=0.5;
    /*L1 = linear predictor of the logistic part of the model;*/

    L1= b0 + kilo*bkilo+u1;
    /*infprob=probability of inflated zeros, which is the logistic
    transform of the linear predictor;*/
    infprob = 1/ (1+exp(-L1));
    /*eta is the linear predictor for the Poisson;*/
    eta= a0+ kilo*akilo+x1*azone1+ x2*azone2+x3*azone3+x4*azone4+
    x5*azone5+x6*azone6+u2;
    /*lamda is the expected Poisson counts;*/
    lamda=exp(eta);
    /*Build the ZIP model;*/
    if y =0 then LL=log(infprob+(1-infprob)*exp(-lamda));
    else LL=log(1-infprob)+y*log(lamda)-lamda-lgamma(y+1);
    model y ~ general(LL);
    random u1 u2 ~ NORMAL([0,0], [V_u1,Cov_u12,V_u2])
     SUBJECT= make;
    run;

      ODS GRAPHICS OFF;
       ODS RTF CLOSE;


    run;
```

The SAS output is

| Moments | | | |
|---|---|---|---|
| N | 2182 | Sum Weights | 2182 |
| Mean | 51.865 | Sum Observations | 113171 |
| Std Deviation | 201.710 | Variance | 40687.203 |
| Skewness | 8.572 | Kurtosis | 93.364 |
| Uncorrected SS | 94608487 | Corrected SS | 88738791.7 |
| Coeff Variation | 388.909 | Std Error Mean | 4.318 |

| Basic Statistical Measures | | | |
|---|---|---|---|
| Location | | Variability | |
| Mean | 51.865 | Std Deviation | 201.710 |
| Median | 5.000 | Variance | 40687 |
| Mode | 0.000 | Range | 3338 |
| | | IQR | 20.000 |

Specifications of the NLMIXED

| Specifications | |
|---|---|
| Data Set | WORK.NEW |
| Dependent Variable | y |
| Distribution for Dependent Variable | General |
| Random Effects | u1 u2 |
| Distribution for Random Effects | Normal |
| Subject Variable | Make |
| Optimization Technique | Dual Quasi-Newton |
| Integration Method | Adaptive Gaussian Quadrature |

| Dimensions | |
|---|---|
| Observations Used | 2182 |
| Observations Not Used | 0 |
| Total Observations | 2182 |
| Subjects | 2182 |
| Max Obs per Subject | 1 |
| Parameters | 13 |
| Quadrature Points | 5 |

| Fit Statistics | |
|---|---|
| -2 Log Likelihood | 16501 |
| AIC (smaller is better) | 16527 |
| AICC (smaller is better) | 16527 |
| BIC (smaller is better) | 16600 |

Estimates of Fixed Effects Parameters

| Parameter | Estimate | Standard Error | DF | Parameter Estimates t Value | Pr > \|t\| | Alpha | Lower | Upper | Gradient |
|---|---|---|---|---|---|---|---|---|---|
| b0 | -1.8338 | 5.5945 | 2180 | -0.33 | 0.7431 | 0.05 | -12.805 | 9.137 | 0.466 |
| bkilo | -3.5359 | 5.0411 | 2180 | -0.70 | 0.4831 | 0.05 | -13.421 | 6.3499 | 0.506965 |
| a0 | 0.08094 | 0.1644 | 2180 | 0.49 | 0.6225 | 0.05 | -0.2414 | 0.4033 | -0.02733 |
| akilo | -0.4379 | 0.02966 | 2180 | -14.76 | <.0001 | 0.05 | -0.4961 | -0.378 | -0.144 |
| azone1 | 3.6721 | 0.1776 | 2180 | 20.68 | <.0001 | 0.05 | 3.3238 | 4.020 | 0.324 |
| azone2 | 3.581 | 0.177 | 2180 | 20.15 | <.0001 | 0.05 | 3.233 | 3.9301 | 0.119 |
| azone3 | 3.5658 | 0.1777 | 2180 | 20.07 | <.0001 | 0.05 | 3.217 | 3.914 | 0.1491 |
| azone4 | 4.0389 | 0.1772 | 2180 | 22.79 | <.0001 | 0.05 | 3.691 | 4.386 | 0.351 |
| azone5 | 2.2206 | 0.1810 | 2180 | 12.27 | <.0001 | 0.05 | 1.865 | 2.575 | -0.086 |
| azone6 | 2.8502 | 0.1790 | 2180 | 15.92 | <.0001 | 0.05 | 2.499 | 3.201 | -0.1569 |

Estimates of the Variance Components of the Random Effects

| Parameter | Estimate | SE | DF | t-Value | p | Alpha | LL | UL | Gradient |
|-----------|----------|-----|-----|---------|---------|-------|---------|--------|----------|
| V_u1 | 0.2996 | 0.7511 | 2180 | 0.40 | 0.6900 | 0.05 | -1.1733 | 1.7725 | -0.76763 |
| Cov_u12 | -0.8925 | 1.1285 | 2180 | -0.79 | 0.4291 | 0.05 | -3.1055 | 1.3205 | -0.65243 |
| V_u2 | 3.1703 | 0.1186 | 2180 | 26.72 | <.0001 | 0.05 | 2.9377 | 3.4030 | -0.06511 |

## Exercises

6.1. The generalized Poisson distribution introduced by Consul & Jain (1973) and discussed in the paper by Janardan et al. (1979) has two parameters and has its variance is larger than the mean. The probability distribution is given by $P_r(Y = y) = (y!)^{-1}\lambda(\lambda + y\theta)^{y-1}exp(-(\lambda + y\theta))$. The mean and variance are given, respectively, by $E(Y) = \lambda/(1-\theta)$ and $var(Y) = \lambda/(1-\theta)^3$. Clearly the distribution reduces to the Poisson when $\theta = 0$. Fit the Poisson and the GPD to the data in Table 6.1 using the method of moments. Using the chi-square goodness of fit, which model fits the data better?

6.2. Suppose that $Y_i(i = 1,2,...,n)$ is a random variable whose probability distribution is Poisson, with mean $\lambda$. Assume a log-link, $log\lambda = \beta_0 + \beta_1 X_i$, where $X$ is a random covariate that has a standard normal distribution. Derive the likelihood equations for estimating the parameters $\beta_0$ and $\beta_1$.

6.3. Show that the elements of Fisher's information matrix are given by

$$-E\left(\frac{\partial^2\ell}{\partial\beta_0^2}\right) = ne^{\beta_0}E\left(e^{\beta_1 X}\right) \equiv i_{00}$$

$$-E\left(\frac{\partial^2\ell}{\partial\beta_0\partial\beta_1}\right) = ne^{\beta_0}E\left(xe^{\beta_1 X}\right) = i_{01}$$

and

$$-E\left(\frac{\partial^2\ell}{\partial\beta_1^2}\right) = ne^{\beta_0}E\left(x^2e^{\beta_1 X}\right) = i_{11}$$

6.4. Given that the moments generating function of the standard normal distribution is

$$E\left(e^{tX}\right) = e^{1/2 t^2}$$

and that differentiation is permitted inside the integration sign, show that

$$i_{00} = n e^{\beta_0 + 1/2 \beta_1^2}$$

$$i_{01} = n \beta_1 e^{\beta_0 + 1/2 \beta_1^2}$$

and

$$i_{11} = n\left(1 + \beta_1^2\right) e^{\beta_0 + 1/2 \beta_1^2}$$

6.5. Show that by inverting Fisher's information matrix, the large sample variance of $\hat{\beta}_1$, the maximum likelihood estimation of $\beta$ is given by

$$n \cdot \mathrm{var}\left(\hat{\beta}_1\right) = e^{\beta_0 - 1/2 \beta_1^2}$$

6.6. Suppose that you are requested to test the hypothesis $H_0 : \beta_1 = 0$ versus $H_1 : \beta_1 = 0.1$. What are the values of $\mathrm{var}(\hat{\beta}_1)$ under both hypotheses when $n = 1500$?

6.7. If you want to test the hypotheses $H_0 : \beta_1 = 0$ versus a two-sided alternative $H_1 : \beta_1 = 0.1$, with power $1-\gamma$ and type I error rate $\alpha$, show that the required sample size is given by

$$n \geq \left(Z_\gamma a_1 + Z_{\alpha/2} a_2\right)^2 / \beta_1^2$$

where

$$a_1^{-1} = e^{1/2 \beta_0 + 1/4 \beta_1^2}$$

$$a_2^{-1} = e^{1/2 \beta_0}$$

6.8. What would be the required sample size for $\alpha = 0.05$, $\gamma = 0.20$, $\beta = \log(0.07)$, and $\beta = 0.1$?

6.9. Suppose that $Y_{ij}$ has Poisson distribution with mean $\log \lambda_{ij} = \beta_0 + \beta_1 x_{ij}$, $i = 1,2,...n$, $j = 1,2,...m$, with $m$ being the cluster size, and $n$ the number of clusters. Assuming that the intracluster correlation is $\rho$, we have a sample inflation factor (design effect) $D = 1 + (m - 1)\rho$. To test the hypotheses stated in Exercise 6.8, develop a sample size approach to estimate the number of clusters, $n$, that takes into account the intracluster correlation.

6.10. Execute the R code for the data of Example 6.3. Comment on the results.

# 7

## Repeated Measures and Longitudinal Data Analysis

### 7.1 Introduction

Experimental designs with repeated measures over time are very common in biological and medical research. This type of design is characterized by repeated observations on a large number of individuals for a relatively small number of time points, where each individual is observed at such points. The repeated measures design has several advantages over completely randomized designs, the most important of which is the ability to control biological heterogeneity between individuals by measuring individuals repeatedly over time. This reduction in heterogeneity makes the repeated measures design more efficient than completely randomized designs.

Longitudinal and repeated measures data are very frequent in almost all scientific fields, including agriculture, biology, medicine, epidemiology, geography, and demography. It is not the aim of this chapter to provide a comprehensive coverage of the subject. The reader is referred to the many excellent texts such as Crowder and Hand (1990), Lindsay (1993), Hand and Crowder (1996), and Diggle et al. (1994), and many articles that have appeared in scientific journals (e.g., *Biometrics* and *Statistics in Medicine*). This chapter describes the statistical methodologies that are used to answer the scientific questions posed. Detailed examples on repeated measures and longitudinal data using the population-averaged generalized estimating equations models and the subject-specific generalized mixed models are provided to illustrate the utility of these models in the context of repeated measures or longitudinal studies that include covariate effects. The following section provides examples of repeated measures experiments that are frequently encountered.

## 7.2  Examples

### 7.2.1  Experimental Studies

#### 7.2.1.1 Liver Enzyme Activity

Frison and Pocock (1992) describe a clinical trial in which 152 patients with heart disease were randomly allocated to treatment using an active drug or a placebo during a 12-month follow-up period. The concentration of the liver enzyme creatinine phosphokinase (CPK) in the patient's serum was measured as an indicator of liver damage arising as a side effect of the treatment. Each patient had three pretreatment measurements that were taken at two months before, one month before, and at the time of randomization. They also had eight posttreatment measurements taken every one and a half months after randomization. The main objective of the study was to see how the enzyme level varied over time before and after treatment.

#### 7.2.1.2 Effect of Mycobacterium Inoculation on Weight

An experiment was conducted in a veterinary microbiology lab with the main objective to determine the effect of mycobacterium inoculation on the weight of immune-deficient mice. Severely immune-deficient beige mice (six to eight weeks old) were randomly allocated to the following groups:

Group 1—Control group where animals did not receive any inoculation.

Group 2—Animals were inoculated intraperitoneally with live mycobacterium paratuberculosis (MPTB) and transplanted with peripheral blood leucocytes (PBL) from humans with Crohn's disease.

Group 3—Animals were inoculated with live MPTB and transplanted with PBL from bovine.

In each group the mice were weighed at baseline (week 0), week 2, and week 4. The question of interest concerns differences between the mean weights among the three groups. The data are given in Table 7.1.

### 7.2.2  Observational Studies

#### 7.2.2.1 Variations in Teenage Pregnancy Rates in Canada

Table 7.2 shows teenage pregnancies rates (per 1000 females aged 15–17 and 18–19). These rates include live births, therapeutic abortions, hospitalized cases of spontaneous and other unspecified abortions, and registered stillbirths with at least 20 weeks gestation at pregnancy termination (Statistics Canada, 1991, vol. 3, no. 4). The Newfoundland data are not included. It is important to detect variations regionally and over time in teenage pregnancy rates.

**TABLE 7.1**

Weights in Grams of Inoculated Immune-Deficient Mice

| | | Time in Weeks | | |
|---|---|---|---|---|
| Group | Mice | 0 | 2 | 4 |
| 1 | 1 | 28 | 25 | 45 |
| 2 | 1 | 40 | 31 | 70 |
| 3 | 1 | 31 | 40 | 44 |
| 4 | 1 | 27 | 21 | 26 |
| 5 | 1 | 27 | 25 | 40 |
| 6 | 2 | 34 | 25 | 38 |
| 7 | 2 | 36 | 31 | 49 |
| 8 | 2 | 41 | 21 | 25 |
| 9 | 2 | 28 | 22 | 10 |
| 10 | 2 | 29 | 24 | 22 |
| 11 | 2 | 31 | 18 | 36 |
| 12 | 2 | 31 | 16 | 5 |
| 13 | 3 | 28 | 28 | 61 |
| 14 | 3 | 27 | 23 | 63 |
| 15 | 3 | 31 | 30 | 42 |
| 16 | 3 | 19 | 16 | 28 |
| 17 | 3 | 20 | 18 | 39 |
| 18 | 3 | 22 | 24 | 52 |
| 19 | 3 | 22 | 22 | 25 |
| 20 | 3 | 28 | 26 | 53 |

**TABLE 7.2**

Teenage Pregnancy Rates per 1000 Females Aged 15–17 and 18–19

| | | East | | | | | West | | | | Territories | |
|---|---|---|---|---|---|---|---|---|---|---|---|---|
| Year | Age Group | PEI | NS | NB | QU | ONT | MAN | SASK | ALTA | BC | Y | NWT |
| 1975 | 15–17 | 41.6 | 43.4 | 40.5 | 12.8 | 39 | 41.7 | 48.5 | 50.5 | 48.6 | 70 | 77.5 |
| | 18–19 | 96.6 | 98.6 | 103.7 | 43.4 | 91.5 | 98.8 | 116.1 | 120.9 | 102 | 110 | 192.5 |
| 1980 | 15–17 | 28 | 34.9 | 28.9 | 12.9 | 32.8 | 34.2 | 46.3 | 43.2 | 41.0 | 65 | 104 |
| | 18–19 | 59.6 | 81.4 | 73.6 | 39.8 | 74.4 | 85.4 | 106.6 | 111.4 | 89.4 | 105 | 165 |
| 1985 | 15–17 | 20.3 | 28.6 | 21.6 | 11.9 | 26 | 35.4 | 37.9 | 34.5 | 29.9 | 50 | 109.3 |
| | 18–19 | 64.5 | 65.8 | 59.5 | 37.7 | 63 | 88.9 | 88.4 | 85.8 | 72 | 77.5 | 186 |
| 1989 | 15–17 | 21.1 | 30.8 | 22.6 | 16.6 | 27.2 | 40.5 | 37.7 | 64.4 | 31.6 | 38.3 | 97.8 |
| | 18–19 | 60.5 | 69.4 | 56.4 | 48.0 | 64.8 | 96.9 | 95.0 | 87.6 | 77.2 | 120 | 220 |

*Note:* ALTA, Alberta; BC, British Columbia; MAN, Manitoba; NB, New Brunswick; NS, Nova Scotia; NWT, North West Territories; ONT, Ontario; PEI, Prince Edward Island; QU, Quebec; SASK, Saskatchewan; Y, Yukon.

Longitudinal surveys are also common in medical research. Laird and Ware (1982), for example, describe a survey in which pulmonary function in about 200 schoolchildren was examined under normal conditions, then during an air pollution alert and on three successive weeks following the alert. The main aim of the study was to determine whether the volume of air exhaled in the first second of a forced exhalation ($FEV_1$) was depressed during the alert. The analysis of repeated categorical measures has been illustrated by Ware et al. (1988). Children were assessed annually at ages 9 to 12 to evaluate the potential effects of air pollution on persistent wheeze. Parents were asked about wheezing by their children during the previous year and responses were grouped into three mutually exclusive categories or states: no wheeze, wheeze with colds, or wheeze apart from colds.

### 7.2.2.2 Number of Tuberculosis Cases in Saudi Arabia

In the Kingdom of Saudi Arabia, there are up to 9 million migrants, mainly from tuberculosis (TB) endemic regions in South/South East Asia and Africa. In addition, 3 million to 6 million pilgrims annually visit the holy cities located in the western region of the Kingdom for performing the holy rituals, Hajj and Umrah. Interestingly, the majority of these pilgrims are coming from TB endemic areas of Asia and Africa. There are differences in the rate of TB infection between different regions of the country. For instance, in Jeddah (sea and airports for pilgrims) the infection rate can reach up to 64 cases per 100,000 compared with 32 per 100,000 in Riyadh (central region). The higher rate in Jeddah may be attributed to the inflow of pilgrims. The data and its graphical representation will be discussed later in this chapter.

## 7.3 Methods for the Analysis of Repeated Measures Data

In the analysis of repeated measures or longitudinal data, the critical feature to recognize is that, since sets of measures are obtained from the same subjects, these measures are likely to be correlated and can rarely be considered as independent. How that dependence is dealt with is a principal distinguishing feature of different methods of analysis. However, before we discuss in detail this fundamental feature, more preliminary examination of the data must be considered. Such examination includes the graphical display of the data, an important step before modeling and data analyses are performed.

Diggle et al. (1994) gave the following simple guidelines for the exploration of repeated measures data using graphical displays:

1. They show as much of the relevant data as possible rather than data summaries.
2. Graphics highlight aggregate patterns of potential scientific interest.
3. Graphics identify both cross-sectional and longitudinal patterns in the data.
4. Graphics identify unusual individuals or unusual observations.

---

## 7.4 Basic Models

Suppose that observations are obtained on $n$ time points for each of $k$ subjects that are divided into $h$ groups, with $k_j$ subjects in the $j$th group.

Let $y_{hij}$ denote the $j$th measurement made on the $i$th subject within the $h$th group. The analysis of data from the repeated measures designs considered in this chapter will be dealt with linear mixed model analysis of variance. As for Chapter 1, the linear model representation of the observations $y_{hij}$ is as follows:

$$y_{hij} = \mu_h + \alpha_{ij} + e_{hij} \tag{7.1}$$

where

- $\mu_h$ is the overall mean response in the $h$th group. The $\mu_h$ is called the fixed effect because it takes one unique value irrespective of the subject being observed and irrespective of the time point.
- $\alpha_{ij}$ represents the departure of $y_{hij}$ from $\mu_h$ for a particular subject; this is called a random effect.
- $e_{hij}$ is the error term representing the discrepancy between $y_{hij}$ and $\mu_h + \alpha_{ij}$.

Further to the above setup, some assumptions on the distributions of the random components of the mixed model (7.1) are needed. These assumptions are

1. $E(\alpha_{ij}) = E(e_{hij}) = 0, V(\alpha_{ij}) = \sigma_\alpha^2, V(e_{ij}) = \sigma_e^2, \mathrm{Cov}(\alpha_{ij}, \alpha_{ij'}) = 0$ and $\mathrm{Cov}(\alpha_{ij}, \alpha_{ij'}) = \sigma_\alpha^2$.
2. $\mathrm{Cov}(e_{hij}, e_{h'ij'}) = 0$ $i \neq i'$ or $j \neq j'$, $h \neq h'$.
3. The random effects $\alpha_{ij}$ is uncorrelated with $e_{hij}$.
4. $\alpha_{ij}$ and $e_{ij}$ are normally distributed.

Consequently, it can be shown that the covariance between any pair of measurements is

$$
\text{Cov}\left(y_{hij}, y_{h'i'j'}\right) =
\begin{bmatrix}
0 & i \neq i' \\
\sigma_\alpha^2 & i = i' \quad j \neq j' \\
\sigma_\alpha^2 + \sigma_e^2 & i = i' \quad \text{and} \quad j = j'
\end{bmatrix}
\tag{7.2}
$$

Therefore, any pair of measurements on the same individual is correlated, and the correlation is given by

$$
\text{Corr}\left(y_{hij}, y_{h'i'j}\right) = \rho = \frac{\sigma_\alpha^2}{\sigma_\alpha^2 + \sigma_e^2}
\tag{7.3}
$$

We have already discussed the definition given in Equation 7.2 in Chapters 1 and 2 (known as the intraclass correlation). The assumptions of common variance of the observations and common correlation among all pairs of observations on the same subject can be expressed as an $n_i \times n_i$ matrix $\Sigma$, where $n_i$ is the number of observations taken repeatedly on the $i$th subject and

$$
\text{Cov}\left(y_{hij}, y_{h'i'j}\right) = \sum = \sigma^2
\begin{bmatrix}
1 & \rho & \dots & \rho \\
\rho & 1 & \dots & \rho \\
\cdot & \cdot & \dots & \cdot \\
\cdot & \cdot & \dots & \cdot \\
\cdot & \cdot & \dots & \cdot \\
\rho & \rho & \dots & 1
\end{bmatrix}
\tag{7.4}
$$

where $\sigma^2 = \sigma_\alpha^2 + \sigma_e^2$.

A matrix of the form Equation 7.4 is said to possess the property of compound symmetry or uniformity or exchangeable correlation structure (Geisser, 1963).

Unfortunately, the data obtained in many repeated measures or longitudinal settings rarely satisfy the assumption of compound symmetry. In such designs, it is common to find that the adjacent observations or successive measurements on adjacent time points are more highly correlated than nonadjacent time points, with the correlation between these measurements decreasing the farther apart the measurement.

It will have become apparent that for analyzing longitudinal data, although the main interest may lie in estimating the effects of risk factors and exposures on the expected value of the response, it often seems necessary to expend more effort to ensure that the model for the variance and covariance among the response is correct. Huber (1967) proposed a heteroscedastic consistent "sandwich" estimator for the parameter covariance matrix. Variants of this covariance estimator are available among many of the software implementations of procedures described in this chapter. At the cost of reduced efficiency—often trivial but sometimes large—the use of this method provides some relief from an excessive concern that the nonsystematic component of the adopted model need be correctly specified in every detail. Summarizing the objectives of analyzing longitudinal data include:

- Measure the average treatment effect over time.
- Assess treatment effects separately at each time point and to test whether treatment interacts with time.
- Identify any covariance patterns in the repeated measurements.
- Determine a suitable model to describe the relationship of a measurement with time.

## 7.5 The Issue of Missing Observations

Missing data is quite common for longitudinal and repeated measures data. It is important to consider the reasons behind missing observations. If the reason that an observation is missing is related to the response variable that is missed, the analysis will be biased. Rubin (1976, 1994) discusses the concept of data that are missing at random in a general statistical setting. Little and Rubin (1987) distinguish between observations that are missing completely at random, where the probability of missing an observation is independent of both the observed responses and the missing responses, and missing at random, where the probability of missing an observation is independent of the responses. Laird (1988) used the term *ignorable missing data* to describe observations that are missing at random.

Diggle (1989) discussed the problem of testing whether dropouts in repeated measures investigations occur at random. Dropouts are a common source of missing observations, and it is important for the analysis to determine if the dropouts occur at random. In the data analysis sections, we will assume that the mechanism for missing observations is ignorable.

## 7.6 Mixed Linear Regression Models

### 7.6.1 Formulation of the Models

A general linear mixed model for the analysis of longitudinal data has been proposed by Laird and Ware (1982):

$$Y_i = X_i\beta + Z_i\alpha_1 + \varepsilon_i \tag{7.5}$$

where $Y_i$ is an $n_i \times 1$ column vector of measurements for subject (cluster) $i$; $X_i$ is an $n_i \times p$ design matrix; $\beta$ is a $p \times 1$ vector of regression coefficients assumed to be fixed; and $Z_i$ is an $n_i \times q$ design matrix for the random effects, $\alpha_i$, which are assumed to be independently distributed across subjects with distribution $\alpha_i \sim N(0,\sigma^2 G)$, where $G$ is an arbitrary covariance matrix. The within-subjects errors, $\varepsilon_i$, are assumed to be distributed $\alpha_i \sim N(0,\sigma^2 R_i)$. It is also assumed that $\varepsilon_i$ and $\alpha_i$ are independent of each other. The fact that $Y_i$, $X_i$, $Z_i$, and $R_i$ are indexed by $i$ means that these matrices are subject specific. Note that the model 7.1 is a special case of 7.5.

Mixed models have the following advantages:

- A single model can be used to estimate overall treatment effects and to estimate treatment effects at each time point. There is no need to calculate mean values across all time points (to obtain the overall treatment effects) or to analyze time points separately (to obtain treatment effects at each time point). Standard errors for treatment effects at individual time points are calculated using information from all time points and are therefore more robust than standard errors calculated from separate time points.

- The presence of missing data causes no problems provided they can be assumed missing at random. This assumption will be assumed to hold throughout this chapter.

- The covariance pattern of the repeated measurements can be determined and properly accounted for.

There are several ways a mixed model can be used to analyze repeated measures data. The simplest approach is to use a random effects model with subject effects considered as random. This will allow for a constant correlation between all observations on the same patient. When the relationship of the response variable with time is of interest, a random coefficients model is appropriate. Here, regression curves are fitted for each patient and the regression coefficients are allowed to vary randomly between the patients. These models are considered in the following sections.

## 7.6.2 Covariance Patterns

A large selection of covariance patterns is available for use in mixed models. Most of the patterns are dependent on measurements being taken at fixed times and some are also easier to justify when the observations are evenly spaced. There are also other patterns where covariances are based on the exact value of time, and these are most useful in situations where the time intervals are irregular.

Some simple covariance patterns for the $R_i$ matrices for a trial with four time points are shown next. In the general pattern, sometimes also referred to as "unstructured," the variances of responses, $\sigma_i^2$, differ for each time period $i$, and the covariances, $\theta_{jk}$, differ between each pair of periods $j$ and $k$. For the first-order autoregressive model, the variances are equal and the covariance decreases exponentially depending on their separation $|j-k|$, so $\theta_{jk} = \rho^{|j-k|}\sigma^2$. This is sometimes an appropriate model when time periods are evenly spaced. It can then be seen as a "natural" model from a time series viewpoint. However, it may be justified empirically in circumstances where the observations are not evenly spaced. For the compound symmetry covariance model, all covariances are equal. Other covariance patterns are possible, but we shall restrict our presentations to these three commonly occurring patterns.

$$\text{General (unstructured; UN)} \quad R_i = \begin{pmatrix} \sigma_1^2 & \theta_{12} & \theta_{13} & \theta_{14} \\ \theta_{12} & \sigma_2^2 & \theta_{23} & \theta_{24} \\ \theta_{13} & \theta_{23} & \sigma_3^2 & \theta_{34} \\ \theta_{14} & \theta_{24} & \theta_{34} & \sigma_4^2 \end{pmatrix}$$

$$\text{First–order autoregressive (AR)} \quad R_i = \sigma^2 \begin{pmatrix} 1 & \rho & \rho^2 & \rho^3 \\ \rho & 1 & \rho & \rho^2 \\ \rho^2 & \rho & 1 & \rho \\ \rho^3 & \rho^2 & \rho & 1 \end{pmatrix}$$

$$\text{Compound symmetry (CS)} \quad R_i = \begin{pmatrix} \sigma^2 & \theta & \theta & \theta \\ \theta & \sigma^2 & \theta & \theta \\ \theta & \theta & \sigma^2 & \theta \\ \theta & \theta & \theta & \sigma^2 \end{pmatrix}$$

There are many covariance patterns available and choosing the most appropriate one is not always easy. The idea is to select the pattern that best fits the true covariance of data, and provide appropriate standard errors for fixed effect estimates. There are two alternative approaches to making a model choice. One is to compare models based on measures of fit that are adjusted for the number of covariance parameters. Another is to use likelihood ratio tests to find whether additional parameters cause a statistically significant improvement in the model.

The likelihood statistics is expected to become larger as more parameters are included in the model. The two statistics below are based on the likelihood but make allowance for the number of covariance parameters fitted. They can be used to make direct comparisons between models that fit the same fixed effects. Akaike's information criterion (AIC) (Akaike, 1973) is given by AIC = $-2 \log (L) + 2 p$, where $p$ is the number of covariates including the intercept. Schwarz's information criterion (SIC) (Schwarz, 1978), takes into account the number of fixed effects, $p$, and the number of observations, $n$, and is given by SIC = $-2 \log (L) + p \log (N)$.

Models with smaller values of AIC and SIC denote better fits. However, it is unclear to us which criterion is preferable. Both criteria are calculated within PROC MIXED and PROC GLIMMIX.

### 7.6.3 Statistical Inference and Model Comparison

Models can be compared statistically using likelihood ratio tests provided that they fit the same fixed effects and their covariance parameters are nested. Nesting of covariance parameters occurs when the covariance parameters in the simpler model can be obtained by restricting some of the parameters in the more complex model (e.g., compound symmetry pattern is nested within a Toeplitz pattern, but it is not nested within a first-order autoregressive pattern). The likelihood ratio test statistic is given by $2(\log (L_1) - \log (L_2)) \sim X^2_{DF}$, where $DF$ is the difference in number of covariance parameters fitted.

If the covariance parameters in the models compared are not nested, statistical comparison using a likelihood ratio test is not valid. The AIC or SIC could be used.

As a simple strategy, we find that in many data sets, especially those with only a few repeated measurements, estimates of overall group differences will differ little between models using different covariance patterns. If obtaining a reliable treatment estimate and standard error is the only objective, a compound symmetry pattern is likely to be robust. A rough check can be made of this by comparing the results with those obtained using a general pattern. If the differences are small, then the compound symmetry pattern can be used with reasonable confidence.

### 7.6.4 Estimation of Model Parameters

The estimation of the parameters of the general normal mixed model (Equation 7.5) is obtained using the method of maximum likelihood, which is described in full detail in Jones (1993) with application given earlier by Potthoff and Roy (1964). The solutions of the likelihood equations are

$$\hat{\beta} = \sum_j \left( X'_j V_j^{-1} X_j \right)^{-1} \left( \sum_j X'_j V_j^{-1} Y_i \right) \tag{7.6}$$

and

$$\text{Cov}\left(\hat{\beta}\right) = \hat{\sigma}^2 \left(\sum X_j' V_j^{-1} X_j\right)^{-1} \tag{7.7}$$

for given $G$ and $R_j$, where $V_j = Z_j G Z_j' + R_j$ is the covariance matrix for the $j$th subject. The maximum likelihood estimator (MLE) of the scale parameter $\sigma^2$ is

$$\hat{\sigma}^2 = \frac{1}{N} \sum_j \left(Y_j - X_j\hat{\beta}\right)' V_j^{-1} \left(Y_j - X_j\hat{\beta}\right) \tag{7.8}$$

where $N = \sum_j n_j$ is the total number of observations on all subjects. When the number of parameters is not small relative to the total number of observations, the estimated variances of the maximum likelihood estimators become seriously biased. To reduce the bias, the restricted maximum likelihood (REML) is used. For repeated measures experiments, Diggle (1988) showed that the REML estimate of $\beta$ is the same as in Equation 7.6; however, the unbiased estimate of $\sigma^2$ is

$$\hat{\sigma}^2 = \frac{1}{n-p} \sum_j \left(Y_j - X_j\hat{\beta}\right)' V_j^{-1} \left(Y_j - X_j\hat{\beta}\right) \tag{7.9}$$

## 7.7 Examples Using the SAS Mixed and GLIMMIX Procedures

Examples 7.1 to 7.5 cover repeated measures data. The first two examples are for normally distributed observations, the third example is for count data, and the last two examples are on repeated binary outcome measurement.

### 7.7.1 Linear Mixed Model for Normally Distributed Repeated Measures Data

Following the approach in Chapter 3, we shall first start with models with two levels. The level 1 model is a linear model that has "time" as an independent variable, and the level 2 model relates the variability in the parameters as random coefficients unrelated to any subject covariates. To be able to extend our argument to a three-level hierarchical model in which subjects within treatment groups are followed over time, we represent the parameters in the level 1 (within-subject) model using $\gamma$ and the parameters in the level 2 (between-subjects) model using $\beta$. Let $y_{ij}$ denote the measurement taken from the $i$th subject at the $j$th time point $t_{ij}$: $y_{ij} = \gamma_{0i} + \gamma_{1i} t_{ij} + e_{1j}$, where $e_{ij} \sim N(0, \sigma^2)$.

Moreover, we assume that $\gamma_{0i} = \beta_0 + u_{0i}$ and $\gamma_{1i} = \beta_1 + u_{1i}$, where

$$u = \begin{pmatrix} u_{0i} \\ u_{1i} \end{pmatrix} \sim BVN \left[ \begin{pmatrix} 0 \\ 0 \end{pmatrix}, \begin{pmatrix} \tau_{00} & \tau_{01} \\ \tau_{10} & \tau_{11} \end{pmatrix} \right]$$

is independent of $e_{ij}$. Here "BVN" denotes bivariate normal.
Therefore

$$y_{ij} = \beta_0 + u_{01} + (\beta_1 + u_{1i})t_{ij} + e_{ij}$$

or

$$y_{ij} = \left( \beta_0 + \beta_1 t_{ij} \right) + \left( u_{0i} + u_{1i}t_{ij} + e_{ij} \right) \tag{7.10}$$

There is a similarity between this model and the unconditional mean model of Chapter 3. The multilevel model (Equation 7.10) has two components: a fixed component, which contains "intercept" and "$t$", and a random component, which contains three random effects.

### Example 7.1: Arrhythmogenic Dose of Epinephrine (ADE) Data

The arrhythmogenic dose of epinephrine (ADE) is determined by infusing epinephrine into a dog until the arrhythmia criterion is reached. The arrhythmia criterion was the occurrence of four intermittent or continuous premature ventricular contractions. Once the criterion has been reached, the infusion required to produce the criteria is recorded. The ADE is then calculated by multiplying the duration of the infusion by the infusion rate. Table 7.3 gives the ADE for six dogs, measured at the 0, ½, 1½, 2, 2½, 3, 3½, 4, and 4½ hours.

The ADE data are analyzed to monitor the levels of ADE across time. Figure 7.1 displays the ADE data over time.

**TABLE 7.3**

Arrhythmogenic Dose of Epinephrine (ADE) Data

| Dog | 0 | ½ | 1 | 1½ | 2 | 2½ | 3 | 3½ | 4 | 4½ |
|-----|------|-----|-----|------|------|------|------|-----|------|-----|
| | | | | | Time | | | | | |
| 1 | 5.7 | 4.7 | 4.8 | 4.9 | 3.88 | 3.51 | 2.8 | 2.6 | 2.5 | 2.5 |
| 2 | 5.3 | 3.7 | 5.2 | 4.9 | 5.04 | 3.5 | 2.9 | 2.6 | 2.4 | 2.5 |
| 3 | 4.0 | 4.6 | 4.1 | 4.58 | 3.58 | 4.0 | 3.6 | 2.5 | 3.5 | 2.1 |
| 4 | 13.7 | 8.9 | 9.6 | 8.6 | 7.5 | 4.0 | 3.1 | 4.1 | 4.08 | 3.0 |
| 5 | 5.0 | 4.0 | 4.1 | 3.9 | 3.4 | 3.39 | 2.95 | 3.0 | 3.1 | 2.0 |
| 6 | 7.1 | 3.0 | 2.4 | 3.3 | 3.9 | 4.0 | 3.0 | 2.4 | 2.1 | 2.0 |

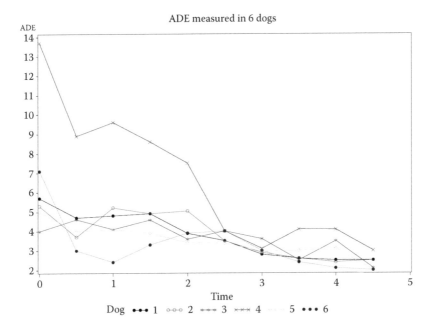

**FIGURE 7.1**
ADE data for six dogs over time.

The SAS code that produced the plot in Figure 7.1 is

```
goptions ftext=swiss;
    axis1 label = ('Time')
    minor=none;
    axis2 label = ('ADE')
    minor=none;
    symbol1 color=black i=join v=dot ;
    symbol2 color=red i=join v=circle ;
    symbol3 color=blue i=join v=star ;
    symbol4 color=green i=join v=x ;
    symbol5 color=yellow i=join v=diamond ;
    symbol6 color=brown i=join v=dot ;
  proc gplot data=ade;
    plot ade*time=dog / frame haxis=axis1 vaxis=axis2;
    title1 'ADE measured in 6 Dogs';
  run;
  quit;
```

Three mixed models will be fitted to these data:

```
data ade; set ade; period=time;

* Model 1: Random coefficient model without exploring the error
covariance structure;
proc mixed covtest data=ade;
 class dog;
 model ade=time/ s ddfm=bw notest;
 random intercept time / type=un sub=dog g;
run;
```

```
* Model 2: Exploring the correlation structures within a subject;
proc mixed covtest data=ade;
  class dog period;
  model ade=time/ s ddfm=bw notest;
  repeated period / type=ar(1) sub=dog r;
run;

* Model 3: Modeling between and within subject covariance;
proc mixed covtest data=ade;
  class dog period;
  model ade=time/ s ddfm=bw notest;
  random intercept time / type=un sub=dog g;
  repeated period / type=ar(1) sub=dog r;
run;
```

The SAS output of the ADE example is listed next:
Model 1: Unstructured Correlation

### Covariance Parameter Estimates

| Cov Parm | Subject | Estimate | Standard Error | Z Value | Pr Z |
|----------|---------|----------|----------------|---------|--------|
| UN(1,1)  | Dog     | 6.9941   | 4.5965         | 1.52    | 0.0641 |
| UN(2,1)  | Dog     | −1.6450  | 1.0973         | −1.50   | 0.1339 |
| UN(2,2)  | Dog     | 0.3775   | 0.2631         | 1.43    | 0.0757 |
| Residual |         | 0.7906   | 0.1614         | 4.90    | <.0001 |

### Fit Statistics

| | |
|---|---|
| -2 Res Log Likelihood | 170.5 |
| AIC (smaller is better) | 178.5 |
| AICC (smaller is better) | 179.2 |
| BIC (smaller is better) | 177.6 |

### Solution for Fixed Effects

| Effect | Estimate | Standard Error | DF | t Value | Pr > \|t\| |
|--------|----------|----------------|----|---------|-----------|
| Intercept | 6.0468 | 1.1005 | 5 | 5.49 | 0.0027 |
| Time | −0.8570 | 0.2633 | 53 | −3.26 | 0.0020 |

Model 2: Autocorrelation Structure

### Covariance Parameter Estimates

| Cov Parm | Subject | Estimate | Standard Error | Z Value | Pr Z |
|----------|---------|----------|----------------|---------|--------|
| AR(1)    | Dog     | 0.8445   | 0.07506        | 11.25   | <.0001 |
| Residual |         | 4.1532   | 1.8760         | 2.21    | 0.0134 |

**Fit Statistics**

| | |
|---|---|
| -2 Res Log Likelihood | 186.4 |
| AIC (smaller is better) | 190.4 |
| AICC (smaller is better) | 190.7 |
| BIC (smaller is better) | 190.0 |

**Solution for Fixed Effects**

| Effect | Estimate | SE | DF | t Value | Pr > \|t\| |
|---|---|---|---|---|---|
| Intercept | 6.5415 | 0.8144 | 5 | 8.03 | 0.0005 |
| Time | −0.9713 | 0.2307 | 53 | -4.21 | <.0001 |

Model 3: Autocorrelation Structure for Within-Subjects and Unstructured Correlation For Between-Subjects Variation

**Covariance Parameter Estimates**

| Cov Parm | Subject | Estimate | Standard Error | Z Value | Pr Z |
|---|---|---|---|---|---|
| UN(1,1) | Dog | 7.1916 | 4.8981 | 1.47 | 0.0710 |
| UN(2,1) | Dog | −1.6627 | 1.1576 | −1.44 | 0.1509 |
| UN(2,2) | Dog | 0.3624 | 0.2759 | 1.31 | 0.0944 |
| AR(1) | Dog | 0.3573 | 0.2222 | 1.61 | 0.1078 |
| Residual | | 0.9634 | 0.3197 | 3.01 | 0.0013 |

**Fit Statistics**

| | |
|---|---|
| -2 Res Log Likelihood | 167.5 |
| AIC (smaller is better) | 177.5 |
| AICC (smaller is better) | 178.6 |
| BIC (smaller is better) | 176.4 |

**Solution for Fixed Effects**

| Effect | Estimate | Standard Error | DF | t Value | Pr > \|t\| |
|---|---|---|---|---|---|
| Intercept | 6.1722 | 1.1358 | 5 | 5.43 | 0.0029 |
| Time | −0.8924 | 0.2691 | 53 | -3.32 | 0.0017 |

Comments on the SAS output of Example 7.1:

1. For model 1, the average value of the intercept parameter is $\hat{\beta}_0 = 6.0468$, and $\hat{\beta}_1 = -0.857$ is the estimate of the average slope. We reject the hypothesis that either of the parameters is 0 in the

population. For the random effects $\hat{\tau}_{00} = 6.9941$, $\hat{\tau}_{11} = 0.3775$, and $\hat{\tau}_{01} = -1.645$. The estimated residual variance is $\hat{\sigma}^2 = 0.7906$. Note that the test results show that there are no significant variations among subjects. The AIC of this model is 178.5.

2. Since the components of variance were not significant, model 2 fits the data with autocorrelation structure for the within-subject covariance. The autocorrelation parameter was significant. The solution for fixed effects of model 2 is not different from model 1 and there is a substantial increase in the within residual variance. The AIC is 190.4.

3. In Model 3 we attempted to model both the between-subjects variations and the within-subject covariance structure. This required the inclusion of a class variable called "period" needed in the "repeated" statement of SAS. Again the fixed effects estimates were the same, with time being the significant factor. However, none of the covariance parameters nor the AR(1) correlation were significant. The AIC is 177.5.

In conclusion, the first model seems to provide adequate representation of the data.

We illustrate the use of the lme4 function in **R** to fit model 1.

```
library(lme4)
library(lme)
library(MASS)
  dog <- c(rep(1,10),rep(2,10),rep(3,10),rep(4,10),rep(5,10),
rep(6,10))
    time <- rep(c(0.0,0.5,1.0,1.5,2.0,2.5,3.0,3.5,4.0,4.5),6)
  ade <- c(5.70,4.70,4.80,4.90,3.88,3.51,2.80,2.60,2.50,2.50,
      5.30,3.70,5.20,4.90,5.04,3.50,2.90,2.60,2.40,2.50,
      4.00,4.60,4.10,4.58,3.58,4.00,3.60,2.50,3.50,2.10,
      13.70,8.90,9.60,8.60,7.50,4.00,3.10,4.10,4.08,3.00,
      5.00,4.00,4.10,3.90,3.40,3.39,2.95,3.00,3.10,2.00,
      7.10,3.00,2.40,3.30,3.90,4.00,3.00,2.40,2.10,2.00)
    ade <- data.frame(dog,time,ade)

        lme1<-glmmPQL(ade~time,random=~1| dog,family=gaussian,
data= ade)
```

**R Output**
```
Linear mixed-effects model fit by maximum likelihood
  Fixed: ade ~ time
(Intercept)      time
  6.0468182    -0.8570303
```

**Random effects:**
```
 Formula: ~1 | dog
      (Intercept)      Residual
StdDev:   1.081084    1.223508
```

The results of fitting the linear mixed models in R may be different than those produced by SAS. For illustration purpose, alternative models will be fitted using lme in Example 7.2.

### Example 7.2: ch7_growth (Data with Subject-Level Covariate)

Potthoff and Roy (1964) present a set of growth data for 11 girls and 16 boys. For each subject, the distance (mm) from the center of the pituitary to the pterygomaxillary fissure was recorded at the ages of 8, 10, 12, and 14. None of the data are missing. The questions posed by the authors were

- Should the growth curves be presented by a second-degree equation in age, or are linear equations adequate?
- Should two separate curves be used for boys and girls, or do both have the same growth curve?
- Can we obtain confidence bands for the expected growth curves?

Before we address these questions, we should emphasize that little is known about the nature of the correlation between the $n_i = 4$ observations on any subject, except perhaps that they are serially correlated. The simplest correlation model is the one in which the correlation coefficient between any two observations $t$ periods of time apart is equal to $\rho^t$, and in which the variance is constant with respect to time. Under this model, the covariance matrix is

$$
w_i \sigma^2 = \begin{bmatrix} 1 & \rho & \rho^2 & \rho^3 \\ \rho & 1 & \rho & \rho^2 \\ \rho^2 & \rho & 1 & \rho \\ \rho^3 & \rho^2 & \rho & 1 \end{bmatrix} \sigma^2
$$

This is the AR(1) correlation structure. We shall explore several covariance structures while modeling these data.

We now consider a model in which we address the possible variation in intercept and slope attributed to the measured subject-level covariate $x_i$. We first consider the model

$$
y_{ij} = \gamma_{0i} + \gamma_{1i} t_{ij} + e_{ij}
$$

$$
\gamma_{0i} = \beta_{00} + \beta_{01} x_i + u_{0i}
$$

$$
\gamma_{1i} = \beta_{10} + \beta_{11} x_i + u_{1i}
$$

where the assumptions on the distributions of $u_i$ and $e_{ij}$ are the same as in Chapter 3. Following the argument of Singer (1998) it is convenient to centralize the covariate $x_i$ around its mean (if it is measured on the continuous scale), so that the model is written as

$$
y_{ij} = \beta_{00} + \beta_{01}(x_i - \bar{x}) + u_{0i} + (\beta_{10} + \beta_{11}(x_i - \bar{x}) + u_{1i})t_{ij} + e_{ij}
$$

Arranging terms we have

$$
y_{ij} = \left[ \beta_{00} + \beta_{01}(x_i - \bar{x}) + \beta_{10} t_{ij} + \beta_{11}(x_i - \bar{x})t_{ij} \right] + \left[ u_{0i} + u_{1i} t_{ij} + e_{ij} \right] \quad (7.11)
$$

Letting $\tilde{x} \equiv x_i - \bar{x}$, model 7.11 again has two components: a fixed component, which contains the "intercept", "$\tilde{x}$", "$t$" and an interaction term $(\tilde{x}) * (t)$; and a random component, which contains intercept, time, and the error term. When a quadratic component of age was added to the model, its $\beta$ coefficient was not significant and thus a linear effect deemed sufficient.

Based on the fitted model, it seems that a linear function in age is adequate and that separate curves be used for boys and girls (significant coefficient of $x_{i1}$).

To address the third question posed by Potthoff and Roy, we follow the approach suggested by Jones (1993). Since for each subject in the study, there is an $X_i$ matrix, let $x$ denote a possible row of $X_i$ for any subject. For example, for a subject who is a girl at age 14, then $x = (1\ 1\ 14)'$ or for a subject who is a boy at age 10, then $x = (1\ 0\ 10)'$. The estimated population mean for a given $x$ vector is

$$\hat{y} = \hat{x}\hat{\beta} \tag{7.12}$$

and has estimated variance

$$V(\hat{y}) = xCov\left(\hat{\beta}\right)x' = \hat{\sigma}^2 x \left(\sum_i x_i' V_i^{-1} x_i\right)^{-1} x' \tag{7.13}$$

By varying the elements of the $x$ vector, estimated population curves can be generated for different values of the covariates. The $(1 - \alpha)100\%$ confidence limits are thus $\hat{y} \pm z_{\alpha/2}\sqrt{var(\hat{y})}$.

```
data growth;
input gender subject age distance @@;
wave=age; age=age-11;
cards;
1   1 8 21.0 1   1 10 20.0 1   1 12 21.5 1   1 14 23.0
1   2 8 21.0 1   2 10 21.5 1   2 12 24.0 1   2 14 25.5
1   3 8 20.5 1   3 10 24.0 1   3 12 24.5 1   3 14 26.0
1   4 8 23.5 1   4 10 24.5 1   4 12 25.0 1   4 14 26.5
1   5 8 21.5 1   5 10 23.0 1   5 12 22.5 1   5 14 23.5
1   6 8 20.0 1   6 10 21.0 1   6 12 21.0 1   6 14 22.5
1   7 8 21.5 1   7 10 22.5 1   7 12 23.0 1   7 14 25.0
1   8 8 23.0 1   8 10 23.0 1   8 12 23.5 1   8 14 24.0
1   9 8 20.0 1   9 10 21.0 1   9 12 22.0 1   9 14 21.5
1 10 8 16.5 1 10 10 19.0 1 10 12 19.0 1 10 14 19.5
1 11 8 24.5 1 11 10 25.0 1 11 12 28.0 1 11 14 28.0
2 12 8 26.0 2 12 10 25.0 2 12 12 29.0 2 12 14 31.0
2 13 8 21.5 2 13 10 22.5 2 13 12 23.0 2 13 14 26.5
2 14 8 23.0 2 14 10 22.5 2 14 12 24.0 2 14 14 27.5
2 15 8 25.5 2 15 10 27.5 2 15 12 26.5 2 15 14 27.0
2 16 8 20.0 2 16 10 23.5 2 16 12 22.5 2 16 14 26.0
2 17 8 24.5 2 17 10 25.5 2 17 12 27.0 2 17 14 28.5
2 18 8 22.0 2 18 10 22.0 2 18 12 24.5 2 18 14 26.5
2 19 8 24.0 2 19 10 21.5 2 19 12 24.5 2 19 14 25.5
2 20 8 23.0 2 20 10 20.5 2 20 12 31.0 2 20 14 26.0
2 21 8 27.5 2 21 10 28.0 2 21 12 31.0 2 21 14 31.5
2 22 8 23.0 2 22 10 23.0 2 22 12 23.5 2 22 14 25.0
2 23 8 21.5 2 23 10 23.5 2 23 12 24.0 2 23 14 28.0
```

```
2 24  8 17.0 2 24 10 24.5 2 24 12 26.0 2 24 14 29.5
2 25  8 22.5 2 25 10 25.5 2 25 12 25.5 2 25 14 26.0
2 26  8 23.0 2 26 10 24.5 2 26 12 26.0 2 26 14 30.0
2 27  8 22.0 2 27 10 21.5 2 27 12 23.5 2 27 14 25.0
```

The following codes show how to fit the data using SAS:

```
ODS RTF;
#We fitted 3 models to the data:


    /* Model 1: Mixed model with unstructured between subject
covariance*/
  proc mixed;
   class gender subject;
   model distance = gender age gender*age / s ddfm=bw notest;
   random intercept age / type=un sub= subject gcorr;
   run;

  /* Model 2: AR(1) structure for within subject variation */
   proc mixed;
   class gender wave;
   model distance = gender age gender*age / s ddfm=bw notest;
   repeated wave / type=ar(1) sub= subject r;
   run;
  /* Model 3: Combining repeated statement with the random
  statement*/
  proc mixed;
   class gender wave subject;
   model distance = gender age gender*age / s ddfm=bw notest;
   random intercept age / type=un sub= subject g;
   repeated wave / type=ar(1) sub= subject r;
   run;
ODS RTF CLOSE;
```

The selected SAS output is given next:
Model 1:

### Covariance Parameter Estimates

| Cov Parm | Subject | Estimate |
|----------|---------|----------|
| UN(1,1)  | Subject | 3.3501   |
| UN(2,1)  | Subject | 0.06814  |
| UN(2,2)  | Subject | 0.03252  |
| Residual |         | 1.7162   |

### Fit Statistics

| | |
|---|---|
| -2 Res Log Likelihood | 432.6 |
| AIC (smaller is better) | 440.6 |
| AICC (smaller is better) | 441.0 |
| BIC (smaller is better) | 445.8 |

### Solution for Fixed Effects

| Effect | Gender | Estimate | Standard Error | DF | t-value | Pr > \|t\| |
|---|---|---|---|---|---|---|
| Intercept | | 24.9688 | 0.4860 | 25 | 51.38 | <.0001 |
| Gender | 1 | −2.3210 | 0.7614 | 25 | −3.05 | 0.0054 |
| Gender | 2 | 0 | . | . | . | . |
| Age | | 0.7844 | 0.08600 | 79 | 9.12 | <.0001 |
| Age*gender | 1 | −0.3048 | 0.1347 | 79 | -2.26 | 0.0264 |
| Age*gender | 2 | 0 | . | . | . | . |

Model 2:

### Covariance Parameter Estimates

| Cov Parm | Subject | Estimate |
|---|---|---|
| AR(1) | Subject | 0.6245 |
| Residual | | 5.2145 |

### Fit Statistics

| | |
|---|---|
| -2 Res Log Likelihood | 444.6 |
| AIC (smaller is better) | 448.6 |
| AICC (smaller is better) | 448.7 |
| BIC (smaller is better) | 451.8 |

### Solution for Fixed Effects

| Effect | Gender | Estimate | Standard Error | DF | t-value | Pr > \|t\| |
|---|---|---|---|---|---|---|
| Intercept | | 25.0610 | 0.4387 | 25 | 57.13 | <.0001 |
| Gender | 1 | −2.4184 | 0.6873 | 25 | −3.52 | 0.0017 |
| Gender | 2 | 0 | . | . | . | . |
| Age | | 0.7693 | 0.1170 | 79 | 6.58 | <.0001 |
| Age*gender | 1 | −0.2854 | 0.1832 | 79 | −1.56 | 0.1233 |
| Age*gender | 2 | 0 | . | . | . | . |

Model 3:

### Covariance Parameter Estimates

| Cov Parm | Subject | Estimate |
|---|---|---|
| UN(1,1) | Subject | 3.6778 |
| UN(2,1) | Subject | 0.1150 |
| UN(2,2) | Subject | 0.08455 |
| AR(1) | Subject | −0.4733 |
| Residual | | 1.1924 |

**Fit Statistics**

| | |
|---|---|
| -2 Res Log Likelihood | 428.8 |
| AIC (smaller is better) | 438.8 |
| AICC (smaller is better) | 439.4 |
| BIC (smaller is better) | 445.3 |

**Solution for Fixed Effects**

| Effect | Gender | Estimate | Standard Error | DF | t-value | Pr > \|t\| |
|---|---|---|---|---|---|---|
| Intercept | | 24.9299 | 0.4876 | 25 | 51.12 | <.0001 |
| Gender | 1 | −2.2800 | 0.7640 | 25 | −2.98 | 0.0063 |
| Gender | 2 | 0 | . | . | . | . |
| Age | | 0.7979 | 0.08707 | 79 | 9.16 | <.0001 |
| Age*gender | 1 | −0.3222 | 0.1364 | 79 | −2.36 | 0.0206 |
| Age*gender | 2 | 0 | . | . | . | . |

Clearly, for the fixed effects estimates, there is little or no difference among the four models. Since none of the models is considered nested within either of the others, model comparisons should be confined to the AIC. Based on the AIC values, model 3 outperforms models 1 and 2.

Note that we subtracted the mean age (11 years) from the age of each subject, so that the intercept would be interpreted as the average response for a randomly selected subject whose age is 11 years.

Now we illustrate the use of the lme function in R to fit the linear mixed effect models. Following is the R code to setup the data for Example 7.2 and fitting alternative mixed-effect models with AR1 correlation structure. In lme1, intercept and age are the random effects where as in lme2 only intercept is specified as the random effect so that lme2 is nested within lme1. The likelihood ratio test is then performed to see if age as a random effect is required in the model and it turns out that age is not required as a random effect. In order to see if a nonzero correlation structure is needed, lme3 was fit with the default independence correlation structure. Comparison of lme2 and lme3 by likelihood ratio test shows that the AR1 correlation structure does not improve the fit of the model. The statistics of fitting the final model are then displayed. Last, the intervals function is used to display the confidence intervals for the parameters of the model.

```
# Setting up the data for Example 7.2
g <- factor(c(rep(1,44),rep(2,64)))
subject <- gl(27, 4, length = 108, labels = 1:27)
age <- rep(c(8,10,12,14),27)-11

distance <-
c(21.0,20.0,21.5,23.0,21.0,21.5,24.0,25.5,20.5,24.0,24.5,26.0,
 23.5,24.5,25.0,26.5,21.5,23.0,22.5,23.5,20.0,21.0,21.0,22.5,
```

```
      21.5,22.5,23.0,25.0,23.0,23.0,23.5,24.0,20.0,21.0,22.0,21.5,
      16.5,19.0,19.0,19.5,24.5,25.0,28.0,28.0,26.0,25.0,29.0,31.0,
      21.5,22.5,23.0,26.5,23.0,22.5,24.0,27.5,25.5,27.5,26.5,27.0,
      20.0,23.5,22.5,26.0,24.5,25.5,27.0,28.5,22.0,22.0,24.5,26.5,
      24.0,21.5,24.5,25.5,23.0,20.5,31.0,26.0,27.5,28.0,31.0,31.5,
      23.0,23.0,23.5,25.0,21.5,23.5,24.0,28.0,17.0,24.5,26.0,29.5,
      22.5,25.5,25.5,26.0,23.0,24.5,26.0,30.0,22.0,21.5,23.5,25.0)
# Changing the reference level as 2
gender <- relevel(g,2)
growth <- data.frame(gender,subject,age,distance)
# Fitting the model with intercept and age as random effects and AR1
correlation structure
lme1 <-lme(distance~gender*age,random=~age|subject,corr=corAR1
(form=~age|subject),data=growth)

# Fitting the model with intercept only as random effect and AR1
correlation structure
 lme2  <-lme(distance~gender*age,random=~1 |subject,corr=corAR1
(),data=growth)
>
> # Likelihood Ratio test to test if age as random effect is required
> anova(lme1,lme2)
   Model df    AIC      BIC    logLik  Test  L.Ratio p-value
lme1   1 9 450.5817 474.3812 -216.2908
lme2   2 7 447.7081 466.2188 -216.8541 1 vs 2 1.126451   0.5694
>
> # Fittinng the model with zero random effects correlation
> lme3 <-lme(distance~gender*age,random=~1|subject,data=growth)
>
> # Likelihood Ratio test of zero random effects correlation
> anova(lme2,lme3)
   Model df    AIC      BIC    logLik  Test    L.Ratio p-value
lme2   1 7 447.7081 466.2188 -216.8541
lme3   2 6 445.7572 461.6236 -216.8786 1 vs 2 0.04913688  0.8246
>
> summary(lme3)
Linear mixed-effects model fit by REML
 Data: growth
     AIC     BIC    logLik
  445.757 461.624 -216.879

Random effects:
 Formula: ~1 | subject
        (Intercept)  Residual
StdDev: 1.816214   1.386382

Fixed effects: distance ~ gender * age
            Value Std.Error DF t-value p-value
(Intercept) 24.969   0.486   79 51.376   0.000
gender1      -2.321   0.761   25 -3.048   0.005
 age          0.784   0.077   79 10.121   0.000
gender1:age -0.3045  0.121   79 -2.511   0.014
 Correlation:
          (Intr) gendr1  age
gender1   -0.638
age        0.000  0.000
```

```
gender1:age  0.000  0.000  -0.638

Standardized Within-Group Residuals:
        Min          Q1        Med         Q3        Max
-3.59804400 -0.45461690 0.01578365 0.50244658 3.68620792

Number of Observations: 108
Number of Groups: 27
> intervals(lme3)
Approximate 95% confidence intervals

Fixed effects:
                lower       est.       upper
(Intercept)  24.0013897 24.9687500 25.9361103
gender1      -3.8891901 -2.3210227 -0.7528554
age           0.6301129  0.7843750  0.9386371
gender1:age  -0.5465118 -0.3048295 -0.0631473
attr(, "label")
[1] "Fixed effects:"

Random Effects:
 Level: subject
                lower   est.  upper
sd((Intercept)) 1.321  1.816  2.497

Within-group standard error:
  lower   est.  upper
  1.186  1.386  1.620
```

It is interesting to note that the estimates of the fixed effect parameters and the log-likelihood statistics are the same as for model 2 in SAS, indicating that fitting a more complicated model did not really improve the fit of the model.

## 7.7.2 Analysis of Longitudinal Binary and Count Data

Examples in this section will focus on the analysis of longitudinal count and binary data. We shall adopt two modeling strategies, the first being a population-average (PA) approach, and the second is known as subject-specific (SS) approach. Both methodologies have been discussed in previous chapters to deal with clustered data.

The PA models use the GEE (GENMOD) methodology of Liang and Zeger (1986, 1993), and the SS models use the generalized linear mixed models (GLIMMIX). We have demonstrated the use of both techniques on cross-sectional continuous, binary, and count data. Here we illustrate both techniques on repeated measures nonnormally distributed outcome variables. We first discuss situations where such data may be available.

### Example 7.3: EPA-CHESS Study (Longitudinal Binary Data)

An asthma study conducted by the Environmental Protection Agency's Community Health and Environment Surveillance System (EPA-CHESS) was described in Korn and Whittemore (1979). Daily records of the absence/presence of an asthma attack were recorded on each participant, in addition to measurements of air pollutants and meteorological variables, such as daily temperature and humidity. Here, the aerometric and meteorological variables are time-varying covariates, since they can change from one time point to another, whereas individual characteristics, such as sex and ethnicity, do not vary with time. One of the objectives of the study was to determine the effects of outside air quality on rates of asthma attacks. Since we are interested in modeling the probability of disease as a function of covariates, a regression model is appropriate. However, the over-time correlation among the binary responses must be properly accounted for.

The issue of robustness of the GEE approach against misspecification of the correlation structure was the subject of investigation by many authors. In particular, if the correlation structure is misspecified as "independence," which assumes that the within cluster responses are independent, the GEE has been shown to be nearly efficient relative to the maximum likelihood in a variety of settings. When the correlation between responses is not too high, Zeger et al. (1988) suggested that this estimator should be nearly efficient. McDonald (1993), focusing on the case of clusters of size $n_i = 2$ (i.e., the bivariate case) concluded that the estimator obtained under specifying independence may be recommended whenever the correlation between the within-cluster pair is nuisance. This may have practical implications since the model can be implemented using standard software packages. In a more recent article, Fitzmaurice (1995) investigated this issue analytically. He has confirmed the suggestions made by Zeger et al. (1988) and McDonald (1993). Furthermore, he showed that when the responses are strongly correlated and the covariate design includes a within-cluster covariate, assuming independence can lead to a considerable loss of efficiency if the GEE is used in estimating the regression parameters associated with that covariate. His results demonstrate that the degree of efficiency depends on both the strength of the correlation between the responses and the covariate design. He recommended that an effort should be made to model the association between responses, even when this association is regarded as a nuisance feature of the data and its correct nature is unknown.

### Example 7.4: The Epilepsy Data

This data have been analyzed by Thall and Vail (1990) and Stukel (1993) using several models for longitudinal count data. It is a randomized clinical trial of the antiepileptic drug Progabide, in which 59 patients were seen, and the count of seizures was taken at baseline and at four follow-up visits. The main objective was to estimate the treatment effect and its possible modification by disease severity (seizure count at baseline), and age. The data are given in Table 7.4.

First, we note from Figures 7.2 and 7.3 that the number of seizures decline as time elapses. From Figure 7.4, it is clear that the two treatments (0, 1) are equally effective.

**TABLE 7.4**

Epilepsy Data

| ID | Treat | Age | Base | Y1 | Y2 | Y3 | Y4 | ID | Treat | Age | Base | Y1 | Y2 | Y3 | Y4 |
|----|-------|-----|------|-----|-----|-----|-----|-----|-------|-----|------|-----|-----|-----|-----|
| 101 | 1 | 18 | 76 | 11 | 14 | 9 | 8 | 104 | 0 | 31 | 11 | 5 | 3 | 3 | 3 |
| 102 | 1 | 32 | 38 | 8 | 7 | 9 | 4 | 106 | 0 | 30 | 11 | 3 | 5 | 3 | 3 |
| 103 | 1 | 20 | 19 | 0 | 4 | 3 | 0 | 107 | 0 | 25 | 6 | 2 | 4 | 0 | 5 |
| 108 | 1 | 30 | 10 | 3 | 6 | 1 | 3 | 114 | 0 | 36 | 8 | 4 | 4 | 1 | 4 |
| 110 | 1 | 18 | 19 | 2 | 6 | 7 | 4 | 116 | 0 | 22 | 66 | 7 | 18 | 9 | 21 |
| 111 | 1 | 24 | 24 | 4 | 3 | 1 | 3 | 118 | 0 | 29 | 27 | 5 | 2 | 8 | 7 |
| 112 | 1 | 30 | 31 | 22 | 17 | 19 | 16 | 123 | 0 | 31 | 12 | 6 | 4 | 0 | 2 |
| 113 | 1 | 35 | 14 | 5 | 4 | 7 | 4 | 126 | 0 | 42 | 52 | 40 | 20 | 23 | 12 |
| 117 | 1 | 27 | 11 | 2 | 4 | 0 | 4 | 130 | 0 | 37 | 23 | 5 | 6 | 6 | 5 |
| 121 | 1 | 20 | 67 | 3 | 7 | 7 | 7 | 135 | 0 | 28 | 10 | 14 | 13 | 6 | 0 |
| 122 | 1 | 22 | 41 | 4 | 18 | 2 | 5 | 141 | 0 | 36 | 52 | 26 | 12 | 6 | 22 |
| 124 | 1 | 28 | 7 | 2 | 1 | 1 | 0 | 145 | 0 | 24 | 33 | 12 | 6 | 8 | 4 |
| 128 | 1 | 23 | 22 | 0 | 2 | 4 | 0 | 201 | 0 | 23 | 18 | 4 | 4 | 6 | 2 |
| 129 | 1 | 40 | 13 | 5 | 4 | 0 | 3 | 202 | 0 | 36 | 42 | 7 | 9 | 12 | 14 |
| 137 | 1 | 33 | 46 | 11 | 14 | 25 | 15 | 205 | 0 | 26 | 87 | 16 | 24 | 10 | 9 |
| 139 | 1 | 21 | 36 | 10 | 5 | 3 | 8 | 206 | 0 | 26 | 50 | 11 | 0 | 0 | 5 |
| 143 | 1 | 35 | 38 | 19 | 7 | 6 | 7 | 210 | 0 | 28 | 18 | 0 | 0 | 3 | 3 |
| 147 | 1 | 25 | 7 | 1 | 1 | 2 | 3 | 213 | 0 | 31 | 111 | 37 | 29 | 28 | 29 |
| 203 | 1 | 26 | 36 | 6 | 10 | 8 | 8 | 215 | 0 | 32 | 18 | 3 | 5 | 2 | 5 |
| 204 | 1 | 25 | 11 | 2 | 1 | 0 | 0 | 217 | 0 | 21 | 20 | 3 | 0 | 6 | 7 |
| 207 | 1 | 22 | 151 | 102 | 65 | 72 | 63 | 219 | 0 | 29 | 12 | 3 | 4 | 3 | 4 |
| 208 | 1 | 32 | 22 | 4 | 3 | 2 | 4 | 220 | 0 | 21 | 9 | 3 | 4 | 3 | 4 |
| 209 | 1 | 25 | 41 | 8 | 6 | 5 | 7 | 222 | 0 | 32 | 17 | 2 | 3 | 3 | 5 |
| 211 | 1 | 35 | 32 | 1 | 3 | 1 | 5 | 226 | 0 | 25 | 28 | 8 | 12 | 2 | 8 |
| 214 | 1 | 21 | 56 | 18 | 11 | 28 | 13 | 227 | 0 | 30 | 55 | 18 | 24 | 76 | 25 |
| 218 | 1 | 41 | 24 | 6 | 3 | 4 | 0 | 230 | 0 | 40 | 9 | 2 | 1 | 2 | 1 |
| 221 | 1 | 32 | 16 | 3 | 5 | 4 | 3 | 234 | 0 | 19 | 10 | 3 | 1 | 4 | 2 |
| 225 | 1 | 26 | 22 | 1 | 23 | 19 | 8 | 238 | 0 | 22 | 47 | 13 | 15 | 13 | 12 |
| 228 | 1 | 21 | 25 | 2 | 3 | 0 | 1 | | | | | | | | |
| 232 | 1 | 36 | 13 | 0 | 0 | 0 | 0 | | | | | | | | |
| 236 | 1 | 37 | 12 | 1 | 4 | 3 | 2 | | | | | | | | |

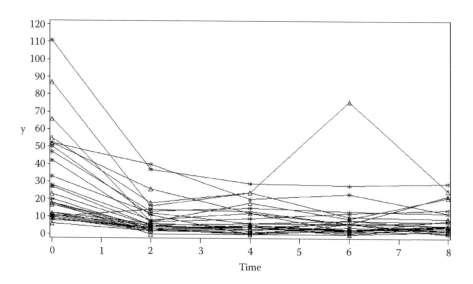

**FIGURE 7.2**
Plot of number of episodes by time for control group.

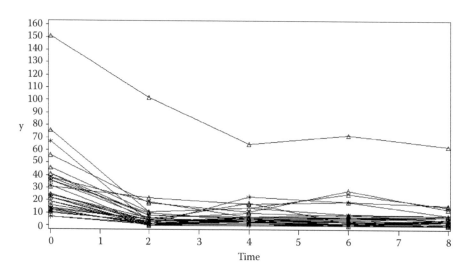

**FIGURE 7.3**
Plot of number of episodes by time for treatment group.

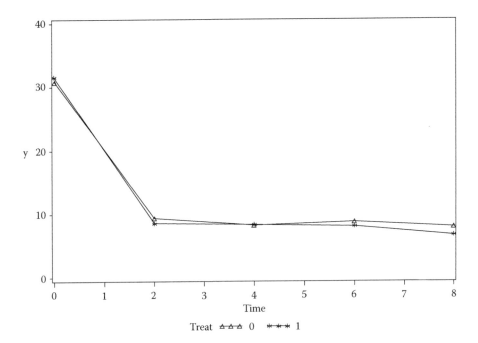

**FIGURE 7.4**
Plot of the average number of episodes for two groups over time.

The data would then be transformed into the longitudinal format so that there would be four observations for each subject corresponding to the four repeated measurements. If the data set is stored in a temporary SAS data set say "epilepsy", transformation may be achieved using the following code.

```
data epilepsy1; set epilepsy;
  period=1; count=y1; output; period=2; count=y2; output;
  period=3; count=y3; output; period=4; count=y4; output;
drop y1-y4;
```

The models are now run as follows:

```
* Population averaged model with exchangeable correlation struc-
ture;
proc genmod data=a;
  class id treat;
  model count = treat base age/ d=poisson;
  repeated subject=id/ corrw type=exch;
run;
```

```
* Subject specific model with exchangeable correlation structure;
proc glimmix data=a;
  class id treat;
  model count = treat base age / s d=poisson;
  random intercept /subject=id type=cs;
run;
```

**The SAS output**
The two models assume that the number of seizures follow Poisson distribution. First we used PROC GENMOD to fit the data, with exchangeable correlation structure. We note that the value of the intrasubject correlation is quite high ($\hat{\rho} = 0.395$). Other correlation structures may be explored, but we shall leave this issue to be discussed in the exercises at the end of the chapter. Note also that we used the baseline count as a covariate. Although this is not a strictly valid approach, it is modeled as such to demonstrate the dependence of the patient's profile on the base line measurement. Again, we shall elaborate on this issue in the exercises.

The GLIMMIX results are not qualitatively different from the PROC GENMOD, although the estimated coefficients and their standard errors differ between the two models. Note that the default correlation in the GLIMMIX is the exchangeable correlation. Both models indicate that there is no significant difference between the treatment and the placebo. Moreover, both models indicate that the baseline counts are significant predictors. Note that the GEE identified "age" as a significant covariate, although this was not the case under the cluster-specific model.

Model 1: Population Averaged Model

#### Criteria For Assessing Goodness Of Fit

| Criterion | DF | Value | Value/DF |
|---|---|---|---|
| Deviance | 232 | 958.4636 | 4.1313 |
| Scaled Deviance | 232 | 958.4636 | 4.1313 |
| Pearson's Chi-Square | 232 | 1180.7884 | 5.0896 |
| Scaled Pearson $X^2$ | 232 | 1180.7884 | 5.0896 |
| Log Likelihood | | 2949.5627 | |

#### Exchangeable Working Correlation

| Correlation | 0.3948 |
|---|---|

#### Analyiss of GEE Parameter Estimates Empirical Standard Error Estimates

| Parameter | | Estimate | Standard Error | 95% Confidence Limits | | Z | Pr > \|Z\| |
|---|---|---|---|---|---|---|---|
| Intercept | | 0.4098 | 0.3630 | -0.3017 | 1.1214 | 1.13 | 0.2589 |
| Treat | 0 | 0.1527 | 0.1711 | -0.1827 | 0.4881 | 0.89 | 0.3722 |
| Treat | 1 | 0.0000 | 0.0000 | 0.0000 | 0.0000 | . | . |
| Base | | 0.0227 | 0.0012 | 0.0202 | 0.0251 | 18.33 | <.0001 |
| Age | | 0.0227 | 0.0116 | 0.0000 | 0.0454 | 1.96 | 0.0496 |

Model 2: Subject Specific Model

### Fit Statistics

| | |
|---|---|
| -2 Res Log Pseudo-Likelihood | 602.13 |
| Generalized Chi-Square | 418.36 |
| Gener. Chi-Square/DF | 1.80 |

### Covariance Parameter Estimates

| Cov Parm | Subject | Estimate | Standard Error |
|---|---|---|---|
| Variance | id | 0.3785 | 0.06921 |
| CS | id | −0.07723 | . |

### Solutions fro Fixed Effecst

| Effect | Treat | Estimate | Standard Error | DF | t Value | Pr > \|t\| |
|---|---|---|---|---|---|---|
| Intercept | | 0.3000 | 0.4000 | 55 | 0.75 | 0.4565 |
| Treat | 0 | 0.2584 | 0.1574 | 177 | 1.64 | 0.1025 |
| Treat | 1 | 0 | . | . | . | . |
| Base | | 0.02694 | 0.002862 | 177 | 9.41 | <.0001 |
| Age | | 0.01400 | 0.01283 | 177 | 1.09 | 0.2767 |

The following R codes read the epilepsy data and run models 1 and 2. First we read the csv data table into the R data file "epilepsy":

```
epilepsy=read.csv(file.choose())
names(epilepsy)
ylim=range(epilepsy$y)
```

Next graphically examine the data using the boxplot:

```
placebo=subset(epilepsy,treat="1")
progabide=subset(epilepsy,treat="2")
boxplot(y~time,data=placebo,
        ylab="Number of seizures",xlab="time", ylim=ylim,
main="Placebo")
boxplot(y~time,data=progabide,
        ylab="Number of seizures",xlab="time", ylim=ylim,
main="Progabide")
```

### Model 1:

```
gee1            <-gee(count~treat+base+age,id=factor(subject),
family=poisson("log"),   corstr="exchangeable",data=epilepsy,
scale.fix=T)
summary(gee1)
```

```
gee(formula = y ~ treat + base + age, id = factor(subject), data =
epilepsy,
  family = poisson("log"), corstr = "exchangeable", scale.fix = T)
```

Summary of Residuals:
```
       Min         1Q      Median         3Q        Max
-41.975570  -5.956965  -3.527219   2.584624  57.926230
```

Coefficients:
```
            Estimate   Naive S.E.  Naive z  Robust S.E.   Robust z
(Intercept)          1.44593075         0.0948597577       15.242826
0.329618256  4.3866829
treat               -0.10982631 0.0340140577  -3.228850  0.125314583
-0.8764048
base                 0.02037982 0.0003712292  54.898223  0.001289145
15.8087921
age                              0.01092135 0.0028634334   3.814076
0.010156894  1.0752651
```

Estimated Scale Parameter: 1
Number of Iterations: 1

## Model 2:

```
glmm2       <-glmmPQL(y~treat+base+age,       random=~1|subject,
family=poisson("log"),correlation=corCompSymm(),
data=epilepsy)
summary(glmm2)
```

Linear mixed-effects model fit by maximum likelihood
 Data: epilepsy
 AIC BIC logLik
  NA  NA    NA

Random effects:
 Formula: ~1 | subject
        (Intercept)  Residual
StdDev:  0.3558148   2.989689

Correlation Structure: Compound symmetry
 Formula: ~1 | subject
 Parameter estimate(s):
        Rho
-0.04848686
Variance function:
 Structure: fixed weights
 Formula: ~invwt
Fixed effects: y ~ treat + base + age
```
                Value   Std.Error   DF   t-value  p-value
(Intercept) 1.2575413 0.25121484 1121  5.005840   0.0000
treat       -0.0908169 0.09522120   55 -0.953746   0.3444
base         0.0263238 0.00179193   55 14.690191   0.0000
age          0.0062869 0.00774824   55  0.811398   0.4206
```
 Correlation:
```
        (Intr) treat  base
treat  -0.289
base   -0.392  0.004
age    -0.937  0.103 0.190
```

```
Standardized Within-Group Residuals:
      Min         Q1        Med        Q3        Max
-1.2543529 -0.5921948 -0.3272269  0.1713517  3.8188977

Number of Observations: 1180
Number of Groups: 59
```

Slight differences in the results of model 2 with those produced by SAS are because of different model fitting methods.

The next example is on longitudinal binary data with covariates measured at the subject level. During the past few decades, repeated binary data have been the subject of a great deal of research in the statistical literature, with special emphasis on the random effects model. In modeling such data, the central problem will be modeling the repeated responses as functions of covariates measured at several levels of hierarchies, and taking into account the interdependencies of the observations and explore several models as we did for continuous and count longitudinal data.

### Example 7.5: Hip Dysplasia in Dogs

The following data are the results of a longitudinal investigation that aimed at assessing the long-term effects of two types of treatments (1 = excision arthroplasty; 2 = triple-pelvic osteotomy) for hip dysplasia in dogs. The owner's assessment was the response variable and was recorded on a binary scale at weeks 1, 3, 6, 10, and 20. Using the generalized estimating equations (GEE) approach we test the treatment effect controlling for laterality and age as possible confounders.

The data in Example 7.5 have been fitted under three different specifications of the correlations among the scores in weeks 1, 3, 6, 10, and 20. The data are entered in a longitudinal format so that the new data will have 6 times the number of observations as in Table 7.5. The SAS programs to fit the models are given next:

```
* Mode 1: Fitting a PA (GEE) model with AR(1) correlation struc-
ture;
proc genmod descending;
 class dog lateral typsurg;
 model score= week lateral typsurg age / dist=binomial;
 repeated subject= dog / type=AR(1);
run;

 * Model 2: Fitting a PA (GEE) model with exchangeable correlation
structure;
proc genmod descending;
 class dog lateral typsurg;
 model score= week lateral typsurg age / dist=binomial;
 repeated subject=dog / type=cs;
run;

 * Model 3: Fitting GLMM, a cluster specific model with AR(1) cor-
relation structure;
proc glimmix;
```

**TABLE 7.5**

Hip Dysplasia in Dogs

| Dog No. | Laterality* | Age | Type of Surgery[+] | Owner's Assessment[‡] | | | | |
|---|---|---|---|---|---|---|---|---|
| | | | | Week 1 | Week 3 | Week 6 | Week 10 | Week 20 |
| 1 | U | 6 | 1 | 1 | 1 | 1 | 1 | 1 |
| 2 | U | 4 | 1 | 1 | 1 | 1 | 1 | 1 |
| 3 | U | 7 | 2 | 1 | 1 | 1 | 1 | 1 |
| 4 | B | 7 | 2 | 1 | 1 | 1 | 1 | 1 |
| 5 | U | 4 | 2 | 1 | 1 | 1 | 1 | 1 |
| 6 | U | 8 | 2 | 1 | 1 | 0 | 1 | 1 |
| 7 | U | 7 | 2 | 1 | 0 | 1 | 1 | 1 |
| 8 | U | 6 | 2 | 1 | 1 | 1 | 0 | 1 |
| 9 | U | 8 | 1 | 1 | 1 | 1 | 1 | 1 |
| 10 | U | 5 | 2 | 1 | 1 | 1 | 1 | 1 |
| 11 | U | 6 | 1 | 1 | 1 | 1 | 1 | 1 |
| 12 | U | 6 | 2 | 0 | 0 | 0 | 1 | 1 |
| 13 | U | 7 | 1 | 1 | 1 | 1 | 1 | 1 |
| 14 | U | 7 | 1 | 1 | 1 | 1 | 1 | 1 |
| 15 | B | 7 | 2 | 1 | 1 | 1 | 1 | 1 |
| 16 | B | 7 | 2 | 1 | 1 | 1 | 0 | 1 |
| 17 | B | 5 | 1 | 1 | 0 | 1 | 0 | 1 |
| 18 | B | 6 | 1 | 0 | 0 | 1 | 1 | 0 |
| 19 | B | 8 | 2 | 1 | 1 | 0 | 1 | 0 |
| 20 | B | 6 | 1 | 1 | 1 | 1 | 0 | 1 |
| 21 | U | 8 | 1 | 1 | 1 | 1 | 1 | 1 |
| 22 | B | 2 | 2 | 1 | 1 | 0 | 0 | 0 |
| 23 | B | 1 | 1 | 0 | 1 | 1 | 1 | 1 |
| 24 | U | 1 | 1 | 1 | 1 | 1 | 1 | 1 |
| 25 | U | 1 | 1 | 1 | 1 | 1 | 1 | 1 |
| 26 | B | 2 | 2 | 1 | 0 | 1 | 0 | 1 |
| 27 | B | 2 | 2 | 0 | 0 | 0 | 0 | 0 |
| 28 | B | 2 | 2 | 0 | 0 | 0 | 0 | 0 |
| 29 | U | 1 | 1 | 1 | 0 | 1 | 1 | 1 |
| 30 | B | 2 | 2 | 1 | 1 | 0 | 1 | 1 |
| 31 | B | 2 | 2 | 0 | 0 | 0 | 0 | 0 |
| 32 | B | 2 | 2 | 0 | 0 | 0 | 1 | 1 |
| 33 | U | 1 | 1 | 1 | 1 | 1 | 1 | 1 |
| 34 | U | 2 | 2 | 1 | 1 | 0 | 1 | 1 |

(*Continued*)

**TABLE 7.5 (CONTINUED)**

Hip Dysplasia in Dogs

| Dog No. | Laterality* | Age | Type of Surgery[†] | Owner's Assessment[‡] | | | | |
|---|---|---|---|---|---|---|---|---|
| | | | | Week 1 | Week 3 | Week 6 | Week 10 | Week 20 |
| 35 | U | 1 | 1 | 1 | 1 | 1 | 1 | 1 |
| 36 | U | 2 | 2 | 1 | 0 | 1 | 1 | 1 |
| 37 | U | 1 | 1 | 0 | 1 | 1 | 1 | 1 |
| 38 | U | 2 | 2 | 1 | 1 | 1 | 1 | 1 |
| 39 | U | 6 | 1 | 1 | 1 | 1 | 1 | 1 |
| 40 | U | 8 | 1 | 0 | 0 | 1 | 0 | 0 |
| 41 | U | 8 | 2 | 1 | 1 | 0 | 0 | 0 |
| 42 | U | 8 | 1 | 0 | 1 | 0 | 0 | 0 |
| 43 | U | 2 | 2 | 1 | 0 | 0 | 1 | 0 |
| 44 | U | 1 | 1 | 0 | 0 | 1 | 1 | 1 |
| 45 | U | 1 | 1 | 1 | 1 | 1 | 1 | 1 |
| 46 | U | 2 | 2 | 0 | 0 | 0 | 1 | 0 |

* U, unilateral; B, bilateral.
[†] 1 = excision arthroplasty, 2 = triple pelvic osteotomy (TPO).
[‡] 1 = good, 0 = poor.

```
  class dog lateral typsurg;
  model score= week lateral typsurg age /s dist=binomial;
  random intercept/ subject= dog type=AR(1);
run;

* Model 4: Fitting GLIM, a cluster specific model compound symme-
try;
proc glimmix;
  class dog lateral typsurg;
  model score= week lateral typsurg age /s dist=binomial;
  random intercept/ subject= dog type=cs;
run;
```

### Model 1: GEE with AR(1) correlation

**Criteria For Assessing Goodness Of Fit**

| Criterion | DF | Value | Value/DF |
|---|---|---|---|
| Deviance | 225 | 248.2603 | 1.1034 |
| Scaled Deviance | 225 | 248.2603 | 1.1034 |
| Pearson's Chi-Square | 225 | 235.1769 | 1.0452 |
| Scaled Pearson $X^2$ | 225 | 235.1769 | 1.0452 |
| Log Likelihood | | -124.1301 | |

Analysis Of GEE Parameter Estimates
Empirical Standard Error Estimates

| Parameter | | Estimate | Standard Error | 95% Confidence Limits | | Z | Pr > \|Z\| |
|---|---|---|---|---|---|---|---|
| Intercept | | 0.7011 | 0.5773 | −0.4305 | 1.8327 | 1.21 | 0.2246 |
| Week | | 0.0148 | 0.0179 | −0.0203 | 0.0499 | 0.82 | 0.4101 |
| Lateral | B | −1.0174 | 0.4505 | −1.9002 | −0.1345 | −2.26 | 0.0239 |
| Lateral | U | 0.0000 | 0.0000 | 0.0000 | 0.0000 | . | . |
| Typsurg | 1 | 0.7000 | 0.4898 | −0.2600 | 1.6600 | 1.43 | 0.1530 |
| Typsurg | 2 | 0.0000 | 0.0000 | 0.0000 | 0.0000 | . | . |
| Age | | 0.0585 | 0.1023 | −0.1419 | 0.2590 | 0.57 | 0.5671 |

**Model 2: GEE with exchangeable correlation**
Correlation = 0.343.

### Solutions for Fixed Effects

| Effect | Lateral | Typsurg | Estimate | Standard Error | Z | Pr > \|Z\| |
|---|---|---|---|---|---|---|
| Intercept | | | 0.5863 | 0.5591 | 1.05 | 0.2943 |
| Week | | | 0.0169 | 0.0193 | 0.87 | 0.3817 |
| Lateral | B | | −1.0059 | 0.4349 | -2.3 | 0.0207 |
| Lateral | U | | 0 | . | . | . |
| Typsurg | | 1 | 0.7987 | 0.4735 | 1.69 | 0.0916 |
| Typsurg | | 2 | 0 | . | . | . |
| Age | | | 0.0667 | 0.0986 | 0.68 | 0.4988 |

**Model 3: GLIMMIX with AR(1) correlation**

### Fit Statistics

| | |
|---|---|
| -2 Res Log Pseudo-Likelihood | 1096.81 |
| Generalized Chi-Square | 149.80 |
| Gener. Chi-Square/DF | 0.67 |

### Covariance Parameter Estimates

| Cov Parm | Subject | Estimate | Standard Error |
|---|---|---|---|
| Variance | Dog | 1.9652 | 0.7513 |
| AR(1) | Dog | 0 | . |

**Solutions for Fixed Effects**

| Effect | Lateral | Typsurg | Estimate | Standard Error | DF | t Value | Pr > \|t\| |
|--------|---------|---------|----------|----------------|-----|---------|-----------|
| Intercept | | | 0.7694 | 0.6858 | 42 | 1.12 | 0.2682 |
| Week | | | 0.02069 | 0.02612 | 183 | 0.79 | 0.4295 |
| Lateral | B | | −1.2576 | 0.5898 | 183 | −2.13 | 0.0343 |
| Lateral | U | | 0 | . | . | . | . |
| Typsurg | | 1 | 0.9772 | 0.5802 | 183 | 1.68 | 0.0938 |
| Typsurg | | 2 | 0 | . | . | . | . |
| Age | | | 0.06153 | 0.1045 | 183 | 0.59 | 0.5566 |

**Comments**

1. All the covariates in the model are measured at the cluster (subject) level, and no time-varying covariates are included in the study. With this type of covariate design one should expect little or no difference between the PA models. The same remark holds for the subject-specific models.
2. All the models indicate that, relative to the unilateral, lateral has significant negative effect on the condition of the dog. However, neither age nor type of surgery have an effect on the subject's condition.

The R code to read the data and fit the four models are given next:

```
dogs <- read.csv(file.choose())
```

We use the Dogs_ Hip_ Displasia file
```
names(dogs)

dogs$dog <- factor(dogs$dog)
dogs$lateral   <-   factor(dogs$lateral,levels=c("u","b"))   #
Changing the order for reference level
dogs$typsurg <- factor(dogs$typsurg,levels=c(2,1))   # Changing
the order for reference level
```

**library(gee)**
**library(MASS)**
**library(nlme)**

```
# Model 1
gee1            <-gee(score~week+lateral+typsurg+age,id=dog,
family=binomial("logit"),corstr="AR-M",Mv=1,data=dogs,
scale.fix=T)
summary(gee1)
```

Coefficients:

|              | Estimate | Naive S.E. | Naive z | Robust S.E. | Robust z |
|--------------|----------|------------|---------|-------------|----------|
| (Intercept)  | 0.701    | 0.513      | 1.366   | 0.577       | 1.214    |
| week         | 0.015    | 0.024      | 0.602   | 0.018       | 0.824    |
| lateral b    | −1.017   | 0.417      | −2.438  | 0.450       | −2.259   |
| typsurg 1    | 0.701    | 0.427      | 1.642   | 0.490       | 1.431    |
| age          | 0.058    | 0.077      | 0.759   | 0.102       | 0.572    |

Estimated Scale Parameter: 1
Number of Iterations: 2

```
# Model 2
gee2=gee(score~week+lateral+typsurg+age,id=dog,
family=binomial("logit"),corstr="exchangeable",data=dogs,
scale.fix=T)
summary(gee2)
```

Coefficients:

|              | Estimate | Naive S.E. | Naive z | Robust S.E. | Robust z |
|--------------|----------|------------|---------|-------------|----------|
| (Intercept)  | 0.586    | 0.579      | 1.011   | 0.559       | 1.049    |
| week         | 0.016    | 0.019      | 0.884   | 0.019       | 0.875    |
| lateral b    | −1.005   | 0.497      | −2.021  | 0.435       | −2.313   |
| typsurg 1    | 0.798    | 0.511      | 1.564   | 0.473       | 1.687    |
| age          | 0.066    | 0.092      | 0.722   | 0.098       | 0.676    |

Estimated Scale Parameter: 1
Number of Iterations: 3
Exchangeable correlation = 0.343

```
# Model 3

glmm3 <-glmmPQL(score~week+lateral+typsurg+age,random=~1|dog,
family=binomial,correlation=corCAR1(),data=dogs)
summary(glmm3)
```

```
Fixed effects: score ~ week + lateral + typsurg + age
            Value Std.Error  DF  t-value p-value
(Intercept) 0.815    0.708  183    1.151   0.251
week        0.023    0.021  183    1.057   0.291
lateralb   -1.429    0.621   42   -2.302   0.026
typsurg1    1.106    0.607   42    1.819   0.076
age         0.075    0.109   42    0.692   0.492
```

Slight differences in the results of Models 3 and 4 with those produced by SAS are because of different model fitting methods.

## 7.8 Two More Examples for Longitudinal Count Data: Fixed Effect Modeling Strategy

**Example 7.6: Regional Variation in the Incidence of Tuberculosis (TB) in Saudi Arabia; Use R.**

In the Kingdom of Saudi Arabia, as of 2013, there are up to 9 million migrants mainly from tuberculosis (TB) endemic regions, in South/South East Asia and Africa. In addition, annually 3 to 6 million pilgrims visit the holy cities located in the western region of the Kingdom for performing the holy rituals, Hajj and Umrah (visiting the Holy city of Mekka). Interestingly, the majority of these pilgrims are coming from TB-endemic areas of Asia and Africa. There are differences in the rate of TB infection between different regions of the country. For instance, in Jeddah (sea and air ports for pilgrims) the infection rate can reach up to 64 cases per 100,000 compared with 32 per 100,000 in Riyadh (central region). The higher rate in Jeddah may be attributed to the inflow of pilgrims.

The data are presented in Figure 7.5 (see Shoukri et al., 2014) prevalence data in each province show the trend over time. From Figure 7.5, there seems to be a clustering of the data in three subgroups. The first is Mekka, the second is the Riyadh region, and all other regions constituted as a third cluster. Regardless of this apparent subgroup clustering, the natural physical grouping of the data in the seven regions will be kept intact, and regions will be explicitly included as six fixed covariates in the proposed modeling strategy. The data are presented as short time series of counts.

Fixed effects have been developed for a variety of different data types and models, including the Poisson regression models for count data (Allison, 1996, 2005a). The fixed-effects Poisson regression model for longitudinal data has been described in detail by Cameron and Trivedi (1998). We provide the R code for fitting both the Poisson regression and the negative binomial models:

```
# We first transform the data from the wide format as given in the
Excel data set, into the long format using the following R code:
tb=read.csv(file.choose())
names(tb)=c
("ID","region","TB.1","TB.2","TB.3","TB.4","TB.5","TB.6",
"TB.7","TB.8","TB.9","TB.10")
tb=reshape(tb,direction="long",idvar="ID",varying=colnames
(tb)[-(1:2)])
```

The above code produces the plot given in Figure 7.5.

```
# Model_1: GLM with Poisson family
model3_glm    =glm(TB~    time+    factor(region),    data=tb,
family="poisson")
summary(model3_glm)
Deviance Residuals:
    Min      1Q    Median      3Q      Max
 -5.6685  -2.3495  -0.2382  2.0344  7.4191
Coefficients:
    Estimate    Std. Error    z value    Pr(>|z|)
(Intercept)    5.667    0.020    279.726    <2e-16
```

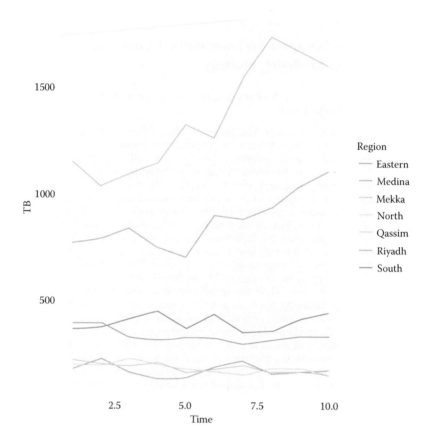

**FIGURE 7.5**
Regional distribution of TB in Saudi Arabia.

```
time            0.027   0.002    14.647     <2e-16
Medina         -0.657   0.029   -22.282    < 2e-16
Mekka           1.392   0.019    72.302    < 2e-16
North          -0.614   0.029   -21.115    < 2e-16
Qassim         -0.606   0.029   -20.918    < 2e-16
Riyadh          0.95    0.020    46.817    < 2e-16
South           0.168   0.023     7.173   7.35e-13
(Dispersion parameter for Poisson family taken to be 1)
Null deviance: 21919.09 on 69 degrees of freedom
Residual deviance:  557.75 on 62 degrees of freedom
AIC: 1115.6
```

Because of the clear over dispersion in the counts, we shall also fit the negative binomial distribution (NBD), which can be regarded as a generalization of the Poisson distribution with an additional parameter (dispersion parameter) allowing us to fit count data with variance exceeding the mean.

```
librarty(MASS)
```

```
#Model_2: GLM with negative binomial family
Model2_glm=glm.nb(formula = TB ~ time + factor(region), link =
"log",data = tb)
summary(model2_glm)
Call:
glm.nb(formula = TB ~ time + factor(region), data = tb, link =
"log",
    init.theta = 67.74920757)
Deviance Residuals:
    Min        1Q     Median        3Q       Max
-1.9290    -0.7649   -0.1871    0.6574    2.0906
Coefficients:
    Estimate      Std.    Error    z value    Pr(>|z|)
(Intercept)    5.795    0.052    111.184     <2e-16
time           0.005    0.006      0.863      0.388
Medina        -0.657    0.062    -10.633     <2e-16
Mekka          1.3888   0.058     24.093     <2e-16
North         -0.613    0.062     -9.957     <2e-16
Qassim        -0.606    0.062     -9.845     <2e-16
Riyadh         0.947    0.058     16.337     <2e-16
South          0.167    0.059      2.822      0.005
```

(Dispersion parameter for Negative Binomial (67.7492) family
taken to be 1)
Null deviance: 2572.888 on 69 degrees of freedom
Residual deviance: 69.065 on 62 degrees of freedom
**AIC: 761.44**
**Theta: 67.7, and its Std. Error: 13.6**

The AIC under the negative binomial model (761.44) is much smaller than the AIC values under the Poisson model (1115.6) indicating that the negative binomial provides a better fit to the data.

**Example 7.7: Number of Traffic Violations in a Large Urban Area**

From 110 locations in a large urban area, traffic violations have been recorded by installed traffic control cameras. The area was divided into two subareas: north (N) and south (S). At the beginning of the study that was conducted by the traffic administration (motor vehicle branch) of the ministry of interior, the number of registered vehicles surrounding a given location was obtained. Traffic violations were recorded in four different time points (w.1, w.2, w.3, w.4). The fixed-effect longitudinal models for counts will be used to model this data. Figure 7.6 shows the boxplot of member of violations over time for each location.

```
traffic=read.csv(file.choose())
names(traffic) = c(ID,
location,"w.1","w.2","w.3","w.4","registered")
library(tidyr)
library(reshape2)
#this code will transform the data from the wide format to the long
format
traffic_long=gather(traffic,time,score,w.1,w.2,w.3,w.4)

#To examine the data we used the boxplot (Figure 7.6) with the R
code:
library(ggplot2)
ggplot(data=traffic_long,aes(x=time,y=score))+geom_boxplot
(aes(fill=location))
```

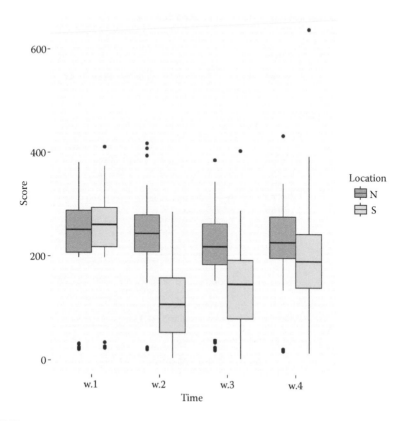

**FIGURE 7.6**
Boxplot for the number of traffic violations by location. Score represents the number of violations.

```
# Now we shall calculate the means and variances of the number of
traffic violations
# for all interactions between location and time
library(HSAUR2)
inter=interaction(traffic_long$location, traffic_long$time)
tapply (traffic_long$score, inter, mean)
  N.w.1   S.w.1   N.w.2   S.w.2   N.w.3   S.w.3   N.w.4   S.w.4
243.61  249.96  239.65  115.02  209.35  140.80  227.17  197.66
tapply (traffic_long$score, inter, var)
  N.w.1   S.w.1   N.w.2   S.w.2    N.w.3   S.w.3   N.w.4    S.w.4
7422.20 6307.27 6104.38 5603.58 7131.440 7728.82 6663.88 10891.14
```

The presence of overdispersion may be due to presence of correlation among the repeated counts, or it may be a true overdispersion. We shall fit three models to this data.

```
library(gee)
library(MASS)
```

```
#Model_1: independence correlation structure
model_1=glm(score~time+location+registered,family=poisson
(link="log"),data=traffic_long)
summary(model_1)
#Deviance Residuals:
     Min       1Q    Median       3Q       Max
-19.4363  -2.5749    0.5486    3.0991   22.1188
```

```
Coefficients:
              Estimate   Std. Error   Pr(>|z|)
(Intercept)     4.934         .014      <2e-16
timew.2         -.337         .009      <2e-16
timew.3         -.347         .009      <2e-16
timew.4         -.151         .009      <2e-16
locationS       -.282    .007      <2e-16
registered    .00001    .000      <2e-16
```

(Dispersion parameter for Poisson family taken to be 1)

Null deviance: 24822 on 439 degrees of freedom
Residual deviance: 17393 on 434 degrees of freedom

**AIC: 20450**

#### #Model_2: exchangeable correlation structure

```
model_2=gee(score~time+location+registered,family=poisson,
id=ID,data=traffic_long,
corstr="exchangeable",scale.fix=TRUE,scale.value=1)
summary(model_2)
```

```
gee(formula = score ~ time + location + registered, id = ID,
  data = traffic_long, family = poisson, corstr = "exchangeable",
  scale.fix = TRUE, scale.value = 1)
```

```
Summary of Residuals:
       Min          1Q      Median          3Q          Max
-244.893997  -37.484830    8.176676   47.130471   409.378921
```

```
Coefficients:
              Estimate   Naive S.E.   Robust S.E.
(Intercept)     4.933         .013          .011
timew.2         -.337         .009          .007
timew.3         -.347         .009          .049
timew.4         -.151         .008          .043
locationS     -.2820          .006          .037
registered    .00001         .0000          .000
```

Estimated Scale Parameter: 1
Number of Iterations: 1

#### #Model_3. Negative binomial is selected because of the over dispersion issue.

```
model_3=glm.nb(score ~ time + location + registered,link=log,
data = traffic_long)
summary(model_3)
```

```
Deviance Residuals:
    Min       1Q    Median       3Q      Max
 -3.9895  -0.3423    0.0760    0.4093   2.2137
```

```
Coefficients:
               Estimate    Std. Error     Pr(>|z|)
(Intercept)      5.047         .010         <2e-16
timew.2          -.035         .078        8.47e-06
timew.3         -.3374         .078        1.69e-05
timew.4          -.119         .078          0.129
locationS        -.340         .055        8.91e-10
registered      .000009        .000        1.10e-12
```

(Dispersion parameter for Negative Binomial (3.0013) family taken to be 1)

Null deviance: 581.62 on 439 degrees of freedom
Residual deviance: 471.33 on 434 degrees of freedom
AIC: 5320.3

Number of Fisher Scoring iterations: 1
          Theta: 3.001, Std. Error: 0.200.   -2 x log-likelihood
= 5306.285

On comparing the three models, we can see that there are no tangible differences among the estimated regression coefficients of the three models. However, the smallest AIC is obtained under the negative binomial model, and that is the model of choice in this situation.

## 7.9 The Problem of Multiple Comparisons in Repeated Measures Experiments

The problem of multiplicity is encountered in randomized controlled clinical trials (RCTs) and observational studies alike. This can arise from testing multiple hypotheses that reflect multiple primary outcome measures, multiple assessment time points, or multiple subgroups. The primary concern is that the multiple testing can elevate the risk of committing a type I error. Specifically, the probability $\alpha_{FWE}$ as familywise type I error (FWE) with $k$ independent tests is as follows:

$$\alpha_{FWE} = 1 - (1 - \alpha)^k$$

where $\alpha$ is a pre-specified significance level for each individual null hypothesis. $H_{0j}$, $j = 1,\ldots,k$, that is, $\alpha = P$ (Reject $H_{0j} \mid H_{0j}$). For instance, the nominal $\alpha$ level of 0.05 is clearly exceeded with $k = 10$ independent tests, $\alpha_{FWE} = 1 - (1 - 0.05)^{10} = 0.401$, where $\alpha_{FWE} = (P$ Reject $H_0 \mid H_0)$ in that $H_0$ is a "global" null hypothesis such that $H_0{:}H_{0j}$ is true for all $j$. Based on concerns about elevated type I error, scientific journals, and guidelines such as the CONSORT Statement (Altman, 2001) advocate the use of multiplicity adjustments.

Several such adjustments are based on the Bonferroni inequality, which in essence states that the probability of occurrence of at least one event among $k$ events $(E_j, j = 1,\ldots k)$ less than or equal to the sum of the probabilities of each of those individual events:

$$P\left(\bigcup_{j=1}^{k} E_j\right) \le \sum_{j=1}^{k} P\left(E_j\right)$$

The so-called Bonferroni adjustment ($\alpha_B$) is based on this inequality. For instance for $k$ tests, the Bonferroni adjustment partitions the nominal alpha-level ($\alpha$) into $k$ equal components and yields $\alpha_B = \alpha/k$ such that $\alpha_B = P$ (Reject $H_{0j} \mid H_{0j}$) for $j = 1,\ldots,k$. Thus, for $k = 10$ tests and a nominal unadjusted $\alpha = 0.05$ $\alpha_B = 0.005$. As a result, the familywise error rate based on the Bonferroni adjustment becomes

$$\alpha_{B-FWE} = 1 - (1 - \alpha_B)^{10} = 1 - (1 - 0.005)^{10} = 0.049$$

In contrast, the Dunn-Šidák adjustment (Ury, 1976), $\alpha_{DS} = 1 - (1 - \alpha)^{1/k}$ yields a familywise error of precisely $\alpha$. In the case of $k = 10$, $\alpha_{DS} = 1 - (1 - 0.05)^{1/10} = 0.005$ which results in the familywise error rate

$$\alpha_{B-FWE} = 1 - (1 - \alpha_{DS})^4 = 1 - (1 - 0.005)^{10} = 0.05$$

Alternatively, sequentially rejective procedures modify the alpha level with each of $k$ successive tests. Holm's (1979) approach, for instance, tests each $H_0$ in *ascending* order of $p$-values. Successively larger $p$-values have less rigorous alpha, $\alpha_{Holm}: \alpha/k, \alpha/(k-1),\ldots, \alpha/(k-(k-1))$. The sequential Holm tests terminate after the first nonsignificant test and no subsequent $H_{0j}$ is rejected.

Hochber's (1988) approach, in contrast, sequences the $k$ tests in *descending* order of $p$-values. Each successive $H_{0j}$ has a more rigorous $\alpha$ threshold. For $k = 10$ the successive $\alpha$ levels $\alpha/(k-9)$, $\alpha/(k-8),\ldots \alpha/(k-1)$ and $\alpha/k$ or 0.05, 0.025, 0.0167, ... and 0.005. The sequential tests terminate after the first significant test and each subsequent $H_{0j}$ is rejected.

**Example 7.8**

The data file is "heart" or the Excel sheet "FDR_R".

In a clinical study on a group of heart patients, a subgroup analyses produced the following $p$-values corresponding to 20 demographic and clinical parameters:

```
DATA heart;
INPUT measure $ RAW_P;
cards;
weight       .006
height       .121
SBP          .001
DBP          .002
HDL          .049
LDL          .031
Platelet     .051
Liver        .116
Kidney       .118
HB           .116
Hip          .006
Family       .001
```

```
Smoking      .059
Diabetes     .007
Gender       .769
Education    .431
Age         .0009
Fat          .073
On_med       .091
Prevsurg     .001
```

As can be seen, 10 of the variables show a significant $p$-value ($p$-value < 0.05). However, because 20 variables were tested, we should expect one or two of the variables to show a significant result purely by chance. Applying the Bonferroni correction, you would divide $P = 0.05$ by the number of tests (20) to get the Bonferroni critical value, so a test would have to have $P < 0.0025$ to be significant. Under that criterion, only the tests for variables 3 (SBP), 4 (DBP), 12 (family history of heart disease), 17 (age), and 20 (previous surgery) are significant.

The SAS Code used to produce the multiplicity adjusted $p$-values is

```
proc print data=heart noobs;
run;

proc multtest inpvalues=heart holm hoc fdr;
run;
```

The R-CODE to produce multiplicity adjusted $p$-values for selected five procedures is:

```
data=read.csv(file.choose())
library(FSA)
names(data)
attach(data)
pvals=c(RAW_P)
BONF=p.adjust(pvals,"bonferroni")
BH=p.adjust(pvals,"BH")
HOLM=p.adjust(pvals,"holm")
Hochberg=p.adjust(pvals,"hochberg")
Hommel=p.adjust(pvals,"hommel")
BY=p.adjust(pvals,"BY")
corrected=cbind(pvals, BONF=round(BONF,3),BH=round(BH,3),
HOLM=round(HOLM,3),Hochberg=round(Hochberg,3),    Hommel=round
(Hommel,3))
Corrected
```

The adjusted $p$-values using six different procedures are given in Table 7.6.

**Controlling the False Discovery Rate: Benjamini–Hochberg Procedure**
An alternative approach to the well-known Bonferroni's correction is to control the false discovery rate (FDR). This is the proportion of "discoveries" (significant results) that are actually false positives. For example, let's say we are using the data from the warehouse to compare clinical parameters for 200 variables between liver tumors and normal liver cells. We are going to do additional study on any variable that shows a significant difference between the normal and tumor cells, and we are willing to accept up to 10% of the parameters with significant results being false positives; we will find out they are false positives when we do the follow-up experiments. In this case, you would set our false discovery rate to 10%.

**TABLE 7.6**

Adjusted *p*-values Using Bonferroni, and Other Procedures

|         | Pvals  | BONF  | BH    | HOLM  | Hochberg | Hommel |
|---------|--------|-------|-------|-------|----------|--------|
| [1,]    | 0.0060 | 0.120 | 0.017 | 0.090 | 0.084    | 0.078  |
| [2,]    | 0.1210 | 1.000 | 0.134 | 0.696 | 0.363    | 0.363  |
| [3,]    | 0.0010 | 0.020 | 0.005 | 0.019 | 0.017    | 0.017  |
| [4,]    | 0.0020 | 0.040 | 0.008 | 0.032 | 0.032    | 0.030  |
| [5,]    | 0.0490 | 0.980 | 0.093 | 0.539 | 0.363    | 0.202  |
| [6,]    | 0.0310 | 0.620 | 0.069 | 0.372 | 0.363    | 0.182  |
| [7,]    | 0.0510 | 1.000 | 0.093 | 0.539 | 0.363    | 0.204  |
| [8,]    | 0.1160 | 1.000 | 0.134 | 0.696 | 0.363    | 0.348  |
| [9,]    | 0.1180 | 1.000 | 0.134 | 0.696 | 0.363    | 0.354  |
| [10,]   | 0.1160 | 1.000 | 0.134 | 0.696 | 0.363    | 0.348  |
| [11,]   | 0.0060 | 0.120 | 0.017 | 0.090 | 0.084    | 0.078  |
| [12,]   | 0.0010 | 0.020 | 0.005 | 0.019 | 0.017    | 0.017  |
| [13,]   | 0.0590 | 1.000 | 0.098 | 0.539 | 0.363    | 0.236  |
| [14,]   | 0.0070 | 0.140 | 0.018 | 0.091 | 0.091    | 0.091  |
| [15,]   | 0.7690 | 1.000 | 0.769 | 0.862 | 0.769    | 0.769  |
| [16,]   | 0.4310 | 1.000 | 0.454 | 0.862 | 0.769    | 0.769  |
| [17,]   | 0.0009 | 0.018 | 0.005 | 0.018 | 0.017    | 0.015  |
| [18,]   | 0.0730 | 1.000 | 0.112 | 0.584 | 0.363    | 0.242  |
| [19,]   | 0.0910 | 1.000 | 0.130 | 0.637 | 0.363    | 0.273  |
| [20,]   | 0.0010 | 0.020 | 0.005 | 0.019 | 0.017    | 0.017  |

One good technique for controlling the false discovery rate was developed in detail by Hockberg and Benjamini (1990), Benjamini and Hochberg (1995) and Benjamini and Yekutieli (2001). We rank the individual *p*-values in order, from smallest to largest. The smallest *p*-value has a rank of $i = 1$, then the next smallest has $i = 2$, and so on. Compare each individual *p*-value to its Benjamini-Hochberg critical value, $(i/m)Q$, where $i$ is the rank, $m$ is the total number of tests, and $Q$ is the false discovery rate we selected. The largest *p*-value that has $p < (i/m)Q$ is significant, and all of the *p*-values smaller than it are also significant, even the ones that are not less than their Benjamini-Hochberg critical value. In Table 7.7 we show the results of the SAS MULTEST procedure.

From Table 7.7, the largest *p*-value with $p < $ the corresponding FDR is parameter number 10 (HB), where the individual *p*-value (0.049) is less than 0.05. Thus the first 10 tests would be significant.

## 7.10 Sample Size Requirements in the Analysis of Repeated Measures

When a trial is being planned, one of the first questions is how many subjects should be enrolled in the study. While this is not a simple matter in

**TABLE 7.7**

Bonferroni and FDR for the Data "heart"

| p-value Adjustment Information | | | |
|---|---|---|---|
| p-value Adjustment | | | Stepdown Bonferroni |
| p-value Adjustment | | | Hochberg |
| p-value Adjustment | | | False Discovery Rate |
| *p*-values | | | |
| Test | Raw | Stepdown Bonferroni | Hochberg | False Discovery Rate |
| 1 | 0.0009 | 0.0180 | 0.0170 | 0.0050 |
| 2 | 0.0010 | 0.0190 | 0.0170 | 0.0050 |
| 3 | 0.0010 | 0.0190 | 0.0170 | 0.0050 |
| 4 | 0.0010 | 0.0190 | 0.0170 | 0.0050 |
| 5 | 0.0020 | 0.0320 | 0.0320 | 0.0080 |
| 6 | 0.0060 | 0.0900 | 0.0840 | 0.0171 |
| 7 | 0.0060 | 0.0900 | 0.0840 | 0.0171 |
| 8 | 0.0070 | 0.0910 | 0.0910 | 0.0175 |
| 9 | 0.0310 | 0.3720 | 0.3630 | 0.0689 |
| 10 | 0.0490 | 0.5390 | 0.3630 | 0.0927 |
| 11 | 0.0510 | 0.5390 | 0.3630 | 0.0927 |
| 12 | 0.0590 | 0.5390 | 0.3630 | 0.0983 |
| 13 | 0.0730 | 0.5840 | 0.3630 | 0.1123 |
| 14 | 0.0910 | 0.6370 | 0.3630 | 0.1300 |
| 15 | 0.1160 | 0.6960 | 0.3630 | 0.1344 |
| 16 | 0.1160 | 0.6960 | 0.3630 | 0.1344 |
| 17 | 0.1180 | 0.6960 | 0.3630 | 0.1344 |
| 18 | 0.1210 | 0.6960 | 0.3630 | 0.1344 |
| 19 | 0.4310 | 0.8620 | 0.7690 | 0.4537 |
| 20 | 0.7690 | 0.8620 | 0.7690 | 0.7690 |

cross-sectional studies, due to the variety of factors that must be taken into account, it becomes even more complicated in longitudinal studies. However, due to an increasing use of longitudinal designs, there is a growing demand for flexible sample size calculations.

The classical method to obtain a sample size is to consider the case where there is no treatment effect, the "null hypothesis," versus any alternative. The sample size is then calculated to ensure that this hypothesis will rarely be accepted if there is a treatment effect of prespecified size.

Suppose that we have two treatment groups and that the effect size is measured by the worthwhile difference ($d$) between the average responses of the two groups. The appropriate formula was given by Diggle et al. (1994, pp. 27–31) to obtain the number of individuals required in this particular type of longitudinal study:

$$m = \frac{2\left(z_n + z_{1-p}\right)^2 \sigma^2 \{1 + (n-1)\rho\}}{nd^2} \tag{7.14}$$

This assumes that only a treatment variable and a random intercept corresponding to a constant correlation over time arc included in the model.

Hence, using the suggestion of Vonesh and Chinchilli (1997, pp. 99–110) so as to include serial correlation rather than a constant correlation, the following sample size calculation is obtained for an average response difference between the groups:

$$m = \frac{2\left(z_n + z_{1-p}\right)^2 \sigma^2 \{1 + (n-1)\rho^{n-1}\}}{n\tau^2} \tag{7.15}$$

Here $\tau$ is the difference for linear time coefficient in the second-degree model for a given meaningful effect size. As an example, suppose that we have $n = 4$ occasions at regular intervals including the baseline, so that $\tau = 0, 1, 2, 3$. Suppose also that a meaningful effect size is $d = 10$, and the residual variance $\sigma^2$ is 200. Excluding the baseline we have $\tau = 2d/3 = 2(10)/3$. The required sample size can then be obtained by using Equation 7.15. If the correlation is assumed 0.2, the required number per group is

$$m = \frac{2 \times (1.64 + .84)^2 \times 200 \times \{1 + (4-1) \times 0.2^3\}}{4 \times \left(\frac{20}{3}\right)^2} = 14$$

The required number is about $14 \times 3 \times 2 = 84$ observations in total, not including baseline measurement, over the time period.

If we examine the sample sizes resulting from this time-average formula, the number of individuals increases with the correlation. This is to be expected, given the assumptions made.

## Exercises

7.1 Situations in which more than two factors would be nested within each other are of frequent occurrence in repeated measures experiments. Example 7.1 illustrates this situation. Pens of animals are randomized into two diet groups. Animals in each pen are approximately of the same age and initial weight. Their weights were measured at weeks 1, 2, and 3 weeks. The data are presented in Table 7.8. The data in Table 7.8 show that pens 1, 2, and 3 are nested within the diet group 1, while pens 4, 5, and 6 are nested in diet group 2 (i.e., different pens within each group) and animals are nested within pens. The main objectives of

**TABLE 7.8**

Data for the Feed Trial

| | | | | Weight | |
|---|---|---|---|---|---|
| Animal | Diet | Pen | Week 1 | Week 2 | Week 3 |
| 1 | 1 | 1 | 2.5 | 3.0 | 4.0 |
| 2 | 1 | 1 | 2.0 | 2.5 | 2.5 |
| 3 | 1 | 1 | 1.5 | 2.0 | 2.0 |
| 4 | 1 | 1 | 1.5 | 2.5 | 3.0 |
| 5 | 1 | 2 | 2.0 | 4.0 | 4.0 |
| 6 | 1 | 2 | 2.5 | 3.5 | 2.5 |
| 7 | 1 | 2 | 1.5 | 2.0 | 2.5 |
| 8 | 1 | 2 | 1.0 | 1.5 | 1.0 |
| 9 | 1 | 3 | 1.5 | 1.5 | 2.0 |
| 10 | 1 | 3 | 2.0 | 2.0 | 2.5 |
| 11 | 1 | 3 | 1.5 | 2.0 | 2.0 |
| 12 | 2 | 4 | 3.0 | 3.0 | 4.0 |
| 13 | 2 | 4 | 3.0 | 6.0 | 8.0 |
| 14 | 2 | 4 | 3.0 | 4.0 | 7.0 |
| 15 | 2 | 4 | 4.0 | 5.0 | 6.0 |
| 16 | 2 | 5 | 3.0 | 3.0 | 5.0 |
| 17 | 2 | 5 | 4.0 | 4.5 | 5.5 |
| 18 | 2 | 5 | 3.0 | 3.0 | 4.0 |
| 19 | 2 | 6 | 3.0 | 7.0 | 9.0 |
| 20 | 2 | 6 | 4.0 | 6.0 | 7.0 |
| 21 | 2 | 6 | 4.0 | 6.0 | 8.0 |
| 22 | 2 | 6 | 4.0 | 6.0 | 7.0 |

the trial were to see if the weights of animals differed between the diet groups, and if such differences were present over time.

7.2 The Familial Polyposis Supplementation Trial (FPST) was a four-year randomized double-blind placebo-controlled trial of high fiber and vitamins C and E to reduce polyps in a high-risk population (part of the data is produced here). Fifty-eight patients were randomized to take either placebo ($k = 22$), vitamin C and E ($k = 16$), or vitamin C and E and high fiber ($k = 20$). Subjects were examined every three months and rectal polyps counted. The study was reported in details in De Cosse et al. (1989). Examine the data carefully (Table 7.9), state the hypotheses, and conduct data analysis using a PA and SS model.

7.3 The data are the results of a feed trial to study the effect of four different diets on the growth rates of rabbits. The R Code produces the following profile plot in Figure 7.7:

**TABLE 7.9**

Data for Familial Polyposis Supplementation Trial

| ID | Treatment | Sex | Baseline Polyp Count | V1 | V2 | V3 | V4 | V5 |
|----|-----------|-----|----------------------|----|----|----|----|----|
| 1 | 0 | 0 | 41 | 36 | 56 | 34 | 46 | 61 |
| 2 | 2 | 1 | 4 | 2 | 2 | 0 | 1 | 1 |
| 3 | 2 | 1 | 26 | 1 | 2 | 0 | 0 | 1 |
| 4 | 2 | 0 | 15 | 13 | 6 | 12 | 6 | 12 |
| 5 | 2 | 0 | 9 | 6 | 2 | 8 | 6 | 4 |
| 6 | 1 | 1 | 6 | 15 | 9 | 4 | 4 | 2 |
| 7 | 1 | 0 | 1 | 7 | 8 | 5 | 3 | 6 |
| 8 | 0 | 1 | 2 | 3 | 1 | 1 | 3 | 3 |
| 9 | 0 | 0 | 2 | 4 | 10 | 9 | 17 | 8 |
| 10 | 1 | 0 | 1 | 2 | 1 | 1 | 1 | 2 |
| 11 | 0 | 0 | 7 | 10 | 31 | 31 | 37 | 11 |
| 12 | 2 | 1 | 25 | 8 | 6 | 8 | 11 | 16 |
| 13 | 1 | 1 | 10 | 6 | 3 | 3 | 7 | 9 |
| 14 | 2 | 0 | 1 | 0 | 0 | 0 | 0 | 0 |
| 15 | 2 | 1 | 4 | 5 | 2 | 1 | 1 | 2 |
| 16 | 2 | 0 | 24 | 21 | 13 | 14 | 9 | 16 |
| 17 | 1 | 1 | 1 | 4 | 4 | 10 | 4 | 7 |
| 18 | 0 | 0 | 3 | 3 | 1 | 1 | 4 | 0 |
| 19 | 2 | 0 | 8 | 1 | 1 | 2 | 2 | 1 |
| 20 | 1 | 1 | 5 | 4 | 6 | 6 | 11 | 16 |
| 21 | 0 | 0 | 8 | 16 | 17 | 22 | 24 | 36 |
| 22 | 1 | 0 | 0 | 0 | 0 | 0 | 0 | 0 |
| 23 | 1 | 0 | 27 | 15 | 10 | 37 | 32 | 30 |
| 24 | 2 | 1 | 3 | 5 | 5 | 6 | 4 | 1 |
| 25 | 0 | 1 | 1 | 6 | 4 | 3 | 11 | 6 |
| 26 | 1 | 1 | 1 | 3 | 3 | 7 | 6 | 7 |
| 27 | 2 | 1 | 10 | 11 | 9 | 9 | 21 | 23 |
| 28 | 0 | 1 | 5 | 1 | 4 | 2 | 0 | 0 |
| 29 | 0 | 0 | 11 | 7 | 4 | 11 | 8 | 8 |

```
library(ggplot2)
rabbitweight=read.csv(file.choose())

diet1 <- subset(rabbitweight, Diet == 1)
diet2 <- subset(rabbitweight, Diet == 2)
diet3 <- subset(rabbitweight, Diet == 3)
diet4 <- subset(rabbitweight, Diet == 4)
add_line <- function(df) {
geom_line(aes(x = Time, y = weight, group = rabbit), data = df)
```

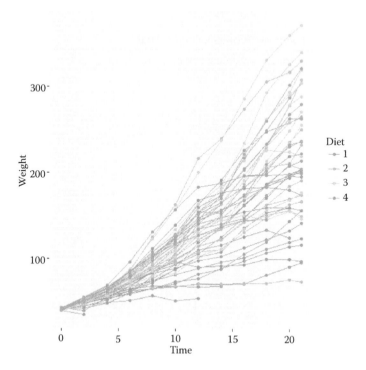

**FIGURE 7.7**
Weight profile of the rabbits over time.

```
}
add_points <- function(df) {
  geom_point(aes(x = Time, y = weight), data = df)
}
  add_line_points <- function(df) {
add_line(df) + add_points(df)
}
(plot1 <- ggplot() + add_line(diet1) + add_points(diet1))
(plot3 <- ggplot(aes(x = Time, y = weight, group = rabbit), data = diet1) +
  geom_line() + geom_point())
## plot the whole dataset at once
(plot5 <- ggplot(aes(x = Time, y = weight, group = rabbit, colour = Diet),
data = rabbitwight +
  geom_line() + geom_point())
```

a. Test the weights for normality using the nonparametric tests in Chapter 2.
b. Use the linear mixed model to analyze the data and to test if there are
   differential effects of treatments on the growth of the animals.

# 8

## Introduction to Time Series Analysis

### 8.1 Introduction

A time series is an ordered sequence of observations. Although the ordering is usually through time, particularly in terms of some equally spaced intervals, the ordering may also be taken through other dimensions, such as space. Time series occur in a wide variety of fields. In economics, interest may be focused on the weekly fluctuations in the stock prices and their relationships to unemployment figures. Agricultural time series analyses and forecasting could be applied to annual crop yields or the prices of produce with regard to their seasonal variations. Environmental changes over time, such as levels of air and water pollution measured at different places can be correlated with certain health indicators. The number of influenza outbreaks in successive weeks during the winter season could be approached as a time series problem by an epidemiologist. In medicine, systolic and diastolic blood pressures followed over time for a group of patients could be useful for assessing the effectiveness of a drug used in treating hypertension. Geophysical time series (Shumway, 1982) are quite important for predicting earthquakes. From these examples one can see the diversity of fields in which time series can be applied. There are, however, some common objectives that must be achieved in collected time series data:

1. *To describe the behavior of the series in a concise way.* This is done by first plotting the data and obtaining simple descriptive measures of the main properties of the series. This may not be useful for all time series because there are series that require more sophisticated techniques, and thus more complex models need to be constructed.

2. *To explain the behavior of the series in terms of several variables.* For example, when observations are taken on more than one variable, it may be feasible to use the variation in one time series to explain the variation in another series.

3. *We may want to predict (forecast) the future values of the series.* This is an important task for the analysis of economic and agricultural time series. It is desirable, particularly if there is sufficient evidence in the

system, to ensure that future behavior will be similar to the past. Therefore, our ability to understand the behavior of the series may provide us with more insight into causal factors and help us make projections into the future.

4. *Controlling the series by generating warning signals of future fluctuations.* For example, if we are measuring the quality of production process, our aim may be to keep the process under control. Statistical quality control provides us with the tools to achieve such an objective by constructing *control charts*. More advanced strategies for control are outlined in Box and Jenkins (1970).

In the following sections we provide examples on modeling and graphing time series. We introduce simple models and the ARIMA models. We restrict our discussion to stationary time series. Other advanced models and methods are available in many books (e.g., Anderson, 1971; Bloomfield, 1976; Cryer, 1986; Diggle, 1990; Kendall and Ord 1990).

**Example 8.1**

An epidemiological time series showing the average somatic cell count (SCC) by month over a number of years is shown in Table 8.1.

A plot of these data in Figure 8.1 shows large fluctuations in both the mean and the variance over time.

The SAS code to read the data and produce Figure 8.1 is given next:

```
data scc;
input year scc @@;
 difscc =dif(scc); time=_n_;
datalines;
84 317 84 292 84 283 84 286 84 314 84 301 84 317 84 344
84 367 84 351 84 321 84 398 85 345 85 310 85 307 85 310
85 340 85 325 85 340 85 370 85 400 85 380 85 345 85 330
86 370 86 360 86 300 86 310 86 389 86 320 86 340 86 400
86 395 86 350 86 400 86 350 87 350 87 420 87 360 87 340
87 335 87 350 87 360 87 395 87 380 87 375 87 402 87 460
88 400 88 385 88 350 88 325 88 345 88 350 88 375 88 410
88 360 88 375 88 370 88 395 89 370 89 335 89 305 89 325
89 310 89 315 89 350 89 370 89 350 89 345 89 355 89 340
90 340 90 345 90 325 90 330 90 360 90 330 90 345 90 350
90 350 90 345 90 325 90 280
;

goptions reset=global gunit=pct cback=white htitle=6 htext=3
ftext=swissb colors=(black);

* Figure 8.1: A time plot of average somatic cell counts per month;
axis1 order=(0 to 90 by 10) label=(angle=0 'Time (Month) ') offset=
(3) minor=(number=1);
axis2 order=(280 to 460 by 20) label=(angle=0 'SCC')
minor=(number=3);

proc gplot data=scc;
  plot scc*time / haxis=axis1 vaxis=axis2;
 symbol1 i=join;
run; quit;
```

**TABLE 8.1**

Average SCC per Farm in Cells (in 1000's) per ml of Milk, 1984–1990

|       | 1984 | 1985 | 1986 | 1987 | 1988 | 1989 | 1990 |
|-------|------|------|------|------|------|------|------|
| Jan   | 317  | 345  | 370  | 350  | 400  | 370  | 340  |
| Feb   | 292  | 310  | 360  | 420  | 385  | 335  | 345  |
| Mar   | 283  | 307  | 300  | 360  | 350  | 305  | 325  |
| Apr   | 286  | 310  | 310  | 340  | 325  | 325  | 330  |
| May   | 314  | 340  | 389  | 335  | 345  | 310  | 360  |
| June  | 301  | 325  | 320  | 350  | 350  | 315  | 330  |
| July  | 317  | 340  | 340  | 360  | 375  | 350  | 345  |
| Aug   | 344  | 370  | 400  | 395  | 410  | 370  | 350  |
| Sept  | 367  | 400  | 395  | 380  | 360  | 350  | 350  |
| Oct   | 351  | 380  | 350  | 375  | 375  | 345  | 345  |
| Nov   | 321  | 345  | 400  | 402  | 370  | 355  | 325  |
| Dec   | 398  | 330  | 350  | 460  | 395  | 340  | 280  |

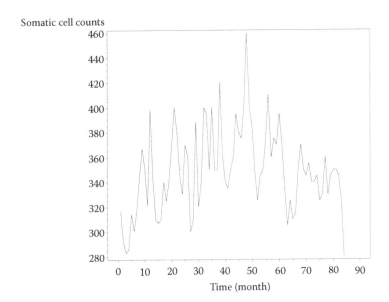

**FIGURE 8.1**

A time series plot of average somatic cell counts per month for 84 months.

## 8.2 Simple Descriptive Methods

This section describes some of the simple techniques that will detect the main characteristics of a time series. From a statistical point of view, the description of a time series is accomplished by a time plot; that is, plotting the observations against time and formulating a mathematical model to characterize the behavior of the series. This models the mechanism that governs the variability of the observations over time. Plotting the data could reveal certain features such as trend, seasonality, discontinuities, and outliers. The term *trend* usually means the upward or downward movement over a period of time, thus reflecting the long-run growth or decline in the time series.

In addition to the effect of trend, most economic time series include seasonal variation. To include both the seasonal variation and the trend effect there are two types of models that are frequently used. The first is the additive seasonal variation model (ASVM) and the second is the multiplicative seasonal variation model (MSVM). If a time series displays additive seasonal variation, the magnitude of the seasonal swing is independent of the mean. On the other hand, in a multiplicative seasonal variation series we see that the seasonal swing is proportional to the mean. The two models are represented by the following equations

$$y_t = T_t + S_t + \varepsilon_t \text{ (ASVM)}$$

$$y_t = (T_t)(S_t) + \varepsilon_t \text{ (MSVM)}$$

where
$y_t$ = observed value at time $t$
$T_t$ = mean trend factor at time $t$
$S_t$ = seasonal effect at time $t$
$\varepsilon_t$ = irregular variation of the time series at time $t$
The two models are illustrated in Figures 8.2 and 8.3.

Note that the MSVM has an additive irregular variation term. If such a term is multiplicative, that is, if $y_t = (T_t)(S_t)(\varepsilon_t)$, then this model can be transformed to ASVM by taking the logarithm of both sides.

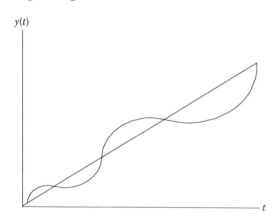

**FIGURE 8.2**
Additive seasonal variation model.

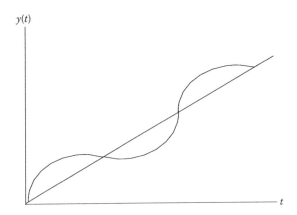

**FIGURE 8.3**
Multiplicative seasonal variation model.

## 8.2.1 Multiplicative Seasonal Variation Model

In this section we will be concerned with simple methods of decomposing MSVM into its trend, seasonal, and random components. The model is somewhat different from the previous model,

$$y_t = (T_t)(S_t) + \varepsilon_t \tag{8.1}$$

and is usually written as

$$y_t = (T_t)(S_t)(C_t)(I_t) \tag{8.2}$$

where $y_t$, $T_t$, and $S_t$ are as previously defined. Here, $C_t$ represents the cyclical effect on the series at time $t$ and $I_t$ is the irregular variation.

How to decompose the multiplicative model is explained in the following example.

### Example 8.2

The data in Table 8.2 represents the number of cases with bovine respiratory disease in particular feedlots reported in eastern Alberta counties over a period of four years in each of the four quarters.

The first step in the analysis of this time series is the estimation of seasonal factors for each quarter. To do this one has to calculate the moving average (MA) in order to remove the seasonal variation from the series. A moving average is calculated by adding the observations for a number of periods in the series and dividing the sum by the number of periods. In this example we have a four-period series since we have quarterly data. If the time series consists of data collected every four months, we have a three-period time series, and hence a three-period moving average should be used. In the example the average for the four observations in the first year is

**TABLE 8.2**

Number of Cases with Bovine Respiratory Disease (BRD) Reported
in Eastern Alberta Counties

| Quarter | Year 1 | Year 2 | Year 3 | Year 4 |
|---------|--------|--------|--------|--------|
| 1 | 21 | 25 | 25 | 30 |
| 2 | 14 | 16 | 18 | 20 |
| 3 | 5 | 7 | 9 | 10 |
| 4 | 8 | 9 | 13 | 15 |

$$\frac{21 + 14 + 5 + 8}{4} = 12$$

The second average is obtained by eliminating the first observation in
year 1 from the average and including the first observation in year 2 in the
new average. Hence

$$\frac{14 + 5 + 8 + 25}{4} = 13$$

The third average is obtained by dropping the second observation in
year 1 and adding the second observation in year 2. This gives

$$\frac{5 + 8 + 25 + 16}{4} = 13.5$$

Continuing in this manner, these moving averages are as found in Table
8.3. Note that since the first average is the average of the observations in
the four quarters, it corresponds to a midpoint between the second and
third quarter.

To obtain the average corresponding to one of the time periods in the
original time series, we calculate a centered moving average. This is
obtained by computing a two-period moving average of the moving
averages previously calculated (Table 8.3).

Note that since the moving average is computed using exactly one
observation from each season, the seasonal variation has been removed
from the data. It is also hoped that this averaging process has removed the
irregular variation $I_t$. This means that the centered moving averages in
column 6 in Table 8.3 represent the trend ($T_t$) and cycle ($C_t$). Now, since

$$y_t = (T_t)(S_t)(C_t)(I_t)$$

then the entries in column 7 of Table 8.3 are computed as:

$$(S_t)(I_t) = \frac{y_t}{(T_t)(C_t)} = \frac{\text{Column (3)}}{\text{Column (6)}}$$

The seasonal coefficients $(S_t I_t)$ are summarized in Table 8.4.

The seasonal effects for each quarter can be computed by summing and
dividing by the number of coefficient. Thus, for quarter 1, this is

$$\hat{S}_1 = \frac{1.818 + 1.667 + 1.655}{3} = 1.713$$

**TABLE 8.3**

Moving Average of the Time Series of Table 8.2

| Year (1) | Quarter (2) | $y_t$ (-3) | Moving Total (-4) | Moving Average (-5) | Centered Moving Average (-6) | $S_t I_t$ (-7) |
|---|---|---|---|---|---|---|
| 1 | 1 | 21 | | | | |
| | 2 | 14 | | | | |
| | | | 48 | 12 | | |
| | 3 | 5 | | | 12.5 | 0.4 |
| | | | 52 | 13 | | |
| | 4 | 8 | | | 13.25 | 0.604 |
| | | | 54 | 13.5 | | |
| 2 | 1 | 25 | | | 13.75 | 1.818 |
| | | | 56 | 14 | | |
| | 2 | 16 | | | 14.125 | 1.133 |
| | | | 57 | 14.25 | | |
| | 3 | 7 | | | 14.25 | 0.491 |
| | | | 57 | 14.25 | | |
| | 4 | 9 | | | 14.5 | 0.621 |
| | | | 59 | 14.75 | | |
| 3 | 1 | 25 | | | 15 | 1.667 |
| | | | 61 | 15.25 | | |
| | 2 | 18 | | | 15.75 | 1.143 |
| | | | 65 | 16.25 | | |
| | 3 | 9 | | | 16.875 | 0.533 |
| | | | 70 | 17.5 | | |
| | 4 | 13 | | | 17.75 | 0.732 |
| | | | 72 | 18 | | |
| 4 | 1 | 30 | | | 18.125 | 1.655 |
| | | | 73 | 18.25 | | |
| | 2 | 20 | | | 18.5 | 1.081 |
| | | | 75 | 18.75 | | |
| | 3 | 10 | | | | |
| | 4 | 15 | | | | |

**TABLE 8.4**

Seasonal Coefficients for Each Quarter by Year

| Quarter 1 | Quarter 2 | Quarter 3 | Quarter 4 |
|---|---|---|---|
| 1.818 | 1.133 | .400 | .604 |
| 1.667 | 1.143 | .491 | .621 |
| 1.655 | 1.081 | .533 | .732 |

Similarly $\hat{S}_2 = 1.119$, $\hat{S}_3 = 0.475$, and $\hat{S}_4 = 0.652$ are the estimated seasonal effects for quarters 2, 3, and 4, respectively.

Once the estimates of the seasonal factors have been calculated, we may obtain an estimate of the trend $T_t$ of the time series. This is done by first estimating the deseasonalized observations.

The deseasonalized observations are obtained by dividing $y_t$ by $S_t$. That is,

$$d_t = \frac{y_t}{S_t}$$

These values should be close to the trend value $T_t$. To model the trend effect, as a first step one should plot $d_t$ against the observation number $t$. If the plot is linear it is reasonable to assume that

$$T_t = \beta_0 + \beta_1 t$$

on the other hand if the plot shows a quadratic relationship then we may assume that

$$T_t = \beta_0 + \beta_1 t + \beta_2 t^2$$

and so on. Table 8.5 gives the deseasonalized observations and Figure 8.4 is a scatter plot of these observations against time.

The estimated trend is found to be

$$\hat{d}_t = \hat{T}_t = 10.05 + 0.685t, \qquad t = 1, 2, \ldots, 16$$

**TABLE 8.5**

Deseasonalized Observations

| Year | Quarter | $t$ | $y_t$ | $S_t$ | $d_t = y_t/S_t$ |
|------|---------|-----|-------|-------|-----------------|
| 1 | 1 | 1 | 21 | 1.713 | 12.26 |
|   | 2 | 2 | 14 | 1.119 | 12.51 |
|   | 3 | 3 | 5 | .475 | 10.53 |
|   | 4 | 4 | 8 | .652 | 12.27 |
| 2 | 1 | 5 | 25 | 1.713 | 14.59 |
|   | 2 | 6 | 16 | 1.119 | 23.24 |
|   | 3 | 7 | 7 | .475 | 14.74 |
|   | 4 | 8 | 9 | .652 | 13.80 |
| 3 | 1 | 9 | 25 | 1.713 | 14.59 |
|   | 2 | 10 | 18 | 1.119 | 16.09 |
|   | 3 | 11 | 9 | .475 | 12.63 |
|   | 4 | 12 | 13 | .652 | 19.94 |
| 4 | 1 | 13 | 30 | 1.713 | 17.51 |
|   | 2 | 14 | 20 | 1.119 | 17.87 |
|   | 3 | 15 | 10 | .475 | 21.05 |
|   | 4 | 16 | 15 | .652 | 23.01 |

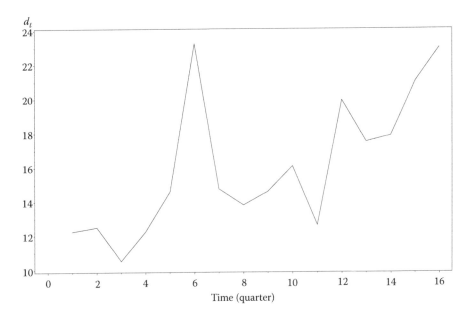

**FIGURE 8.4**
Deseasonalized observations over time for the BRD data.

To compute the cyclical effect, recall that

$$y_t = (T_t)(S_t)(C_t)(I_t)$$

hence,

$$(C_t)(I_t) = \frac{y_t}{(\hat{T}_t)(\hat{S}_t)}$$

We summarize these computations in Table 8.6.

$$C_t I_t = \frac{\text{Column 4}}{\text{Column 7}}$$

Once $(C_t)(I_t)$ has been obtained, a three-quarter moving average may remove the effect of irregular variation. The results are summarized in Table 8.7.

The previous example shows how a time series can be decomposed into its components. Most econometricians use the trend and seasonal effect in their forecast of time series, ignoring the cyclical and irregular variations. Clearly, irregular ups and downs cannot be predicted, however, cyclical variation can be forecasted and is treated in the same manner as the seasonal effects shown in Table 8.4. In our example, the average effect of the cycle at each period is as found in Table 8.8.

It should be noted that the estimated cycles are useful if a well-defined repeating cycle of reasonable fixed duration can be recognized. In many real-life data this may not be possible. In order to obtain reliable estimates of the cyclical effect, data with several cycles should be available. Since

**TABLE 8.6**

Computations of Cyclical Effect and Irregular Variation

| Year (1) | Quarter (2) | t (3) | $y_t$ (4) | $T_t = 10.05+$ <br> .685t (5) | $S_t$ (6) | $(T_t)(S_t)$ (7) | $(C_t)(I_t)$ (7) |
|---|---|---|---|---|---|---|---|
| 1 | 1 | 1 | 21 | 10.74 | 1.713 | 18.39 | 1.14 |
|   | 2 | 2 | 14 | 11.42 | 1.119 | 12.78 | 1.10 |
|   | 3 | 3 | 5 | 12.11 | .475 | 5.75 | 0.87 |
|   | 4 | 4 | 8 | 12.79 | .652 | 8.34 | 0.96 |
| 2 | 1 | 5 | 25 | 13.48 | 1.713 | 23.08 | 1.08 |
|   | 2 | 6 | 16 | 14.16 | 1.119 | 15.85 | 1.64 |
|   | 3 | 7 | 7 | 14.85 | .475 | 7.05 | 0.99 |
|   | 4 | 8 | 9 | 15.53 | .652 | 10.13 | 0.89 |
| 3 | 1 | 9 | 25 | 16.22 | 1.713 | 27.78 | 0.90 |
|   | 2 | 10 | 18 | 16.9 | 1.119 | 18.91 | 0.95 |
|   | 3 | 11 | 9 | 17.59 | .475 | 8.35 | 0.72 |
|   | 4 | 12 | 13 | 18.27 | .652 | 11.91 | 1.09 |
| 4 | 1 | 13 | 30 | 18.96 | 1.713 | 32.47 | 0.92 |
|   | 2 | 14 | 20 | 19.64 | 1.119 | 21.98 | 0.91 |
|   | 3 | 15 | 10 | 20.33 | .475 | 9.65 | 1.04 |
|   | 4 | 16 | 15 | 21.01 | .652 | 13.70 | 1.10 |

**TABLE 8.7**

Estimated Cyclical Effect

| Year | Quarter | t | $(C_t)(I_t)$ | Three-Period Moving <br> Average $C_t$ |
|---|---|---|---|---|
| 1 | 1 | 1 | 1.14 | |
|   | 2 | 2 | 1.10 | 1.037 |
|   | 3 | 3 | 0.87 | 0.977 |
|   | 4 | 4 | 0.96 | 0.970 |
| 2 | 1 | 5 | 1.08 | 1.230 |
|   | 2 | 6 | 1.64 | 1.24 |
|   | 3 | 7 | 0.99 | 1.173 |
|   | 4 | 8 | 0.89 | 0.927 |
| 3 | 1 | 9 | 0.90 | 0.913 |
|   | 2 | 10 | 0.95 | 0.857 |
|   | 3 | 11 | 0.72 | 0.920 |
|   | 4 | 12 | 1.09 | 0.910 |
| 4 | 1 | 13 | 0.92 | 0.973 |
|   | 2 | 14 | 0.91 | 0.957 |
|   | 3 | 15 | 1.04 | 1.017 |
|   | 4 | 16 | 1.10 | |

**TABLE 8.8**

Cycle's Effect for Different Periods

| Quarter 1 | Quarter 2 | Quarter 3 | Quarter 4 |
|-----------|-----------|-----------|-----------|
| 1.23 | 1.04 | 0.98 | 0.97 |
| 0.92 | 1.24 | 1.17 | 0.93 |
| 0.97 | 0.86 | 0.92 | 0.91 |
|      | 0.96 | 1.02 |      |

cyclical fluctuations have a duration of two to seven years or more, more than 25 years of data may be needed to estimate the cycle effect and make accurate forecasts. For these reasons, the cyclical variation in time series cannot be accurately predicted. In such situations, forecasts are based on the trend and seasonal factors only. Having obtained $T_t$ and $S_t$, the forecast of a future observation is given by

$$\hat{y}_t = (\hat{T}_t)(\hat{S}_t)$$

### 8.2.2 Additive Seasonal Variation Model

For this type of model we shall assume, for simplicity, that the series is composed of trend, seasonal effect, and error component, so that

$$y_t = T_t + S_t + I_t$$

As before the trend effect can be modeled either linearly, $T_t = \beta_0 + \beta_1 t$; quadratically, $T_t = \beta_0 + \beta_1 t + \beta_2 t^2$; or exponentially, $T_t = \beta_0 \beta_1^t$ (which may be linearized through the logarithmic transformation).

The seasonal pattern may be modeled by using dummy variables. Let $L$ denote the number of periods or seasons (quarter, month, etc.) in the year. $S_t$ can be modeled as follows:

$$\hat{S}_t = \gamma_1 X_{1t} + \gamma_2 X_{2t} + \ldots + \gamma_{L-1} X_{L-1\,t}$$

where

$$X_{1t} = \begin{cases} 1 & \text{if period } t \text{ is season 1} \\ 0 & \text{otherwise} \end{cases}$$

$$X_{2t} = \begin{cases} 1 & \text{if period } t \text{ is season 2} \\ 0 & \text{otherwise} \end{cases}$$

$$X_{L-1\,t} = \begin{cases} 1 & \text{if period } t \text{ is season } L-1 \\ 0 & \text{otherwise} \end{cases}$$

For example, if $L = 4$ (quarterly data) we have

$$y_t = T_t + S_t + I_t$$
$$= T_t + \gamma_1 X_{1t} + \gamma_2 X_{2t} + \gamma_3 X_{3t} + I_t \tag{8.3}$$

Similarly, if $L = 12$ (monthly data) we have

$$y_t = T_t + S_t + I_t$$
$$= T_t + \sum_{i=1}^{11} \gamma_i X_{it} + I_t \tag{8.4}$$

Clearly $T_t$ can be represented by either a linear, quadratic, or exponential relationship.

### Example 8.2 (Continued)

We now decompose an additive time series (BRD occurrence in eastern Alberta) to estimate the trend and seasonal effects under the model

$$y_t = (\beta_0 + \beta_1 t) + \gamma_1 X_{1t} + \gamma_2 X_{2t} + \gamma_3 X_{3t} \tag{8.5}$$

The following SAS program reads the data and runs the regression model.

The fitted series is given by

$$\hat{y}_t = 5.6875 + 0.5563t + 15.6688X_{1,t} + 6.8625X_{2,t} - 2.9438X_{3,t}$$

from which $\hat{T}_t = 5.6875 + 0.5563t$ and $\hat{S}_t = 15.6688X_{1,t} + 6.8625X_{2,t} - 2.9438X_{3,t}$.

```
data brd;
input y x1 x2 x3 @@;
timw=_n_;
cards;
21 1 0 0 14 0 1 0 5 0 0 1 8 0 0 0
25 1 0 0 16 0 1 0 7 0 0 1 9 0 0 0
25 1 0 0 18 0 1 0 9 0 0 1 13 0 0 0
30 1 0 0 20 0 1 0 10 0 0 1 15 0 0 0
;
proc reg;
model y = time x1 x2 x3/dw;
output out=new(keep=time y x1 x2 x3 yhat yresid) p=yhat r=yresid;
run;

data new; set new; that=5.6875+0.5563*time; shat=15.6688*x1
+6.8625*x2-2.9438*x3;

proc print data=new; run;
```

The estimated components of the series are summarized in Table 8.9 and Figure 8.5 Visual inspection of the time series plot in Figure 8.5, of the observed and estimated data points under the assumed model shows that the estimation procedure is quite efficient.

**TABLE 8.9**

Estimated Trend and Seasonal Effect of the Series: BRD Data

| Year | Quarter | $t$ | $y_t$ | $\hat{T}_t$ | $S_t$ | $\hat{y}_t = T_t + S_t$ | $e_t = y_t - \hat{y}_t$ |
|------|---------|-----|-------|-------------|-------|-------------------------|-------------------------|
| 1 | 1 | 1 | 21 | 6.25 | 15.67 | 21.92 | −0.92 |
|   | 2 | 2 | 14 | 6.80 | 6.86 | 13.66 | 0.34 |
|   | 3 | 3 | 5 | 7.36 | −2.94 | 4.41 | 0.58 |
|   | 4 | 4 | 8 | 7.91 | 0.00 | 7.91 | 0.09 |
| 2 | 1 | 5 | 25 | 8.47 | 15.67 | 24.14 | 0.86 |
|   | 2 | 6 | 16 | 9.03 | 6.86 | 15.89 | 0.11 |
|   | 3 | 7 | 7 | 9.58 | −2.94 | 6.64 | 0.36 |
|   | 4 | 8 | 9 | 10.14 | 0.00 | 10.14 | −1.14 |
| 3 | 1 | 9 | 25 | 10.69 | 15.67 | 26.36 | −1.36 |
|   | 2 | 10 | 18 | 11.25 | 6.86 | 18.11 | −0.11 |
|   | 3 | 11 | 9 | 11.81 | −2.94 | 8.86 | 0.14 |
|   | 4 | 12 | 13 | 12.36 | 0.00 | 12.36 | 0.64 |
| 4 | 1 | 13 | 30 | 12.92 | 15.67 | 28.59 | 1.41 |
|   | 2 | 14 | 20 | 13.48 | 6.86 | 20.34 | −0.34 |
|   | 3 | 15 | 10 | 14.03 | −2.94 | 11.09 | −1.09 |
|   | 4 | 16 | 15 | 14.59 | 0.00 | 14.59 | 0.41 |

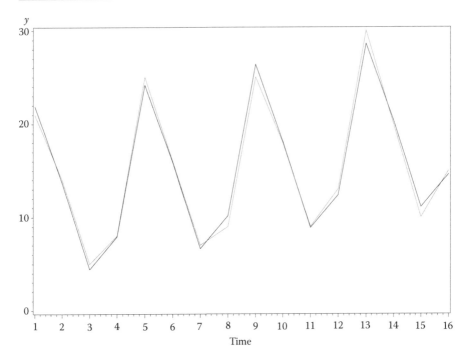

**FIGURE 8.5**

Observed and predicted values of $y$ from the additive model plotted over time.

### 8.2.3 Detection of Seasonality: Nonparametric Test

Seasonality may be tested using a test based on ranks. The test is a simple adaptation of the nonparametric analysis of variance procedure. After removing a linear trend, if desired, we rank the values within each year from 1 (smallest) to 12 (largest) for monthly data. In general, let the years represent $c$ columns and the months $r$ (=12) rows. Then each column represents a permutation of the integers 1, 2, ..., 12. Summing across each row gives the monthly score $M_j$, $j = 1, 2, ..., 12$. Under the null hypothesis $H_0$: no seasonal pattern, the test statistic

$$T = 12 \sum_{j=1}^{r} \left( M_j - \frac{c(r+1)}{2} \right)^2 / cr(r+1)$$

$$= 12 \left[ \sum_{j=1}^{r} M_j^2 / cr(r+1) + \frac{c(r+1)}{4} \sum_{j=1}^{r} M_j / r \right]$$

(8.6)

is approximately distributed as chi-square with $(r-1)$ degrees of freedom.

#### Example 8.3

The data from this example were kindly provided by Dr. J. Mallia of the Ontario Veterinary College. Cyanonsis is one of the leading causes of condemnation of poultry in Canada. To investigate seasonal patterns in the proportion of turkeys condemned, we use Equation 8.6. The data are summarized in Table 8.10 and plotted in Figure 8.6.

**TABLE 8.10**

Number of Turkeys Condemned (per 100,000) Because of Cyanosis

| Month | 1987 | 1988 | 1989 | 1990 | 1991 | 1992 | 1993 |
|-------|------|------|------|------|------|------|------|
| Jan | 643.0 | 1168.7 | 1173.7 | 1140.4 | 691.2 | 1154.4 | 556.7 |
| Feb | 508.6 | 1422.4 | 1492.3 | 1446.4 | 370.9 | 683.0 | 489.3 |
| Mar | 646.2 | 1748.4 | 1600.5 | 1002.7 | 454.3 | 535.6 | 466.2 |
| Apr | 849.1 | 1226.9 | 1141.0 | 999.5 | 393.9 | 351.6 | 448.9 |
| May | 710.2 | 1061.0 | 861.0 | 485.1 | 374.0 | 430.2 | 302.1 |
| June | 653.0 | 905.6 | 706.3 | 416.9 | 253.2 | 371.5 | 260.3 |
| July | 542.2 | 875.7 | 537.7 | 562.6 | 428.2 | 317.1 | 215.6 |
| Aug | 502.6 | 943.0 | 583.3 | 483.7 | 429.5 | 425.2 | 272.9 |
| Sept | 789.5 | 1228.2 | 810.8 | 490.4 | 393.7 | 332.5 | 286.0 |
| Oct | 409.5 | 1286.0 | 750.0 | 670.5 | 387.9 | 327.0 | 270.8 |
| Nov | 836.4 | 1434.8 | 1137.6 | 605.6 | 587.0 | 427.6 | 373.3 |
| Dec | 792.4 | 860.3 | 1178.7 | 618.5 | 618.5 | 381.8 | 259.6 |

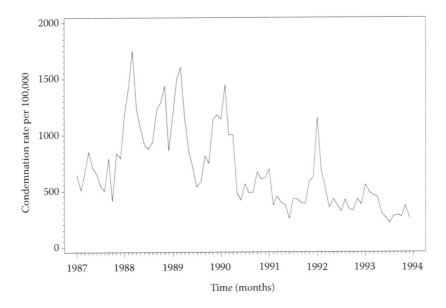

**FIGURE 8.6**
Time series plot of proportion of turkeys condemned.

In Table 8.11 we provide the ranks, $M_j$ and $M_j^2$.

$$\sum_{j=1}^{12} M_j = 546, \ \sum_{j=1}^{12} M_j^2 = 27,750, \ \text{and } T = 31.94$$

Since $\chi^2_{0.05,11} = 19.67$, the null hypothesis of no seasonal pattern is not supported by the data.

**TABLE 8.11**

Ranked Values

| Month | 1987 | 1988 | 1989 | 1990 | 1991 | 1992 | 1993 | $M_j$ | $M_j^2$ |
|---|---|---|---|---|---|---|---|---|---|
| 1 | 1 | 6 | 9 | 11 | 12 | 12 | 12 | 63 | 3969 |
| 2 | 4 | 10 | 11 | 12 | 2 | 11 | 11 | 61 | 3721 |
| 3 | 6 | 12 | 12 | 10 | 9 | 10 | 10 | 69 | 4761 |
| 4 | 12 | 7 | 8 | 9 | 6 | 4 | 9 | 55 | 3025 |
| 5 | 8 | 5 | 6 | 3 | 3 | 9 | 7 | 41 | 1681 |
| 6 | 7 | 3 | 3 | 1 | 1 | 5 | 3 | 23 | 529 |
| 7 | 5 | 2 | 1 | 5 | 7 | 1 | 1 | 22 | 484 |
| 8 | 3 | 4 | 2 | 2 | 8 | 7 | 5 | 31 | 961 |
| 9 | 9 | 8 | 5 | 4 | 5 | 3 | 6 | 40 | 1600 |
| 10 | 2 | 9 | 4 | 8 | 4 | 2 | 4 | 33 | 1089 |
| 11 | 11 | 11 | 7 | 6 | 10 | 8 | 8 | 61 | 3721 |
| 12 | 10 | 1 | 10 | 7 | 11 | 6 | 2 | 47 | 2209 |

The following SAS code is used to read the data in Table 8.10 and produce Figure 8.6:

```
data condemn;
input month:monyy5. rate @@;
rate=rate*100000;
t=_n_; year=year(month);
cards;
jan87 .006430 feb87 .005086 mar87 .006462 apr87 .008491 may87
.007102
jun87 .006530 jul87 .005422 aug87 .005026 sep87 .007895 oct87
.004095
nov87 .008364 dec87 .007924 jan88 .011687 feb88 .014224 mar88
.017484
apr88 .012269 may88 .010610 jun88 .009056 jul88 .008757 aug88
.009430
sep88 .012282 oct88 .012860 nov88 .014348 dec88 .008603 jan89
.011737
feb89 .014923 mar89 .016005 apr89 .011410 may89 .008610 jun89
.007063
jul89 .005377 aug89 .005833 sep89 .008108 oct89 .007500 nov89
.011376
dec89 .011787 jan90 .011404 feb90 .014464 mar90 .010027 apr90
.009995
may90 .004851 jun90 .004169 jul90 .005626 aug90 .004837 sep90
.004904
oct90 .006705 nov90 .006056 dec90 .006185 jan91 .006912 feb91
.003709
mar91 .004543 apr91 .003939 may91 .003740 jun91 .002532 jul91
.004282
aug91 .004295 sep91 .003937 oct91 .003879 nov91 .005870 dec91
.006185
jan92 .011544 feb92 .006830 mar92 .005356 apr92 .003516 may92
.004302
jun92 .003715 jul92 .003171 aug92 .004252 sep92 .003325 oct92
.003270
nov92 .004276 dec92 .003818 jan93 .005567 feb93 .004893 mar93
.004662
apr93 .004489 may93 .003021 jun93 .002603 jul93 .002156 aug93
.002729
sep93 .002860 oct93 .002708 nov93 .003733 dec93 .002596
;

ODS GRAPHICS ON;
ODS RTF;
goptions reset=global gunit=pct cback=white htitle=6 htext=3
ftext=swissb colors=(black);

*Figure 8.6: Time series plot of proportion of turkeys condemned;
axis1 order=('01jan87'd to '01jan94'd by year) label=(angle=0
'Time (Months)') offset=(3) minor=(number=11);
axis2 order=(0 to 2000 by 500) label=(angle=90 'Condemnation rate
per 100,000');

proc gplot data=condemn;
 plot rate*month / haxis=axis1 vaxis=axis2;
 symbol1 i=join;
```

```
 format month year4.;
run; quit;
ODS GRAPHICS OFF;
ODS RTF CLOSE;
```

The R-code that produces the same time series plot:

```
library(tseries)
condemn=read.csv(file.choose())
names(condemn)
tseriescondemn=ts(rate,frequency=12,star=c(87,1))
plot.ts(tseriescondemn)
# To decomposing the seasonal data into its components, we use the
R-statement.
condemn_components= decompose(tseriescondemn)
```

Note that the output is quite long and is omitted. Plotting will be most informative:

```
plot(condemn_components)
```

The components are plotted in Figure 8.7.

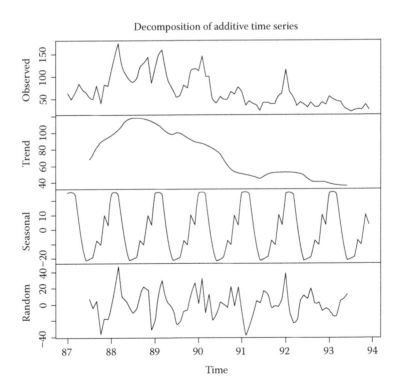

**FIGURE 8.7**
Decomposing the condemnation series into its components.

### 8.2.4 Autoregressive Errors: Detection and Estimation

One of the main characteristics of a true series is that adjacent observations
are likely to be correlated. One way to detect such correlation is to plot the
residuals $e_t = y_t - \hat{y}_t$ against time. This is illustrated using the additive model,
where the residuals (from Table 8.9) are plotted in time order as Figure 8.8.

From the plot one can see that the residuals have the signs −, +, +, +, +, +, +,
−, −, −, +, +, +, −, −, +. This shows a tendency for residuals to be followed by
residuals of the same sign, an indication of possible autocorrelation. Usually,
the Durbin-Watson statistic (1951) given by

$$d = \frac{\sum_{t=2}^{n}(e_t - e_{t-1})^2}{\sum_{t=1}^{n} e_t^2} \tag{8.7}$$

is used to test for the significance of this correlation. The sample autocorrelation is given in the SAS output and is shown in Figure 8.12. The DW option
in the model statement of PROC REG would compute the Durbin-Watson D
statistic to test that the autocorrelation is zero. The relevant output is shown
below.

| Durbin-Watson D | 1.489 |
|---|---|
| Number of Observations | 16 |
| 1st Order Autocorrelation | 0.201 |

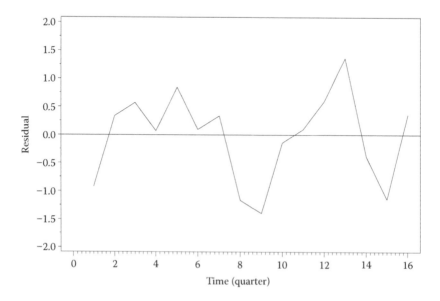

**FIGURE 8.8**
Plot of residuals over time.

When autocorrelation is present, ignoring its effect would produce unrealistically small standard errors for the regression estimates in the fitted model 8.5. Therefore one has to account for the effect of this correlation to produce accurate estimates of the standard error. Our approach to modeling this autocorrelation at present will still be at a descriptive level. More rigorous treatment of the autocorrelation structure will be presented in the next section.

The simplest autocorrelation structure that we shall examine here is called the first-order *autoregression process*. This model assumes that successive errors are linearly related through the relationship:

$$\varepsilon_t = \rho \varepsilon_{t-1} + u_t \tag{8.8}$$

It is assumed that $\{u_t \; ; \; t = 1, 2, \ldots, n\}$ are independent and identically distributed $N(0,\sigma^2)$. Under the above specifications we have the model

$$y_t = \beta_0 + \beta_1 X_{1t} + \beta_2 X_{2t} + \ldots + \beta_k X_{kt} + \varepsilon_t \tag{8.9}$$

where $\varepsilon_t = \rho \varepsilon_{t-1} + u_t$ and $\rho = \text{corr}(\varepsilon_t, \varepsilon_{t-1})$.
Note that

$$\rho y_{t-1} = \rho \beta_0 + \rho \beta_1 X_{1\,t-1} + \rho \beta_2 X_{2\,t-1} + \ldots + \rho \beta_k X_{k\,t-1} + \varepsilon_{t-1} \tag{8.10}$$

Subtracting Equation 8.10 from Equation 8.9 we have:

$$y_t - \rho y_{t-1} = \beta_0(1 - \rho) + \beta_1(X_{1t} - \rho X_{1\,t-1}) + \beta_2(X_{2t} - \rho X_{2\,t-1}) + \ldots$$
$$+ \beta_k(X_{kt} - \rho X_{k\,t-1}) + u_t \tag{8.11}$$

The preceding equation has $u_t$ as an error term that satisfies the standard assumptions of inference in a linear regression model. The problem now is that the left-hand side of Equation 8.11 has a transformed response variable that depends on the unknown parameter $\rho$. A commonly used procedure to estimate the model parameters is to use a procedure known as the Cochran and Orcutt procedure, which is outlined in the following four steps:

1. Estimate the parameters of model 8.9 using least squares or PROC REG from SAS, and compute the residuals $e_1, e_2, \ldots e_n$.

2. From $e_1, e_2, \ldots, e_n$, evaluate the moment estimator of $\rho$ as

$$\hat{\rho} = \frac{\displaystyle\sum_{t=2}^{n} e_t e_{t-1}}{\displaystyle\sum_{t=1}^{n} e_{t-2}^2} \tag{8.12}$$

3. Substitute $\hat{\rho}$ in place of $\rho$ in model 8.11, which has an error term that satisfies the standard assumptions and compute the revised least square estimates.

4. From the least square estimates obtained in step 3, compute the revised residuals and return to step 2; find an updated estimate of $\rho$ using Equation 8.12. We iterate between step 2 and step 4 until the least squares estimate has an insignificant change between successive iterations.

### 8.2.5 Modeling Seasonality and Trend Using Polynomial and Trigonometric Functions

It is desirable, in many applications of time series models, to estimate both the trend and seasonal components in the series. This can be done quite effectively by expressing the series $y_t$ as a function of polynomials in $t$ and a combination of sine and cosine functions.

Therefore, a suggested additive model is given by

$$y_t = Q(t) + F(t) + e_t$$

where

$$Q(t) = \sum_{j=0}^{p} \beta_j t^j$$

models the trend component and

$$F(t) = \sum_{j=1}^{q} \left[ a_j \sin(2\pi j t/L) + b_j \cos(2\pi j t/L) \right]$$

models the seasonal components, and $L$ is the number of seasons in a year. Thus $L = 4$ for quarterly data and $L = 12$ for monthly data. We may fit one of the following models that may be suitable for modeling additive seasonal variation:

Model 1: $p = q = 1$

$$y_t = \beta_0 + \beta_1 t + a_1 \sin(2\pi t/L) + b_1 \cos(2\pi t/L)$$

Model 2: $p = 1$, $q = 2$

$$y_t = \beta_0 + \beta_1 t + a_1 \sin(2\pi t/L) + b_1 \cos(2\pi t/L) + a_2 \sin(4\pi t/L) + b_2 \cos(4\pi t/L)$$

Multiplicative time series may be modeled by extending either model 1 or model 2. For model 1, a time series with multiplicative seasonal variation becomes

$$y_t' = y_t + c_1 t \sin(2\pi t/L) + c_2 t \cos(2\pi t/L)$$

whereas model 2 becomes

$$y_t'' = y_t' + d_1 t \sin(4\pi t/L) + d_2 t \cos(4\pi t/L)$$

**Example 8.3 (Continued)**

The condemnation rate series showed significant seasonal effect. Shown next is how to model both seasonality and trend in the series. Several models were fitted using SAS PROC REG. The best model is found to be

$$\log(\text{rate}) = y_t = \beta_0 + \beta_1 t + \beta_2 t^2 + \beta_3 t^3 + a_1 \sin(2\pi t/12) + b_1 \cos(2\pi t/12)$$

The estimated coefficients are

$$\hat{\beta}_0 = 6.193 \ (.118), \qquad \hat{\beta}_1 = .072 \ (.012), \qquad \hat{\beta}_2 = -.002 \ (.0003)$$

$$\hat{\beta}_3 = 14 \times 10^{-6} (2.5 \times 10^{-6}), \qquad \hat{a}_1 = .199(.040), \qquad \hat{b}_1 = .198(.039)$$

$$R^2 = 0.769$$

and the root mean square error is .255.

Note the PROC REG does not account for the correlation in the series. To account for such correlation, PROC GENMOD was used. The generalized estimating equation (GEE) approach gave similar coefficient estimates with empirical standard errors that are robust against misspecification of the correlation structure (assume AR(1)).

The following SAS program will use the data condemn, generate the additional variables, run the regression and GEE models, and produce

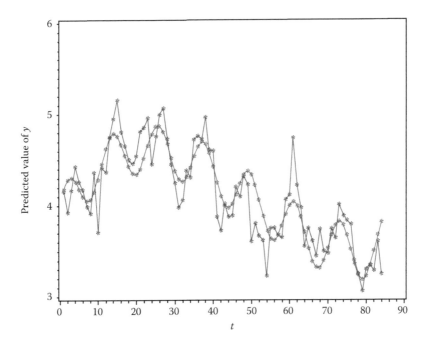

**FIGURE 8.9**
Plot of the rate series with predicted series from the regression with polynomial and trigonometric components.

Figure 8.9. As seen in Figure 8.9, the postulated model fits the observed data quite closely.

```
data condemn; set condemn;
y=log(rate); t2=t**2; t3=t**3;
sin=sin((2*t*22)/(7*12)); cos=cos((2*t*22)/(7*12));

proc reg data=condemn noprint;
 model y = t t2 t3 sin cos;
 output out=new(keep=t month y yhat) p=yhat; run;

proc genmod data=condemn;
 class year;
 model y = t t2 t3 sin cos / dist=n link=id dscale;
 repeated subject=year / type=ar(1);
 run; quit;

axis2 order=(5 to 8 by 1) label=(angle=90 'Log condemnation rate
per 100,000');
*Fig 5.7: Plot of the rate series with predicted series from
regression with polynomial and trigonometric components;
proc gplot data=new;
symbol1 color=red i=join;
symbol2 color=blue i=join;
plot(yhat y ) * month / overlay haxis=axis1 vaxis=axis2;
format month year4.;
run; quit;
```

## 8.3 Fundamental Concepts in the Analysis of Time Series

In order to establish a proper understanding of time series models, we introduce some of the necessary fundamental concepts. Such concepts include a simple introduction to stochastic processes, and autocorrelation and partial autocorrelation functions.

### 8.3.1 Stochastic Processes

As before, $y_t$ denotes an observation made at time $t$. It is assumed that for each time point $t$, $y_t$ is a random variable and hence its behavior can be described by some probability distribution. We need to emphasize an important feature of time series models which is that observations made at adjacent time points are statistically correlated. Our main objective is to investigate the nature of this correlation. Therefore, for two time points $t$ and $s$, the joint behavior of $(y_t, y_s)$ is determined from their bivariate distribution. This is generalized to the collection of observations $(y_1, y_2, ..., y_n)$ where their behavior is described by their multivariate joint distribution.

A stochastic process is a sequence of random variables $\{..., y_{-2}, y_{-1}, y_0, y_1, y_2, ...\}$. We shall denote this sequence by $\{y_t: t = 0, \pm1, \pm2, ...\}$. For a given real-valued process we define the following:

Mean function of the process: $\mu_t = E(y_t)$
Variance function of the process: $\sigma_t^2 = E(y_t - \mu_t)^2$
Covariance function between $y_t$ and $y_s$: $\gamma(t,s) = E[(y_t - \mu_t)(y_s - \mu_s)]$
Correlation function between $y_t$ and $y_s$: $\rho(t,s) = \dfrac{\gamma(t,s)}{\sqrt{\sigma_t^2 \sigma_s^2}} = \dfrac{\gamma(t,s)}{\sqrt{\gamma(t,t)\gamma(s,s)}}$
From this definition it is easily verified that

$$\rho(t,t) = 1$$

$$\rho(t,s) = \rho(s,t)$$

$$|\rho(t,s)| \leq 1$$

Values of $\rho(t,s)$ near $\pm 1$ indicate strong dependence, whereas values near zero indicate weak linear dependence.

### 8.3.2 Stationary Series

The notion of stationarity is quite important in order to make statistical inferences about the structure of the time series. The fundamental idea of stationarity is that the probability distribution of the process does not change with time. Here we introduce two types of stationarity. The first is "strict" or "strong" stationarity and the other is "weak" stationarity.

The stochastic process $y_t$ is said to be strongly stationary if the joint distribution of $y_{t_1}, ..., y_{t_n}$ is the same as the joint distribution of $y_{t_{1-k}}, ..., y_{t_{n-k}}$ for all the choices of the points $t_1, ..., t_n$ and all the time lags $k$. To illustrate this concept we examine two cases: $n = 1$ and $n = 2$. For $n = 1$, the stochastic process $y_t$ is strongly stationary if the distribution of $y_t$ is the same as that of $y_{t-k}$ for any $k$. This implies

$$E(y_t) = E(y_{t-k})$$

and

$$V(y_t) = V(y_{t-k})$$

are constant or independent of $t$. For $n = 2$, the process is strongly stationary if the bivariate distribution of $(y_t, y_s)$ is the same as the bivariate distribution of $(y_{t-k}, y_{s-k})$, from which we have

$$\gamma(t,s) = Cov(y_t, y_s) = Cov(y_{t-k}, y_{s-k})$$

Setting $k = s$, we obtain

$$\gamma(t,s) = Cov(y_{t-s}, y_0)$$
$$= \gamma(0, |t - s|)$$

hence

$$\gamma(t, t - k) = \gamma_k$$

and

$$\rho(t, t - k) = \rho_k$$

A process is said to be weakly stationary if

- $\mu_t = \mu$ for all $t$
- $\gamma(t, t\text{-}k) = \gamma(0, k)$ for all $t$ and $k$

All the series that will be considered in this chapter are stationary unless otherwise specified.

### 8.3.3 Autocovariance and Autocorrelation Functions

For a stationary time series $\{y_t\}$, we have already mentioned that $E(y_t) = \mu$, and $V(y_t) = E(y_t - \mu)^2$ (which are constant) and $Cov(y_t, y_s)$ is a function of the time difference $|t - s|$. Hence we can write

$$Cov(y_t, y_{t+k}) = E[(y_t - \mu)(y_{t+k} - \mu)] = \gamma_k$$

and

$$\rho_k = Corr(y_t, y_{t+k}) = \frac{Cov(y_t, y_{t+k})}{\sqrt{V(y_t)V(y_{t+k})}} = \frac{\gamma_k}{\gamma_0}$$

The functions $\gamma_k$ and $\rho_k$ are called the autocovariance and autocorrelation functions, respectively. Since the values of $\mu$, $\gamma_k$, and $\rho_k$ are unknown, the moment estimators of these parameters are as follows:

1. $\bar{y} = \frac{1}{n} \sum_{i=1}^{n} y_i$ is the sample mean estimator of $\mu$. It is unbiased and has variance given by

$$V(\bar{y}) = \frac{1}{n^2} \sum_{t=1}^{n} \sum_{s=1}^{n} Cov(y_t, y_s)$$

From the strong stationarity assumption,

$$Cov(y_t, y_s) = \gamma(t - s)$$

Hence, letting $k = t - s$

$$V(\bar{y}) = \frac{1}{n^2} \sum_{t=1}^{n} \sum_{s=1}^{n} \gamma(t - s)$$

$$= \frac{\gamma_0}{n} \sum_{k=-(n-1)}^{n-1} \left(1 - \frac{|k|}{n}\right) \rho_k$$

$$= \frac{\gamma_0}{n} \left[1 + 2 \sum_{k=1}^{n-1} \left(1 - \frac{k}{n}\right) \rho_k\right]$$

When $\rho_k = 0$ for $k = 2, 3, \ldots, n - 1$ then, for large $n$

$$V(\bar{y}) \cong \frac{\gamma_0}{n}\left[1 + 2\left(\frac{n-1}{n}\right)\rho_1\right] \cong \frac{\gamma_0}{n}[1 + 2\rho_1]$$

2. $\hat{\gamma}_k = \frac{1}{n}\sum_{t=1}^{n-k}(y_t - \bar{y})(y_{t+k} - \bar{y})$

  is the moment estimate of the autocovariance function.

A natural moment estimator for the autocorrelation function is defined as

$$\hat{\rho}_k = \frac{\hat{\gamma}_k}{\hat{\gamma}_0} = \frac{\sum_{t=1}^{n-k}(y_t - \bar{y})(y_{t+k} - \bar{y})}{\sum_{t=1}^{n}(y_t - \bar{y})^2}, \qquad k = 0, 1, 2, \ldots \qquad (8.13)$$

A plot of $\hat{\rho}_k$ versus $k$ is sometimes called a sample correlogram. Note that $\hat{\rho}_k = \hat{\rho}_{-k}$, which means that the sample autocorrelation function is symmetric around $k = 0$.

For a stationary Gaussian process, Bartlett (1946) showed that for $k > 0$ and $k + j > 0$,

$$Cov(\hat{\rho}_k, \hat{\rho}_{k+j}) \cong \frac{1}{n}\sum_{t=-\infty}^{\infty}\left(\rho_t\rho_{t+j} + \rho_{t+j+k}\rho_{t-k} - 2\rho_k\rho_t\rho_{t-k-j} - 2\rho_{k+j}\rho_t\rho_{t-k} + 2\rho_k\rho_{k+j}\rho_t^2\right)$$

$$(8.14)$$

For large $n$, $\hat{\rho}_k$ is approximately normally distributed with mean $\rho_k$ and variance

$$V(\hat{\rho}_k) \cong \frac{1}{n}\sum_{t=-\infty}^{\infty}(\rho_t^2 + \rho_{t+k}\rho_{t-k} - 4\rho_k\rho_t\rho_{t-k} + 2\rho_k^2\rho_t^2)$$

For processes with $\rho_k = 0$ for $k > l$, Bartlett's approximation becomes

$$V(\hat{\rho}_k) \cong \frac{1}{n}\left(1 + 2\rho_1^2 + 2\rho_2^2 + \ldots + 2\rho_l^2\right) \qquad (8.15)$$

In practice $\rho_i$ ($i = 1, 2, \ldots, l$) are unknown and are replaced by their sample estimates $\hat{\rho}_i$; the large sample variance of $\hat{\rho}_k$ is approximated by replacing $\rho_i$ with $\hat{\rho}_i$ in Equation 8.15.

## 8.4 Models for Stationary Time Series

In this section we consider models based on an observation made by Yule (1927) that time series in which successive values are autocorrelated can be modeled as a linear combination (or linear filter) of a sequence of uncorrelated random variables. Suppose that $\{a_t;\ t = 0, \pm1, \pm2, \dots\}$ are a sequence of identically distributed uncorrelated random variables with $E(a_t) = 0$ and $V(a_t) = \sigma^2$, and $cov(a_t, a_{t-k}) = 0$ for all $k \neq 0$. Such a sequence is commonly known as a *white noise*. With this definition of white noise, we introduce the linear filter representation of the process $y_t$.

A general linear process $y_t$ is one that can be presented as

$$y_t = a_t + \psi_1 a_{t-1} + \psi_2 a_{t-2} + \dots$$

$$= \sum_{j=0}^{\infty} \psi_j a_{t-j}, \quad \psi_0 = 1$$

For the infinite series of the right-hand side of the preceding equation to be meaningful, it is assumed that

$$\sum_{j=1}^{\infty} \psi_j^2 < \infty$$

### 8.4.1 Autoregressive Processes

As their name implies, autoregressive processes are regressions on themselves. To be more specific, the $p$th order autoregressive process $y_t$ satisfies,

$$y_t = \phi_1 y_{t-1} + \phi_2 y_{t-2} + \dots + \phi_p y_{t-p} + a_t \tag{8.16}$$

In this model, the present value $y_t$ is a linear combination of its $p$ most recent values plus an "innovation" term $a_t$, which includes everything in the series at time $t$ that is not explained by the past values. It is also assumed that $a_t$ is independent of $y_{t-1}, y_{t-2}, \dots$.

Before we examine the general autoregressive process, we first consider the first-order autoregressive model, which is denoted by AR(1).

### 8.4.2 The AR(1) Model

Let $y_t$ be a stationary series such that

$$y_t = \phi y_{t-1} + a_t \tag{8.17}$$

Most textbooks write this model as

$$y_t - \mu = \phi(y_{t-1} - \mu) + a_t$$

where $\mu$ is the mean of the series. However, we shall use Equation 8.17 assuming that the mean has been subtracted from the series. The requirement $|\phi| < 1$ is a necessary and sufficient condition for stationarity.

From Equation 8.17, $V(y_t) = \phi^2 V(y_{t-1}) + V(a_t)$ or $\gamma_0 = \phi^2 \gamma_0 + \sigma_a^2$ from which

$$\gamma_0 = \frac{\sigma_a^2}{1 - \phi^2} \tag{8.18}$$

Multiplying both sides of Equation 8.17 by $y_{t-k}$ and taking the expectation, the result is

$$E(y_t y_{t-k}) = \phi E(y_{t-1} y_{t-k}) + E(y_{t-k} a_t)$$

By the stationarity of the series and the independence of $y_{t-1}$ and $a_t$,

$$\gamma_k = \phi \gamma_{k-1}, \qquad k = 1, 2, \ldots$$

For $k = 1$,

$$\gamma_1 = \phi \gamma_0 = \phi \frac{\sigma_a^2}{1 - \phi^2}$$

For $k = 2$,

$$\gamma_2 = \phi \gamma_1 = \phi \left( \phi \frac{\sigma_a^2}{1 - \phi^2} \right) = \phi^2 \frac{\sigma_a^2}{1 - \phi^2} = \phi^2 \gamma_0$$

By mathematical induction one can show that

$$\gamma_k = \phi^k \gamma_0$$

or

$$\rho_k = \frac{\gamma_k}{\gamma_0} = \phi^k \tag{8.19}$$

Note that since $|\phi| < 1$, the autocorrelation function is exponentially decreasing in $k$. For $0 < \phi < 1$, all $\rho_k$ are positive (Figure 8.10). For $-1 < \phi < 0$, $\rho_1 < 0$ and the sign of successive autocorrelations, as can be in Figures 8.10 and 8.11, alternates (positive if $k$ is even and negative if $k$ is odd) (Figure 8.11).

### 8.4.3 AR(2) Model (Yule's Process)

The second-order autoregressive process AR(2) is a stationary series $y_t$ that is a linear combination of the two preceding observations and can be written as

$$y_t = \phi_1 y_{t-1} + \phi_2 y_{t-2} + a_t \tag{8.20}$$

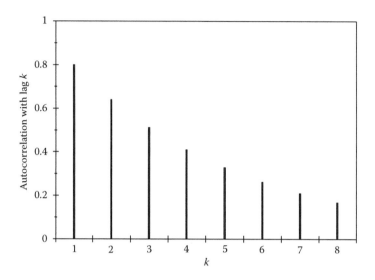

**FIGURE 8.10**
Autocorrelation plot for a $\phi$ of 0.8 and $k = 1, 2, ..., 8$.

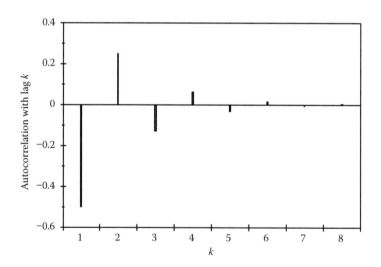

**FIGURE 8.11**
Autocorrelation plot for a $\phi$ of –0.5 and $k = 1, 2, ..., 8$.

To ensure stationarity, the coefficients $\phi_1$ and $\phi_2$ must satisfy

$$\phi_1 + \phi_2 < 1$$
$$\phi_2 - \phi_1 < 1 \quad \text{and}$$
$$-1 < \phi_2 < 1$$

These conditions are called the stationarity conditions for the AR(2) model.

To derive the autocorrelation function for the AR(2), we multiply both sides of Equation 8.20 by $y_{t-k}$ ($k = 1, 2, \ldots$) and take the expectations. Under the assumptions of independence of $y_t$ and $a_t$, and the stationarity of the series we have

$$E(y_t y_{t-k}) = \phi_1 E(y_{t-1} y_{t-k}) + \phi_2 E(y_{t-2} y_{t-k}) + E(a_t y_{t-k})$$

from which

$$\gamma_k = \phi_1 \gamma_{k-1} + \phi_2 \gamma_{k-2} \tag{8.21}$$

On dividing by $\gamma_0$ we get

$$\rho_k = \phi_1 \rho_{k-1} + \phi_2 \rho_{k-2} \tag{8.22}$$

Equation 8.22 is called the Yule-Walker equation. For $k = 1$

$$\rho_1 = \phi_1 \rho_0 + \phi_2 \rho_{-1}$$

Since $\rho_0 = 1$ and $\rho_{-1} = \rho_1$ we have

$$\rho_1 = \frac{\phi_1}{1 - \phi_2}$$

For $k = 2$

$$\rho_2 = \phi_1 \rho_1 + \phi_2$$

or

$$\rho_2 = \phi_2 + \frac{\phi_1^2}{1 - \phi_2}$$

Note also that the variance of the AR(2) process can be written in terms of the model parameters. In fact, from Equation 8.20 we have

$$V(y_t) = \phi_1^2 V(y_{t-1}) + \phi_2^2 V(y_{t-2}) + 2\phi_1 \phi_2 Cov(y_{t-1}, y_{t-2}) + \sigma_a^2$$

or

$$\gamma_0 = \phi_1^2 \gamma_0 + \phi_2^2 \gamma_0 + 2\phi_1 \phi_2 \gamma_1 + \sigma_a^2 \tag{8.23}$$

Setting $k = 1$ in Equation 8.21 we have

$$\gamma_1 = \phi_1 \gamma_0 + \phi_2 \gamma_{-1}$$
$$= \phi_1 \gamma_0 + \phi_2 \gamma_1$$

which gives

$$\gamma_1 = \phi_1 \frac{\gamma_0}{1 - \phi_2} \tag{8.24}$$

Substituting in Equation 8.23,

$$\gamma_0 = (\phi_1^2 + \phi_2^2)\gamma_0 + 2\phi_1^2\phi_2 \frac{\gamma_0}{1 - \phi_2} + \sigma_a^2$$

Hence

$$\gamma_0 = \frac{\sigma_a^2(1 - \phi_2)}{(1 - \phi_2)(1 - \phi_1^2 - \phi_2^2) - 2\phi_1^2\phi_2} \tag{8.25}$$

It should be noted that for $-1 < \phi_2 < 0$, the AR(2) process tends to exhibit sinusoidal behavior, regardless of the value of $\phi_1$. When $0 < \phi_2 < 1$, the behavior of the process will depend on the sign of $\phi_1$. For $\phi_1 < 0$, the AR(2) process tends to oscillate and the series shows ups and downs.

### 8.4.4 Moving Average Processes

Another type of stochastic model that belongs to the class of linear filter models is called a moving average process. This is given as

$$y_t = a_t - \theta_1 a_{t-1} - \theta_2 a_{t-2} - .. - \theta_q a_{t-q}$$

This series is called a *moving average* of order $q$ and is denoted by MA($q$).

### 8.4.5 First-Order Moving Average Process MA(1)

Here we have

$$y_t = a_t - \theta a_{t-1} \tag{8.26}$$

$$E(y_t) = 0$$

$$\gamma_0 = V(y_t) = \sigma_a^2 + \theta^2 \sigma_a^2 = \sigma_a^2(1 + \theta^2)$$

Moreover,

$$Cov(y_t, y_{t-1}) = E(y_t y_{t-1})$$
$$= E[(a_t - \theta a_{t-1})(a_{t-1} - \theta a_{t-2})]$$
$$= E(a_t a_{t-1}) - \theta \left[ E(a_{t-1}^2) + E(a_t a_{t-2}) \right] + \theta^2 E(a_{t-1} a_{t-2})$$

Since $a_1, a_2, \ldots$ are independent with $E(a_t) = 0$ for all $t$, then

$$\gamma_1 = Cov(y_t, y_{t-1}) = -\theta \sigma_a^2$$

$$Cov(y_t, y_{t-k}) = 0, \qquad k = 2, 3, \ldots$$

Furthermore, the autocorrelation function is

$$\rho_1 = \frac{\gamma_1}{\gamma_0} = -\frac{\theta}{1 + \theta^2}, \quad (\rho_k = 0 \text{ for } k = 2, 3, \ldots) \tag{8.27}$$

Note that if $\theta$ is replaced by $1/\theta$ in Equation 8.27, we get exactly the same autocorrelation function. This lack of uniqueness of MA(1) models must be rectified before we estimate the model parameters.

Rewriting Equation 8.26 as

$$a_t = y_t + \theta a_{t-1}$$
$$= y_t + \theta(y_{t-1} + \theta a_{t-2})$$
$$= y_t + \theta y_{t-1} + \theta^2 a_{t-2}$$

and continuing this substitution,

$$a_t = y_t + \theta y_{t-1} + \theta^2 y_{t-2} + \ldots$$

or

$$y_t = -(\theta y_{t-1} + \theta^2 y_{t-2} + \ldots) + a_t \tag{8.28}$$

If $|\theta| < 1$, we see that the MA(1) model can be inverted into an infinite-order AR process. It can be shown (see Box and Jenkins, 1970) that there is only one invertible MA(1) model with the given autocorrelation function $\rho_1$.

### 8.4.6 Second-Order Moving Average Process MA(2)

An MA (2) process is defined by

$$y_t = a_t - \theta_1 a_{t-1} - \theta_2 a_{t-2} \tag{8.29}$$

The autocovariance functions are given by

$$\gamma_1 = Cov(y_t, y_{t-1})$$
$$= E[(a_t - \theta_1 a_{t-1} - \theta_2 a_{t-2})(a_{t-1} - \theta_1 a_{t-2} - \theta_2 a_{t-3})]$$
$$= -\theta_1 \sigma_a^2 + \theta_1 \theta_2 \sigma_a^2$$
$$= (-\theta_1 + \theta_1 \theta_2)\sigma_a^2$$

$$\gamma_2 = Cov(y_t, y_{t-2})$$
$$= E[(a_t - \theta_1 a_{t-1} - \theta_2 a_{t-2})(a_{t-2} - \theta_1 a_{t-3} - \theta_2 a_{t-4})]$$
$$= -\theta_2 \sigma_a^2$$

and

$$\gamma_0 = V(y_t)$$
$$= \sigma_a^2 + \theta_1^2 \sigma_a^2 + \theta_2^2 \sigma_a^2$$
$$= (1 + \theta_1^2 + \theta_2^2)\sigma_a^2$$

Therefore, for an MA(2) process:

$$\rho_1 = \frac{\gamma_1}{\gamma_0} = \frac{-\theta_1 + \theta_1 \theta_2}{1 + \theta_1^2 + \theta_2^2} \tag{8.30}$$

$$\rho_2 = \frac{\gamma_2}{\gamma_0} = -\frac{\theta_2}{1 + \theta_1^2 + \theta_2^2} \tag{8.31}$$

$$\rho_k = 0, \qquad k = 3, 4, \ldots$$

### 8.4.7 Mixed Autoregressive Moving Average Processes

In modeling time series we are interested in constructing a parsimonious model. One type of such a model is obtained from mixing an AR($p$) with an MA($q$). The general form of this is given by

$$y_t = (\phi_1 y_{t-1} + \phi_2 y_{t-2} + \ldots + \phi_p y_{t-p}) + (a_t - \theta_1 a_{t-1} - \theta_2 a_{t-2} - \ldots - \theta_q a_{t-q}) \tag{8.32}$$

The process $y_t$ defined in Equation 8.32 is called the mixed autoregressive moving average process of orders $p$ and $q$, or ARMA($p, q$).

An important special case of the ARMA($p, q$) is ARMA(1,1) which can be obtained from Equation 8.32 for $p = q = 1$. Therefore an ARMA(1,1) is

$$y_t = \phi y_{t-1} + a_t - \theta a_{t-1} \tag{8.33}$$

For stationarity we assume that $|\phi|<1$ and for invertibility we require that $|\theta|<1$. When $\phi = 0$, Equation 8.33 is reduced to an MA(1) process, and when $\theta = 0$, it is reduced to an AR(1) process. Thus the AR(1) and MA(1) may be regarded as special processes of the ARMA(1, 1).

To obtain the autocovariance for the ARMA(1,1), we multiply both sides of Equation 8.33 by $y_{t-k}$ and take the expectations

$$E(y_t y_{t-k}) = \phi E(y_{t-1} y_{t-k}) + E(a_t y_{t-k}) - \theta E(a_{t-1} y_{t-k})$$

from which

$$\gamma_k = \phi \gamma_{k-1} + E(a_t y_{t-k}) - \theta E(a_{t-1} y_{t-k}) \tag{8.34}$$

For $k = 0$

$$\gamma_0 = \phi \gamma_1 + E(a_t y_t) - \theta E(a_{t-1} y_t) \tag{8.35}$$

and

$$E(a_t y_t) = \sigma_a^2$$

Noting that

$$E(a_{t-1} y_t) = \phi E(a_{t-1} y_{t-1}) + E(a_t a_{t-1}) - \theta E(a_{t-1}^2)$$
$$= \phi \sigma_a^2 - \theta \sigma_a^2 = (\phi - \theta)\sigma_a^2$$

and substituting in Equation 8.35 we see that

$$\gamma_0 = \phi \gamma_1 + \sigma_a^2 - \theta(\phi - \theta)\sigma_a^2 \tag{8.36}$$

and from Equation 8.34 when $k = 1$,

$$\gamma_1 = \phi \gamma_0 + E(a_t y_{t-1}) - \theta E(a_{t-1} y_{t-1}) \tag{8.37}$$
$$= \phi \gamma_0 + E(a_t y_{t-1}) - \theta \sigma_a^2$$

But,

$$E(a_t y_t) = \phi E(a_t y_{t-1}) + E(a_t^2) - \theta E(a_t a_{t-1})$$

or

$$a_a^2 = \phi E(a_t y_{t-1}) + a_a^2$$

so,

$$E(a_t y_{t-1}) = 0 \tag{8.38}$$

Substituting Equation 8.38 in Equation 8.37,

$$\gamma_1 = \phi \gamma_0 - \theta \sigma_a^2 \tag{8.39}$$

and using this in Equation 8.36 we have

$$\gamma_0 = \phi^2 \gamma_0 - \phi\theta\sigma_a^2 + \sigma_a^2 - \phi\theta\sigma_a^2 + \theta^2\sigma_a^2$$

from which

$$\gamma_0 = \frac{1 + \theta^2 - 2\phi\theta}{1 - \phi^2}\sigma_a^2$$

Thus,

$$\gamma_1 = \phi \left[\frac{1 + \theta^2 - 2\phi\theta}{1 - \phi^2}\right]\sigma_a^2 - \theta\sigma_a^2$$

$$= \frac{(\phi - \theta)(1 - \phi\theta)}{1 - \phi^2}\sigma_a^2$$

From Equation 8.34, we have

$$\gamma_k = \phi\gamma_{k-1}$$

Hence, the ARMA(1,1) has the following autocorrelation function:

$$\rho_k = \begin{cases} 1 & k = 0 \\ \dfrac{(\phi - \theta)(1 - \phi\theta)}{1 + \theta^2 - 2\phi\theta} & k = 1 \\ \phi\rho_{k-1} & k \geq 2 \end{cases} \tag{8.40}$$

### 8.4.8 ARIMA Models

A series $y_t$ is said to follow an autoregressive integrated moving average (ARIMA) model of order $d$ if the $d$th difference denoted by $\nabla^d y_t$ is a stationary ARMA model. The notation used for this model is ARIMA($p, d, q$), where $p$ is the order of the autoregressive component, $q$ is the order of the moving average component, and $d$ is the number of differences performed to produce a stationary series. Fortunately, for practical reasons we can take $d = 1$ or 2. An ARIMA($p,1,q$) process can be written as

$$w_t = \phi_1 w_{t-1} + \phi_2 w_{t-2} + \ldots + \phi_p w_{t-p} + a_t - \theta_1 a_{t-1} - \theta_2 a_{t-2} \ldots - \theta_q a_{t-q} \tag{8.41}$$

where

$$w_t = y_t - y_{t-1}$$

As an example, the ARIMA(0,1,1) or IMA(1,1) is given by

$$w_t = a_t - \theta_1 a_{t-1} \quad \text{or}$$

$$y_t = y_{t-1} + a_t - \theta_1 a_{t-1} \tag{8.42}$$

which means that the first difference ($d = 1$) would produce a stationary MA(1) series as long as $|\theta_1| < 1$.

### Example 8.1 (Continued)

Using the somatic cell count data, we will fit two models: AR(1) and MA (1). This will be accomplished by running the PROC ARIMA in SAS/ETS. The data was differenced once so as to stabilize the mean.

The following SAS/PROC ARIMA code produces the plots of the autocorrelation function and the partial correlation function and forecasts for the AR(1) and MA(1) model (also see Figure 8.12).

```
* fitting an AR(1) model;
proc arima data=scc;
 identify var=scc(1) nlags=20;
 estimate p= (1) (12) noint method=ml;
 forecast lead=12 id=time interval=month;
run; quit;

* fitting an MA(1) model;
 proc arima data=scc;
  identify var=scc(1) nlags=20;
  estimate q= (1) (12) noint method=ml ;
 forecast lead=12 id=time interval=month;
run; quit;
```

Note that $p(1)$ is replaced by $q(1)$ to fit an MA(1) seasonal model.

The maximum likelihood and the least squares estimates of the parameters of the AR(1) and MA(1) models are shown below. Figure 8.12 is produced by SAS by the "ODS GRAPHICS" facility in SAS.

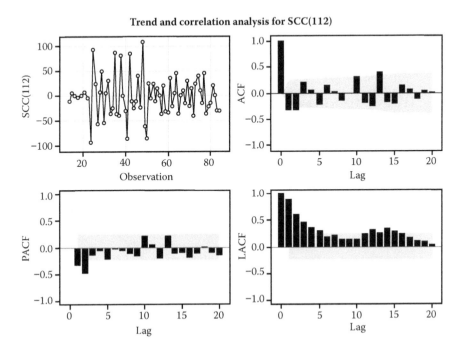

**Trend and correlation analysis for SCC(112)**

**FIGURE 8.12**
Autocorrelation and partial autocorrelation functions.

AR(1)

|  | Parameter | Estimate | Std. Error |
|---|---|---|---|
| Least Square | μ | -0.309 | 2.854 |
|  | AR1,1 | -0.211 | 0.110 |
| MLE | μ | -0.301 | 2.855 |
|  | AR1,1 | -0.210 | 0.110 |

MA(1)

|  | Parameter | Estimate | Std. Error |
|---|---|---|---|
| Least Square | μ | -0.01804 | 1.71859 |
|  | MA1,1 | 0.49064 | 0.09976 |
| MLE | μ | -0.02744 | 1.73327 |
|  | MA1,1 | 0.48592 | 0.10055 |

### Remarks

The autocorrelation function (ACF) lists the estimated autocorrelation coefficients at each lag. The value of the ACF at lag 0 is always 1. The dotted lines provide an approximate 95% confidence limit for the autocorrelation estimate at each lag. If none of the autocorrelation estimates fall outside the strip defined by the two dotted lines (and no outliers in the data) one may assume the absence of serial correlation. In effect the ACF is a measure of how important the sequence of distant observations $y_{t-1}, y_{t-2}, \ldots$ are to the current time series value $y_t$.

The partial autocorrelation function (PACF) is the ACF at lag $p$ accounting for the effects of all intervening observations. Thus the PACF at lag 1 is identical to the ACF at lag 1, but they are different at higher lags.

The time series plot of the log-cell counts is given in Figure 8.13. Taking the differenced with one unit lag we produce the time series plot of the differenced series in Figure 8.14.

The following R code produces time series analysis using the log-SCC (Figure 8.13), the autocorrelations, and the ACF plot (Figure 8.14), and to fit the AR(1) model by conditional least squares method.

```
library(quantmod);library(tseries);
library(timeSeries);library(forecast);library(xts)
ssc=read.csv(file.choose())
y=ssc$SCC
# Compute the log SCC
ly = diff(log(y),lag=1)


# Plot log returns
plot(ly,type='l', main='log somatic cell count plot')
#Summary of the ARIMA model using the determined (p,d,q)
parameters
fit = arima(ly, order = c(1, 1, 0),include.mean=FALSE)
summary(fit)


# plotting a acf plot of the residuals
acf(fit$residuals,main="Residuals plot")


# Forecasting the log scc
arima.forecast = forecast(fit, h = 1,level=99)
summary(arima.forecast)
plot(arima.forecast, main = "ARIMA Forecast")
```

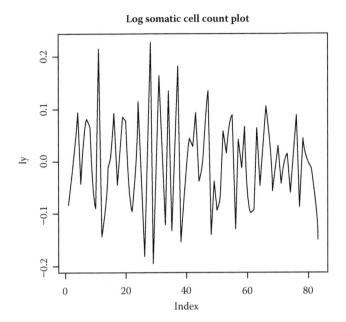

**FIGURE 8.13**
Time series plot of the log-SCC.

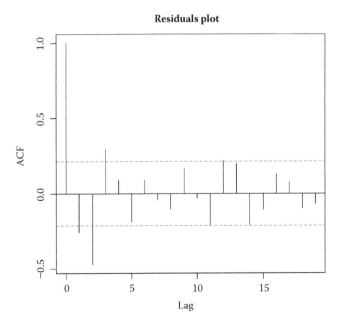

**FIGURE 8.14**
ACF plot of the differenced series.

## 8.5 Forecasting

One of the most important objectives of time series analysis is to forecast the future values of the series. The term *forecasting* is used more frequently in recent time series literature than the term *prediction*. However, most forecasting results are derived from a general theory of linear prediction developed by Kolmogorov (1939, 1941), Kalman (1960), and Whittle (1983), and many others.

Once a good time series model has become available, it can be used to make inferences about future observations. What we mean by a good model is that identification estimation and diagnostics have been completed. Even with good time series models the reliability of the forecast is based on the assumption that the future behaves like the past. However, the nature of the stochastic process may change in time and the current time series model may no longer be appropriate. If this happens the resulting forecast may be misleading. This is particularly true for forecasts with a long lead time.

Let $y_n$ denote the last value of the time series and suppose we are interested in forecasting the value that will be observed $l$ time periods ($l > 0$) in the future and that we are interested in forecasting the future value $y_{n+l}$. We denote the forecast of $y_{n+l}$ by $y_n(l)$, where the subscript denotes the forecast origin and the number in parentheses denotes the lead time. Box and Jenkins (1970) showed that the "best" forecast of $y_{n+l}$ is given by the expected value of $y_{n+l}$ at time $n$, where "best" is defined as that forecast that minimizes the mean square error:

$$E[y_{n+l} - \hat{y}_n(l)]^2$$

It should be noted that this expectation is in fact a conditional expectation since, in general, it will depend on $y_1, y_2, \ldots, y_n$. This expectation is minimized when

$$\hat{y}_n(l) \cong E(y_{n+l})$$

Following is how to obtain the forecast for the time series models AR(1), AR(2), and MA(1).

### 8.5.1 AR(1) Model

$$y_t - \mu = \phi(y_{t-1} - \mu) + a_t \tag{8.43}$$

Consider the problem of forecasting one time unit into the future. Replacing $t$ with $t + 1$ in the equation $y_t - \mu = \phi(y_{t-1} - \mu) + a_t$ we have

$$y_{t+1} - \mu = \phi(y_t - \mu) + a_{t+1} \tag{8.44}$$

Conditional on $y_1, y_2, ..., y_{t-1}, y_t$, the expectation of both sides of Equation 8.44 is

$$E(y_{t+1}) - \mu = \phi\{E(y_t/y_t, y_{t-1}, ..., y_1) - \mu\} + E(a_{t+1}/y_t, y_{t-1}, ..., y_1)$$

Since

$$E(y_t/y_t, y_{t-1}, ..., y_1) = y_t$$

$$E(a_{t+1}/y_t, y_{t-1}, ..., y_1) = E(a_{t+1}) = 0$$

and

$$E(y_{t+1}) = \hat{y}_t(1)$$

then

$$\hat{y}_t(1) = \mu + \phi(y_t - \mu) \tag{8.45}$$

For a general lead time $l$, replace $t$ with $t + l$ in equation $y_t - \mu = \phi(y_{t-1} - \mu) + a_t$ and taking the conditional expectation we get

$$\hat{y}_t(l) = \mu + \phi(y_t(l - 1) - \mu) \qquad l \geq 1 \tag{8.46}$$

It is clear now that Equation 8.46 is recursive in $l$. It can also be shown that

$$\hat{y}_t(l) = \mu + \phi^l(y_t - \mu) \qquad l \geq 1 \tag{8.47}$$

Since $|\phi| < 1$, we may simply have $\hat{y}_t(l) \cong \mu$ For large $l$. Now let us consider the one-step ahead, the forecast error, $e_t(1)$. From Equations 8.44 and 8.45

$$e_t(1) = y_{t+1} - \hat{y}_t(1)$$
$$= \mu + \phi(y_t - \mu) + a_{t+1} - [\mu + \phi(y_t - \mu)] \tag{8.48}$$
$$= a_{t+1}$$

This means that the white noise $a_{t+1}$ can now be explained as a sequence of one-step-ahead forecast errors. From Equation 8.48

$$V[e_t(1)] = V(a_{t+1}) = \sigma_a^2$$

It can be shown (see Abraham and Ledolter, 1983, p. 241) that for the AR(1)

$$V[e_t(l)] = \frac{1 - \varphi^{2l}}{1 - \varphi^2} \sigma_a^2 \tag{8.49}$$

and for large $l$,

$$V[e_t(l)] \cong \frac{\sigma_a^2}{1 - \varphi^2} \qquad (8.50)$$

### 8.5.2 AR(2) Model

Consider the AR(2) model

$$y_t - \mu = \phi_1(y_{t-1} - \mu) + \phi_2(y_{t-2} - \mu) + a_t$$

Setting $t = t+l$, the preceding equation is written as

$$y_{t+l} - \mu = \phi_1(y_{t+l-1} - \mu) + \phi_2(y_{t+l-2} - \mu) + a_{t+l}$$

For the one-step ahead forecast (i.e., $l = 1$)

$$y_{t+1} = \mu + \phi_1(y_t - \mu) + \phi_2(y_{t-1} - \mu) + a_{t+1} \qquad (8.51)$$

From the observed series, $y_t$ and $y_{t-1}$ are the last two observations in the series. Therefore, for given values of the model parameters, the only unknown quantity on the right-hand side of Equation 8.51 is $a_{t+1}$. Therefore, conditional on $y_t, y_{t-1}, ..., y_1$ we have

$$E(y_{t+1}) = \mu + \phi_1(y_t - \mu) + \phi_2(y_{t-1} - \mu) + E(a_{t+1})$$

By assumption, $E(a_{t+1}) = 0$, and hence the forecast of $y_{t+1}$ is

$$\hat{y}_t(1) = E(y_{t+1})$$
$$= \mu + \phi_1(y_t - \mu) + \phi_2(y_{t-1} - \mu)$$

where $\mu$, $\varphi_1$, and $\varphi_2$ are replaced by their estimates. In general we have

$$\hat{y}_t(l) = E(y_{t-l})$$
$$= \mu + \phi_1(\hat{y}_t(l - 1) - \mu) + \phi_2(y_t(l - 2) - \mu) \qquad l > 3 \qquad (8.52)$$

The forecast error is given by

$$e_t(l) = y_t(l) - \hat{y}_t(l)$$

For $l = 1$

$$V[e_t(l)] = \sigma_a^2 \left[1 + \psi_1^2 + \psi_2^2 + ... + \psi_{l-1}^2\right] \qquad (8.53)$$

where

$$\psi_1 = \phi_1, \qquad \psi_2 = \phi_1^2 + \phi_2$$

$$\psi_j = \phi_1 \psi_{j-1} + \phi_2 \psi_{j-2} \qquad j > 2$$

For $l = 1$

$$V[e_t(1)] = \sigma_a^2 \left[ 1 + \phi_1^2 + \left( \phi_1^2 + \phi_2 \right)^2 \right] \tag{8.54}$$

(see Abraham and Ledolter, 1983, p. 243).

### 8.5.3 MA(1) Model

An MA(1) time series model is forecast in a similar manner:

$$y_t = \mu + a_t - \theta a_{t-1}$$

First, we replace $t$ by $t + l$ so that

$$y_{t+l} = \mu + a_{t+l} - \theta a_{t+l-1} \tag{8.55}$$

Conditional on the observed series we have

$$\hat{y}_t(l) = \mu + a_{t+l} - \theta a_{t+l-1} \tag{8.56}$$

because for $l > 1$ both $a_{t+l}$ and $a_{t+l-1}$ are independent of $y_t, y_{t-1}, \ldots, y_1$. Hence

$$\hat{y}_t(l) = \begin{cases} \mu & l > 1 \\ \mu - \theta a_t & l = 1 \end{cases}$$

$$V[\hat{y}_t(l)] = \begin{cases} \sigma_a^2 (1 + \theta^2) & l > 1 \\ \sigma_a^2 & l = 1 \end{cases} \tag{8.57}$$

These results allow constructing $(1-\alpha)100\%$ confidence limits on the future observations $y_{t+l}$ as

$$\hat{y}_t(l) \pm Z_{1-\alpha/2} \sqrt{V[e_t(l)]} \tag{8.58}$$

#### Example 8.1 (Continued)

Again, using the SCC data of Example 8.1, we can compute forecasts for the variable SCC by employing the time series programs in SAS. The following estimates and 95% upper and lower confidence limits are obtained for the AR(1) model (also see Figure 8.15).

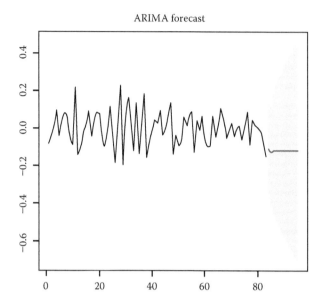

**FIGURE 8.15**
ARIMA forecast based for the logarithm of SCC.

Forecasts for Variable SCC

| Obs | Forecast | Std Error | 95% Confidence Limits | |
|-----|----------|-----------|-----------------------|----------|
| 85 | 289.1262 | 31.4242 | 227.5360 | 350.7164 |
| 86 | 286.8251 | 40.0251 | 208.3773 | 365.2729 |
| 87 | 286.9365 | 47.8325 | 193.1866 | 380.6865 |
| 88 | 286.5386 | 54.3920 | 179.9324 | 393.1449 |
| 89 | 286.2483 | 60.2682 | 168.1247 | 404.3718 |
| 90 | 285.9352 | 65.6151 | 157.3319 | 414.5385 |
| 91 | 285.6269 | 70.5591 | 147.3337 | 423.9202 |
| 92 | 285.3176 | 75.1784 | 137.9708 | 432.6645 |
| 93 | 285.0086 | 79.5298 | 129.1330 | 440.8842 |
| 94 | 284.6995 | 83.6553 | 120.7382 | 448.6608 |
| 95 | 284.3904 | 87.5866 | 112.7238 | 456.0569 |
| 96 | 284.0812 | 91.3489 | 105.0407 | 463.1218 |

The R code and the resulting output of the forecast of the SSC series are displayed below. The ARIMA forecast is illustrated by the shaded region in the right-hand side of Figure 8.15.

```
# Forecasting the log scc
arima.forecast = forecast(fit, h = 12,level=95)
(summary(arima.forecast))
plot(arima.forecast, main = "ARIMA Forecast")
```

```
Forecasts:
  Point Forecast    Lo 95   Hi 95
84   -0.1089440 -0.3459437 0.1280557
85   -0.1269400 -0.3975498 0.1436699
86   -0.1188621 -0.4429690 0.2052449
87   -0.1224880 -0.4825905 0.2376145
88   -0.1208604 -0.5176157 0.2758949
89   -0.1215910 -0.5502350 0.3070530
90   -0.1212631 -0.5802722 0.3377461
91   -0.1214103 -0.6086045 0.3657839
92   -0.1213442 -0.6353040 0.3926156
93   -0.1213739 -0.6607196 0.4179718
94   -0.1213605 -0.6849727 0.4422516
95   -0.1213665 -0.7082327 0.4654997
```

## 8.6 Forecasting with Exponential Smoothing Models

If we have a time series that can be modeled using an additive model with constant level and no seasonality, then we can use simple exponential smoothing models (ESMs) to make forecasts. This approach does not require any assumptions about the correlations between successive observations of the series.

Since one of the main objectives of the analysis of time series data is to forecast future observations, it is very important that the forecast is done with minimum error. An important approach to efficient forecasts is to construct forecasts from weighted averages of the observations, with the most recent observations being given the largest weights. Therefore, we consider forecasts of the form

$$y_t(k) = \sum_{j=0}^{t-1} w_j y_{n-j} \tag{8.59}$$

The sum $\sum_{j=0}^{t-1} w_j = 1$, $w_0 \geq w_1, w_2, \dots \geq t_{t-1}$ and $k$ is the number of steps ahead of a forecast. One possible choice of the weights is

$$w_j = \lambda(1-\lambda)^j \ 0 < \lambda < 1 \tag{8.60}$$

Substituting from Equation 8.60 into Equation 8.56 we get

$$y_n(k) = \lambda y_n + (1-\lambda)y_n(k-1) \tag{8.61}$$

This simple formula is called *exponential smoothing*.

A closer look at Equation 8.61 will reveal that the values of $\lambda$ (which should be determined by model user) depend on $k$.

The ESM is a very simple model and as can be seen from the recursive scheme in Equation 8.61, for any $0<\lambda<1$, the weights attached to the observations decrease exponentially as we go back in time, hence the name "exponential smoothing." If $\lambda$ is small (i.e., close to 0), the more weight is given to observations from the distant time point. If $\lambda$ is large (i.e., close to 1), more weight is given to the more recent data points.

Note that Equation 8.62 may be written in a simpler and intuitive way as

$$S_t = \lambda y_{t-1} + (1 - \lambda)S_{t-1} \tag{8.62}$$

where $y_t$ is the value of observation from the given series at time $t$ and $S_t$ is the value of the smoothed observation at time $t$. Therefore, Equation 8.62 may be written as

$$S_t = S_{t-1} + \lambda(y_{t-1} - S_{t-1})$$
$$= S_{t-1} + \lambda \text{ (forecast error)}$$

Hence, the forecast is

$$S_{t+1} = S_t + \lambda(y_t - S_t) \tag{8.63}$$

The quality of the forecast is assessed by the magnitude of the sum square errors of the forecast. The optimal smoothing depends on the values of the smoothing parameter $\lambda$. An objective way to obtain values is to estimate it from observed data. The initial values of $\lambda$ may be obtained by minimizing the mean square error of predictions where the errors are specified by $e_t = y_t - S_t$.

Remark: To use simple exponential smoothing in R, we need to install two packages "**tseries**" and "**forecast**".

### Example 8.3: Nosocomial Time Series Data

The quality department in a large regional tertiary care hospital keeps track of the number of patients with nosocomial infection, the length of follow up being from the time of admission to detecting the infection (see Figure 8.16). The data is given in the Excel file "nosocomial", and the time series plot is shown in Figure 8.16. Forecasting through the exponential smoothing with R for different smoothing parameters is given next (also see Figure 8.17).

```
nosocomial=read.csv(file.choose())
names(nosocomial)
attach(nosocomial)
library(tseries)
library(forecast)
incidence=NI
plot(ts(NI))


#modeling simple exponential smoothing
fit1 <- ses(incidence, alpha=0.2, initial="simple", h=5)
summary(fit1)
```

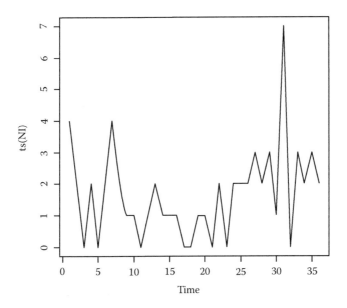

**FIGURE 8.16**
Incidence of nosocomial infection (NI).

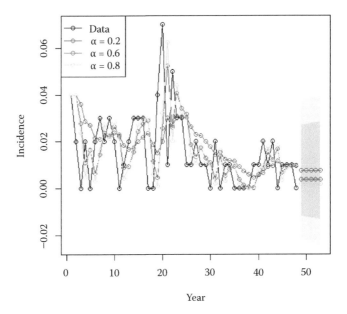

**FIGURE 8.17**
Simple exponential smoothing (SES) using different values for the smoothing parameter.

```
fit2 <- ses(incidence, alpha=0.6, initial="simple", h=5)
summary(fit2)
fit3 <- ses(incidence, alpha=0.8, initial="simple", h=5)
summary(fit3)
plot(fit1, plot.conf=FALSE, ylab="incidence",
xlab="Year", main="", fcol="white", type="o")
lines(fitted(fit1), col="blue", type="o")
lines(fitted(fit2), col="red", type="o")
lines(fitted(fit3), col="green", type="o")
lines(fit1$mean, col="blue", type="o")
lines(fit2$mean, col="red", type="o")
lines(fit3$mean, col="green", type="o")
legend("topleft",lty=1, col=c(1,"blue","red","green"),
 c("data", expression(alpha == 0.2), expression(alpha == 0.6),
  expression(alpha == 0.8)),pch=1)
```

The R exponential smoothing for different values of the smoothing parameter $\alpha$ are depicted in Figure 8.17.
plot(fit3)

Figure 8.18 shows the forecast in solid line and the 95% confidence interval in gray.

To test whether there is an evidence of significant departure from zero of the correlations at lags "1 to 10", we can use the Ljung-Box test (which is explained in Section 8.7):

```
Box.test(fit3$residuals, lag=10, type="Ljung-Box")
 Box-Ljung test.
data: fit3$residuals
X-squared = 26.194, df = 10, p-value = 0.003488
```

**Forecast from simple exponential smoothing**

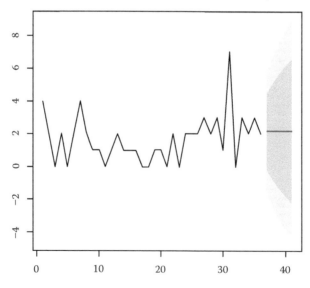

**FIGURE 8.18**
Forecast using the smoothing parameter $\alpha = 0.8$.

Here the Ljung-Box test statistic is 26.194, and the *p*-value is 0.003488, so there is sufficient evidence of nonzero autocorrelation in the in-sample forecast errors at lags "1 to 10".

## 8.7 Modeling Seasonality with ARIMA: Condemnation Rates Series Revisited

The ARIMA models presented in this chapter assume that seasonal effects are removed or that the series is nonseasonal. However, in many practical situations it is important to model and quantify the seasonal effects. For example, it might be of interest to poultry producers to know which months of the year the condemnation rates are higher. This is important in order to avoid potential losses if the supply falls short of the demand due to excess condemnation.

Box and Jenkins (1970) extend their ARIMA models to describe and forecast time series with seasonal variation. Modeling of time series with seasonal variation using ARIMA is quite complicated, and detailed treatment of this topic is beyond the scope of this chapter. However, we shall outline the steps of modeling seasonality in the condemnation rates data of Example 8.3, shown in Table 8.12.

**Example 8.3 (Continued)**

First we examine the monthly means of the series (Table 8.12).

The data suggest that condemnation rates tend to be relatively high in the winter and low in the summer. Figure 8.19 shows the sample series plot.

SAS CODE for the time series plot of the condemnation time series:

```
ODS GRAPHICS ON;
ODS RTF;
goptions reset=global gunit=pct cback=white htitle=6 htext=3
ftext=swissb colors=(black);

*Figure 8.6: Time series plot of proportion of turkeys condemned;
axis1 order=('01jan87'd to '01jan94'd by year) label=(angle=0
'Time (Months)') offset=(3) minor=(number=11);
axis2 order=(0 to 2000 by 500) label=(angle=90 'Condemnation rate
per 100,000');
```

**TABLE 8.12**

Monthly Means of Condemnation Rates per 100,000 Birds

| Jan | Feb | Mar | Apr | May | Jun | Jul | Aug | Sep | Oct | Nov | Dec |
|-----|-----|-----|-----|-----|-----|-----|-----|-----|-----|-----|-----|
| 932.58 | 916.13 | 921.99 | 772.99 | 603.37 | 509.54 | 497.01 | 520.03 | 618.73 | 585.96 | 771.76 | 672.83 |

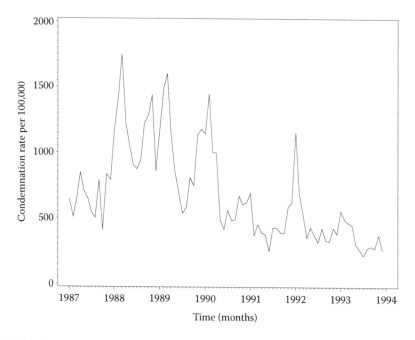

**FIGURE 8.19**
Time series plot of the condemnation rates.

```
proc gplot data=condemn;
 plot rate*month / haxis=axis1 vaxis=axis2;
 symbol1 i=join;
 format month year4.;
run;
ODS GRAPHICS OFF;
ODS RTF CLOSE;
```

Results of the ARIMA procedure in SAS:

```
ODS GRAPHICS ON;
ODS RTF;
Rate=rate*100000;
/* FITTING MA seasonal model*/
 proc arima data=condemn;
  identify var=rate(1,12)nlag=12;
  estimate q=(1,3,8)(12);
run;
/* Fitting AR seasonal model */

proc arima  data=condemn;
  identify var=rate(1) noprint;
  estimate p=(1,3,8)(12);
 run;

ODS GRAPHICS OFF;
ODS RTF CLOSE;
```

The sample autocorrelations beyond lag 13 are insignificant. Moreover, we can see that the autocorrelations are not tailing off, which indicates that the series is nonstationary, and that there is a sinusoidal pattern confirming a seasonal pattern in the series. Note that an AR(1) series $y_t = \mu + \phi y_{t-1} + \rho_t$ is nonstationary if $\phi = 1$. To confirm the nonstationarity of our series, we fitted an AR(1) model. The MLE of $\phi$ was $\hat{\phi} = 0.961$ with SE $= .03$, and hence a value of $1.0$ for $\phi$ seems acceptable. To remove nonstationarity, we formed the series $w_t = y_t - y_{t-1}$. Figure 8.20 is the time series plot of $w_t$.

In modeling seasonality we decided to choose between two models: The first is an MA with one nonseasonal component and several seasonal components, and the second is an AR with similar structure.

Since the series is monthly, it has 12 periods. Accordingly, the series is constructed with 12 periods of differencing. The ACF plot of this series is given next. This plot shows significant autocorrelations at lags 1, 8, and 12.

The candidate model for this situation is an additive AR model:

$$y_t = \mu + \phi_1 y_{t-1} + \phi_2 y_{t-3} + \phi_3 y_{t-8} + \phi_4 y_{t-12} + a_t$$

or an additive MA model:

$$y_t = \mu + a_t - \theta_1 a_{t-1} - \theta_2 a_{t-3} - \theta_3 a_{t-8} - \theta_4 a_{t-12}$$

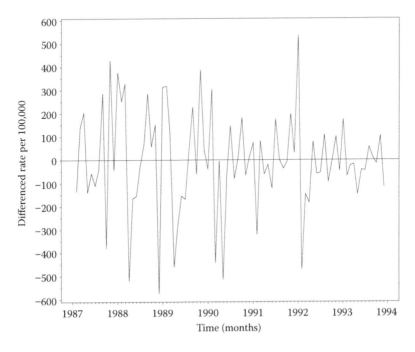

**FIGURE 8.20**
The time series plot after differencing.

Our choice of an additive structure is for simplicity. The fitted seasonal AR model is

$$y_t = -10.03 - .26y_{t-1} - .16y_{t-3} + .22y_{t-8} - .29y_{t-12}$$

$$(18.76) \quad (.12) \quad (.12) \quad (.12) \quad (.13)$$

and the fitted seasonal MA model is

$$y_t = -9.43 + .33\varepsilon_{t-1} + .0.23\varepsilon_{t-3} - .16\varepsilon_{t-8} + .63\varepsilon_{t-12}$$

$$(7.95) \quad (.11) \quad (.11) \quad (.11) \quad (.10)$$

The numbers in parentheses the standard errors of the estimates. Parameters of both models were estimated via conditional least squares.

As a diagnostic tool to check the model adequacy, use the $Q$ statistic of Box and Pierce (1970), later modified by Ljung and Box (1978). They suggested a test to contrast the null hypothesis

$$H_0 : \rho_1 = \rho_2 = \dots = \rho_k = 0$$

against the general alternative

$$H_1 : \quad \text{Not all } \rho_i = 0$$

Based on the autocorrelation between residuals, they suggested the statistic

$$Q = Q(k) = n(n+2) \sum_{j=1}^{k} r_j^2 / (n-j)$$

where $n$ is the length of the series after any differencing and $r_j$ is the residual autocorrelation at lag $j$. Box and Pierce (1970) showed that under $H_0$, $Q$ is asymptotically distributed as chi-squared with $(k - p - q)$ degrees of freedom. Typically, the statistic is evaluated for several choices of $k$. Under $H_0$, for large $n$

$$E[Q(k_2) - Q(k_1)] = k_2 - k_1$$

so that different sections of the correlogram can be checked for departures from $H_0$.

Table 8.13 gives the Box-Pierce-Ljung (BPL) statistics (see Box and Pierce, 1970) for the condemnation rates using the AR and for the models given in Table 8.14.

**TABLE 8.13**

BPL Statistic Using AR Seasonal Model

| Lag(k) | Q(k) | df | Prob. | Autocorrelations $r_j$ | | | | | |
|--------|------|-----|--------|--------|--------|--------|--------|--------|--------|
| 6 | 2.57 | 2 | 0.2762 | −0.062 | −0.142 | −0.046 | −0.088 | −0.015 | 0.011 |
| 12 | 3.09 | 8 | 0.9287 | 0.019 | 0.033 | −0.036 | −0.008 | 0.001 | −0.057 |
| 18 | 5.14 | 14 | 0.9838 | 0.093 | −0.076 | −0.001 | −0.078 | 0.041 | 0.012 |
| 24 | 18.31 | 20 | 0.5673 | −0.067 | 0.022 | 0.170 | 0.050 | 0.181 | −0.231 |

Note: $p = 4, q = 0, \text{df} = k - 4$.

**TABLE 8.14**

BPL Statistic Using MA Seasonal Model

| Lag(k) | Q(k) | df | Prob. | Autocorrelations $r_j$ | | | | | |
|--------|------|----|-------|------|------|------|------|------|------|
| 6 | 0.88 | 2 | 0.6435 | 0.018 | −0.099 | 0.006 | −0.028 | 0.022 | −0.015 |
| 12 | 3.32 | 8 | 0.9129 | 0.031 | 0.048 | −0.095 | −0.022 | −0.036 | 0.120 |
| 18 | 5.63 | 14 | 0.9750 | 0.053 | −0.121 | −0.012 | −0.067 | 0.056 | −0.004 |
| 24 | 12.50 | 20 | 0.8978 | −0.035 | 0.035 | 0.163 | 0.090 | 0.151 | −0.072 |

*Note:* $p = 0$, $q = 4$, df $= k - 4$.

Note that all the $Q$ statistics of the two models are nonsignificant. This means that the models captured the autocorrelations in the data and that the residual autocorrelations are nonsignificant. Therefore, both models provide good fit to the condemnation rates times series process. However, we might argue that the MA seasonal model gives a better fit due to the fact that it has a smaller Akaike information criterion (AIC = 978.07 for the MA, and AIC = 982.25 for the AR).

## 8.8 Interrupted Time Series (Quasi-Experiments)

An important design to establish causality attributed to intervention is the interrupted time-series experiment. In Figure 8.21 we show the effect of intervention (in this case launching campaign to advocate regular hand washing) on the incidence of nosocomial infection in a large tertiary care hospital.

By studying the pattern of the graph, paying particular attention to what happens to the series precisely at the point of shifting from no intervention to intervention, one may learn the effect of introducing this type of intervention on the quality of care.

The data to evaluate the impact of the intervention come from the hospital data warehouse collections of data gathered routinely across time for administrative purposes. An example of this type of archival time-series experiment appears as Figure 8.21. Here we show a graph of the incidence density of nosocomial infection in four consecutive years.

The simple logic of the interrupted time-series experiment is this: If the graph of the dependent variable shows an abrupt shift in level of direction precisely at the point of intervention, then the intervention may be considered if change in the dependent variable is tangible.

Graphs of time series data can change level and direction for many reasons, some related to the intervention and others not. Separating potential reasons for effects into those essentially related to the intervention and those only

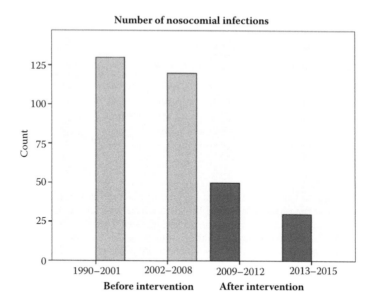

**FIGURE 8.21**
Effect of intervention on the incidence of nosocomial infection.

accidentally related is the principal task in analyzing time-series experiments. In the pretest-posttest design, a group of persons is observed, a treatment is applied to them, and then they are observed a second time. An increase (or a decrease) of the group average from the first observation to the second is taken to be the effect of the independent variable (the intervention) on the dependent variable that was observed. The pretest–posttest design is not a reliable approach to analyze because it requires many sources of variation and correlation in the dependent variable whether pre- or postintervention.

The graph of a real time-series will show a good deal of fluctuation quite apart from any effect an intervention might have on it. Consider, for example, a graph of the series in Figure 8.21.

### Example 8.4: Modeling Interrupted Time Series Using AUTREG in SAS and R

To model the effect of intervention on the incidence of nosocomial infection we use the following SAS code:

```
ODS RTF;
PROC AUTOREG DATA=infection;
MODEL incidence=time;
RUN;
```

```
/* intervention effect*/
PROC AUTOREG DATA=infection;
MODEL incidence = time T1 T2;
run;
ODS RTF CLOSE;
```

The SAS output of AUTOREG for the interrupted time series (nosoco-mial incidence) is given next (also see Figure 8.22):

Parameter Estimates

| Variable | DF | Estimate | Standard Error | t Value | Approx Pr > \|t\| |
|----------|----|----------|----------------|---------|-------------------|
| Intercept | 1 | 0.0136 | 0.005457 | 2.50 | 0.0164 |
| time | 1 | 0.000743 | 0.000382 | 1.95 | 0.0580 |
| T1 | 1 | -0.0227 | 0.007487 | -3.03 | 0.0041 |
| T2 | 1 | -0.000848 | 0.000540 | -1.57 | 0.1237 |

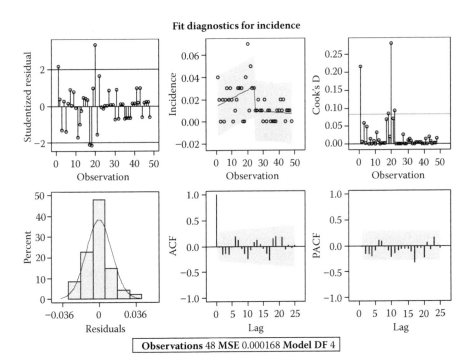

**FIGURE 8.22**
SAS Fit diagnostics for incidence.

The R code to fit ARIMA for the interrupted time series is

```
data=read.csv(file.choose())
library(forecast)
# suppose that we read the CSv Excel file into data
attach(data)
names(data)
dat=data.frame(t1,t2,time)
x=model.matrix(~t1+t2+time,dat)
x=data.frame(x)
new.data=data.frame(incidence,x)
attach(new.data)
fit.model=auto.arima(incidence,xreg=x)
summary(fit.model)
plot(forecast(fit.model,xreg=x,h=20))
#RESULTS
#Coefficients:
      X. Intercept    t1       t2         time
        0.0136      -0.02 3   -8e-04      7e-04
  s.e.  0.0052       0.007    5e-04       4e-04
#sigma^2 estimated as 0.0001677: log likelihood=142.61
AIC=-275.22    BIC=-265.87
```

The results of the analyses of interrupted time series are similar in SAS and R.

## 8.9 Stationary versus Nonstationary Series

A key distinction has to be drawn between two types of series: those that are stationary and those that are nonstationary. The distinction makes all the difference in the world when it comes to inspecting graphs and analyzing whether an intervention had an effect. The difference between stationary and nonstationary time series is the difference between graphs that fluctuate around some fixed levels across time and graphs that wander about, changing their level and region haphazardly.

It is far easier to detect an intervention effect in a stationary process than in a nonstationary one. Intervention effects are seen as displacements of the curve, and as you can see, nonstationary processes are, by definition, curves that are subject to displacements of level and slope at random points in time. The trick with nonstationary time series is to distinguish the random displacements from the one true deterministic displacement caused by the intervention.

## Exercises

8.1. The following table shows the average air temperature at a capital city in a Middle East country from 1960 to 1969.

| Year | Jan | Feb | Mar | Apr | May | Jun | Jul | Aug | Sep | Oct | Nov | Dec |
|------|------|------|------|------|------|------|------|------|------|------|------|------|
| 1960 | 27.1 | 27 | 28 | 29.2 | 32 | 35 | 42 | 49.1 | 38 | 36 | 32 | 26.9 |
| 1961 | 27.1 | 27.1 | 28.1 | 29.1 | 34.1 | 36.1 | 42 | 49.2 | 38.1 | 35.6 | 31 | 27.2 |
| 1962 | 28.1 | 27.2 | 27.9 | 29.1 | 35.1 | 37.2 | 41 | 48.1 | 39.1 | 35.5 | 29.2 | 27.6 |
| 1963 | 27.2 | 27.1 | 26.9 | 28.2 | 35.2 | 37 | 43 | 47.2 | 41.2 | 36.5 | 29.7 | 28.6 |
| 1964 | 26.8 | 27.9 | 30.1 | 30.1 | 36.1 | 37 | 44 | 47 | 41.1 | 34.2 | 28.9 | 28.6 |
| 1965 | 27.0 | 26.8 | 30.1 | 30.1 | 37.1 | 37 | 45 | 47 | 41.6 | 34.1 | 29.6 | 27.5 |
| 1966 | 27.0 | 27.3 | 30.2 | 28.1 | 37.2 | 38 | 45 | 48 | 41.7 | 34.2 | 28.6 | 25.6 |
| 1967 | 28.1 | 28.3 | 29.9 | 29.2 | 38 | 39 | 46.1 | 49 | 42.6 | 35.1 | 27.6 | 27.6 |
| 1968 | 26.8 | 28.2 | 29.9 | 31.2 | 39 | 39 | 47 | 48 | 41.2 | 36.1 | 30.1 | 28.6 |
| 1969 | 26.9 | 29.9 | 31.9 | 33.2 | 39 | 39 | 46.1 | 46 | 41.7 | 36 | 30.1 | 29.6 |

a. Plot the data.

b. Using a nonparametric test for seasonality, what do you conclude?

c. Calculate the overall January average, the overall February average, and so on. Subtract each individual value from the corresponding monthly average. Test for seasonality of the resulting data.

d. Assuming a multiplicative model, evaluate the seasonal coefficients.

e. After fitting a linear trend model, what is the predicted temperature in the month of February 1961?

8.2. Suppose that you are given the series

63, 54, 75, 50, 52, 54, 6.6, 5.5, 6.0, 50, 70, 50, 60, 65, 40, 66, 60, 62, 42, 62, 70, 44, 62, 49, 71, 50, 63, 55, 75, 54

a. Plot the series.

b. What is the approximate value of the first lag autocorrelation coefficient $\rho_1$?

c. Plot $y_t$ against $y_{t+1}$. What is the approximate value of $\rho_1$ from the scatter plot?

d. Calculate and plot the sample ACF $\hat{\rho}_k$ for $k = 0, 1, 2, 3, 4, 5$.

8.3. Consider a stationary time series with theoretical autocorrelation function

$$\rho_k = \alpha \pi^{k-1}, \; k = 1, 2, \ldots$$

What is $\mathrm{var}(\hat{\rho}_k)$ using Bartlett's approximation?

8.4. Find the ACF and plot $\rho_k$ for $k = 0, 1, 2, 3,$ and 4 for each of the following models:

a. $y_t - 0.5y_{t-1} = a_t$

b. $y_t - 1.5y_{t-1} + 0.5y_{t-2} = a_t$

8.5. For the AR (2) model:

$$y_t - 0.6y_{t-1} - 0.3y_{t-2} = a_t$$

Calculate $var(y_t)$ assuming that $\sigma_a^2 = 1$.

8.6. Find the AR representation of the MA(1) process

$$y_t = a_t - 0.4a_{t-1}$$

# 9

## Analysis of Survival Data

## 9.1 Introduction

Time-to-event or survival studies involve observing units until an event is experienced. In the case of medical studies, the units are humans or animals, and failure may be broadly defined as the occurrence of a prespecified event. Events of this nature include time to death, and disease remission, occurrence, or recurrence.

Although survival data analysis is similar to the analysis of other types of data previously discussed (continuous, binary, time series) in that information is collected on the response variable as well as any covariates of interest, it differs in one important aspect: the anticipated event may not occur for each subject under study. Not all subjects will experience the outcome during the course of observation, resulting in the absence of a failure time for that particular individual. This situation is referred to as *censoring* in the analysis of survival data, and a study subject for which no failure time is available is referred to as *censored*.

Unlike the other types of analysis previously discussed, censored data analysis requires special methods to compensate for the information lost by not knowing the time of failure of all subjects. In addition, survival data analysis must account for highly skewed data. Often one or two individuals will experience the event of interest much sooner or later than the majority of individuals under study, giving the overall distribution of failure times a skewed appearance and preventing the use of the normal distribution in the analysis. Thus the analysis of survival data requires techniques that are able to incorporate the possibility of skewed and censored observations.

## 9.2 Fundamental Concept in Survival Analysis

The preceding section referred to censoring in a broad sense, defining censored survival data as data for which the true failure time is not known. This general definition may be broken down for three specific situations, resulting in three types of censoring:

*Type I censoring*—Subjects are observed for a fixed period of time, with exact failure times recorded for those who fail during the observation period. Subjects not failing during the observation period are considered censored. Their failure times become the time at which they were last observed or the time at which the study finished.

*Type II censoring*—Subjects are observed until a fixed number of failures occur. As with type I censoring, those failing during the observation period are considered uncensored and have known failure times. Those not failing are considered censored and have failure times that become the time at which they were last observed or the time at which the longest uncensored failure occurred.

*Random censoring*—Often encountered in clinical trials, random censoring occurs due to the accrual of patients gradually over time, resulting in unequal times under study. The study takes place over a fixed period of time, resulting in exact failure times for those failing during the period of observation and censored failure times for those lost to follow up or not failing before study termination. All failure times reflect the period under study for that individual. The three censoring situations are further elaborated in Table 9.1.

If the survival time is denoted by the random variable $T$, then the following definitions are useful in the context of survival analysis:

*Cumulative distribution function (CDF)*—Denoted by $F(t)$, this quantity defines the probability of failure before time $t$:

$$F(t) = \Pr(\text{individual fails before time } t)$$

$$= \Pr(T < t)$$

*Probability density function (PDF)*—Denoted by $f(t)$, this quantity is the derivative of the CDF and defines the probability that an individual fails in a small interval per unit time:

**TABLE 9.1**

Summary Information for Three Types of Censoring

| Characteristics | Type I | Type II | Random |
|---|---|---|---|
| Duration | Study continues for a fixed period of time | Study continues until a fixed number/proportion of failures | Study continues for a fixed period of time; unequal periods of observation possible |
| Uncensored failure time | Equal to the exact failure time, which is known | Equal to the exact failure time, which is known | Equal to the exact failure time, which is known |
| Censored failure time | Equal to the length of the study | Equal to the largest uncensored failure time | Calculated using time of study completion and time of subject enrollment |
| Lost to follow-up failure time | Calculated using time at which subject is lost and time at which study starts | Calculated using time at which subject is lost and time at which study starts | Calculated using time at which subject is lost and time of subject enrollment |

*Source:* Reproduced with permission from International Biometrics Society.

$$f(t) = \lim_{\Delta t \to 0} \frac{\Pr(\text{an individual dies in}(t, t + \Delta t))}{\Delta t}$$

$$= \frac{d}{dt} F(t).$$

As with all density functions, $f(t)$ is assumed to have the following properties:

- The area under the density curve equals one.
- The density is a nonnegative function such that

$$f(t) > 0 \text{ for all } t > 0$$
$$= 0 \text{ for all } t \le 0$$

*Survival function*—Denoted by $S(t)$. This function gives the probability of survival longer than time $t$:

$$S(t) = \Pr(\text{an individual survives longer than } t)$$
$$= \Pr(T > t)$$

The survival function is assumed to have the following properties:

- The probability of survival at time zero is one, $S(t) = 1$ for $t = 0$.
- The probability of infinite survival is zero, $S(t) = 0$ for $t = \infty$.
- The survival function is nonincreasing.

*Hazard function*—Denoted by $h(t)$, this function gives the probability an individual fails in a small interval of time conditional on their survival at the beginning of the interval:

$$h(t) = \lim_{\Delta t \to 0} \frac{Pr(\text{an individual dies in } (t, t + \Delta t)|T > t)}{\Delta t}$$

In terms of the previously defined quantities, the hazard function may be written as

$$h(t) = \frac{f(t)}{S(t)}$$

In practice, the hazard function is also referred to as the instantaneous failure rate or the force of mortality. It represents the failure risk per unit of time during a lifetime.

The cumulative hazard function is written as $H(t)$ and is the integral of the hazard function:

$$H(t) = \int_0^t h(x)dx$$

The quantities $F(t), f(t), S(t),$ *and* $h(t)$ may be defined for any continuous random variable; however, $S(t)$ and $h(t)$ are usually seen in the context of survival data since they are particularly suited to its analysis. Notice as well that given any one of the four quantities, the other three are easily obtained. Thus specifying the survival function, for instance, also determines what the cumulative distribution function, probability density function, and hazard function are. The following relationships hold:

$$f(t) = -\frac{dS(t)}{dt}$$
$$S(t) = \exp(-H(t))$$
$$f(t) = h(t)\exp[-H(t)]$$
$$h(t) = -\frac{d}{dt}\ln(S(t))$$

In the following section we introduce five examples of survival data.

## 9.3 Examples

### 9.3.1 Cystic Ovary Data

This study examined the effectiveness of hormonal therapy for treatment of cows with cystic ovarian disease. Two groups of cows were randomized to hormonal treatment and one group to placebo. The time of cyst disappearance was then recorded with the possibility of censored data due to not all cysts disappearing. The data are shown in Table 9.2.

### 9.3.2 Breast Cancer Data

An increase in breast cancer incidence in recent years has resulted in a substantial portion of health care dollars being directed towards research in this area. The research studies have focused on early detection through mass screening and recurrence prevention through effective treatments. The focus of one such investigation was to determine which prognostic measures were predictive of breast cancer recurrence in female patients. The data set contains the following variables:

| | |
|---|---|
| id | Patient identification number |
| censor | Breast cancer recurrence (0 = no, 1 = yes) |
| time | Time until breast cancer recurrence in months |
| pag | Proliferative AgNOR index (0 = low, 1 = high) |
| mag | Mean AgNOR count (0 = low, 1 = high) |
| age | Age at the start of the study in years |
| tsize | Size of original tumor (small = 0, large = 1) |

### 9.3.3 Ventilating Tube Data

One-third of pediatric visits arise due to inflammation of the middle ear, also known as otitis media, resulting in a substantial health care burden. In addition, concerns have surfaced relating to long-term language, behavior, and speech development. Unsuccessful treatment with various drug therapies often leads to surgical intervention, in which tubes are placed in the ear. It has been shown that ventilating tubes are successful in preventing otitis media as long as the tubes are in place and unblocked. Lee and Lindgren (1996) studied

**TABLE 9.2**

Cystic Ovary Data

| Treatment 1 | Treatment 2 | Placebo |
|---|---|---|
| 4, 6, 8, 8,9, 10, 12 | 7, 12, 15, 16, 18, 22* | 19, 24, 18*, 20*, 22*, 27*, 30* |

*Censored observation.

the time of tube failure (displacement or blockage) for 78 children. Each child was randomly assigned to be treated with placebo or prednisone and sulfamethoprim. The children were observed from February 1987 to January 1990, resulting in the possibility of censored failure times for children not experiencing tube failure before the completion of the study. In addition, it is anticipated that the failure times of the two ears from one child will correlated. The analysis of these data will be discussed under the topic of correlated survival data analysis. The data set contains the following variables.

| | |
|---|---|
| child | Four-digit ID |
| treat | Treatment group (2 = medical, 1 = control) |
| ear | Left or right ear (1 = right, 2 = left) |
| time | Time, in months |
| status | Event (1 = failed, 0 = censored) |

### 9.3.4 Age at Culling of Dairy Cows

Individual cow data were obtained at 72 Ontario farms from April 1993 to October 1994. The information included date of birth, calving and culling dates, and lactation number. During a two and a half year study, the producer recorded all occurrences of clinical diseases to cows. Culling age was calculated as the date of birth to the date the cow was culled. The definition of culling (the event of interest) was restricted to the removal of cows by slaughter. During the study, cows that were in the herd and removed for other reasons (e.g., sold) were considered lost to follow-up and their unobserved culling age was treated as censored observation. The three covariates believed to be prognostic indicators for culling were parity, total milk production in the previous parity, and presence or absence of clinical mastitis in the period prior to the lactation in which the cow was culled.

### 9.3.5 Model for End-Stage Liver Disease and Its Effect on Survival of Liver Transplanted Patients

The model for end-stage liver disease (MELD) was developed to predict short-term mortality in patients with cirrhosis. In 2002, the New York State Committee on Quality Improvement in Living Liver Donation prohibited live liver donation for potential recipients with Model for End-Stage Liver Disease (MELD) scores greater than 25. Despite the paucity of evidence to support this recommendation, many centers in North America remain reluctant to offer a living donor (LD) to patients with moderate to high MELD scores. It has since become the standard tool to prioritize patients for liver transplantation. The main objective is to assess the value of pretransplant MELD in the prediction of posttransplant survival.

These five examples will be used throughout the chapter to demonstrate methods of survival data analysis using SAS and R.

## 9.4 Estimating Survival Probabilities

Two distinct methodologies exist for the analysis of survival data: nonparametric approaches in which no distributional assumptions are made for the previously defined probability density function $f(t)$, and parametric approaches in which distributional restrictions are imposed. Each methodology will be discussed separately.

## 9.5 Nonparametric Methods

### 9.5.1 Methods for Noncensored Data

Estimates of the survival function, probability density function, and hazard function exist for the specific case of noncensored data. They are given as follows:

- Estimate of the survival function for noncensored data:

$$\hat{S}(t) = \frac{\text{number of patients surviving longer than } t}{\text{total number of patients}}$$

- Estimate of the probability density function for noncensored data:

$$\hat{f}(t) = \frac{\text{number of patients dying in the interval beginning at time } t}{(\text{total number of patients})(\text{interval width})}$$

- Estimate of the hazard function for noncensored data:

$$\hat{h}(t) = \frac{\text{number of patients dying in the interval beginning at time } t}{(\text{number of patients surviving at } t)(\text{interval width})}$$

$$= \frac{\text{number of patients dying per unit time in the interval}}{\text{number of patients surviving at } t}$$

It is also possible to define the average hazard rate, which uses the average number of survivors at the interval midpoint to calculate the denominator of the estimate:

$$\hat{h}^{\bullet}(t) = \frac{\text{number of patients dying per unit time in the interval}}{(\text{number of patients surviving at } t) - .5(\text{number of deaths in the interval})}$$

The estimate given in $\hat{h}^{\bullet}(t)$ results in a smaller denominator and thus a larger hazard rate. The $\hat{h}^{\bullet}(t)$ is used primarily by actuaries.

Obviously different methods are required in the presence of censored data. These methods are now discussed.

### 9.5.2 Methods for Censored Data

The Kaplan-Meier (1958) estimate of the survival function in the presence of censored data, also known as the product-limit estimate, is given by

$$\hat{S}(k) = p_1 \times p_2 \times p_3 \times \cdots \times p_k$$

where $k \geq 2$ years and $p_i$ denotes the proportion of patients surviving the *i*th year conditional on their survival until the $(i - 1)$th year. In practice, $\hat{S}(k)$ is calculated using the following formula:

$$\hat{S}(t) = \Pi_{t_{(r)} \leq t} \frac{n - r}{n - r + 1} \tag{9.1}$$

where the survival times have been placed in ascending order so that $t_{(1)} \leq t_{(2)} \leq \cdots \leq t_{(n)}$ for $n$ the total number of individuals under study and $r$ runs through the positive integers such that $t_{(r)} \leq t$ and $t_{(r)}$ is uncensored.

We use Table 9.3 to calculate the product-limit (PL) survival estimate:

The last column of the table is filled in after calculation of $\frac{n - r}{n - r + 1}$. Notice that the estimate of the survival function is available only for the noncensored times $t$. These computations are illustrated in Example 9.1 using cystic ovary data.

#### Example 9.1: Cystic Ovary Data

The Kaplan-Meier (product-limit) estimates of the survival function for the cystic ovary data are calculated separately for each treatment group as illustrated in Table 9.4.

The following SAS code reads the data and produces the Kaplan-Meier survival estimates. The "censor" variable takes value 1 if failure and 0 if not. Testing the equality of two or more survival curves is achieved by the log-rank test statistic, which is asymptotically distributed as a chi-square

**TABLE 9.3**

Illustration for the Calculation Using the PL Survival Estimates

| Ordered Survival Times (Censored and Uncensored) | Rank | Rank (*r*) (Uncensored Observations) | Number in Sample (*n*) | $\frac{n-r}{n-r+1}$ | $\hat{S}(t)$ |
|---|---|---|---|---|---|
| $t_{(1)}$ | 1 | 1 | N | | |
| $t_{(2)}^*$ | 2 | / | $n - 1$ | / | / |
| $t_{(3)}$ | 3 | 3 | $n - 2$ | | |
| . | . | . | . | | |
| $t_{(n)}$ | n | n | 1 | | |

**TABLE 9.4**

KM Survival Estimate for Cystic Ovary Data

| Treatment | Ordered Survival Times | Rank | Rank Uncensored Observations | Number in Sample | $\dfrac{n-r}{n-r+1}$ | $\hat{S}(t)$ |
|---|---|---|---|---|---|---|
| 1 | 4 | 1 | 1 | 7 | 6/7 | 0.85 |
| | 6 | 2 | 2 | 6 | 5/6 | (0.83)(0.85) = 0.71 |
| | 8 | 3 | 3 | 5 | 4/5 | (0.80)(0.71) = 0.56 |
| | 8 | 4 | 4 | 4 | 3/4 | (0.75)(0.56) = 0.42 |
| | 9 | 5 | 5 | 3 | 2/3 | (0.66)(0.42) = 0.28 |
| | 10 | 6 | 6 | 2 | 1/2 | (0.50)(0.28) = 0.13 |
| | 12 | 7 | 7 | 1 | 0 | 0 |
| 2 | 7 | 1 | 1 | 6 | 5/6 | 0.83 |
| | 12 | 2 | 2 | 5 | 4/5 | (0.80)(0.83) = 0.66 |
| | 15 | 3 | 3 | 4 | 3/4 | (0.75)(0.66) = 0.49 |
| | 16 | 4 | 4 | 3 | 2/3 | (0.66)(0.49) = 0.33 |
| | 18 | 5 | 5 | 2 | 1/2 | (0.50)(0.33) = 0.16 |
| | 22* | 6 | / | 1 | / | / |
| 3 | 18* | 1 | / | 7 | / | / |
| | 19 | 2 | 2 | 6 | 5/6 | 0.83 |
| | 20* | 3 | / | 5 | / | / |
| | 22* | 4 | / | 4 | / | / |
| | 24 | 5 | 5 | 3 | 2/3 | (0.66)(0.83) = 0.55 |
| | 27* | 6 | / | 2 | / | / |
| | 30* | 7 | / | 1 | / | / |

*Censored.

random variable with degrees of freedom equal to the number of groups being compared minus one (Peto and Peto, 1972).

```
data cyst;
input cow treat time censor @@;
datalines;
1   1   4 1   2 1   6 1   3 1   8 1   4 1   8 1
5   1   9 1   6 1 10 1   7 1 12 1   8 2   7 1
9   2 12 1 10 2 15 1 11 2 16 1 12 2 18 1
13 2 22 0 14 3 19 1 15 3 24 1 16 3 18 0
17 3 20 0 18 3 22 0 19 3 27 0 20 3 30 0
;

proc lifetest data=cyst plots=(s) graphics;
time time*censor(0);
strata treat;
symbol1 v=none color=black line=1;
symbol2 v=none color=black line=2;
symbol3 v=none color=black line=3;
run;
```

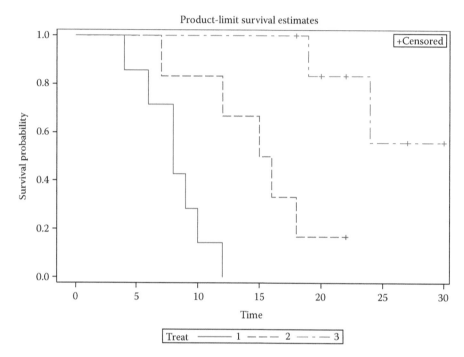

**FIGURE 9.1**
The survival curves of the treatment groups.

**TABLE 9.5**

SAS Output Comparing the Survivor Rates in the Three Groups

| Test of Equality over Strata | | | |
| --- | --- | --- | --- |
| **Test** | **Chi-Square** | **DF** | **Pr > Chi-Square** |
| **Log-Rank** | 21.2401 | 2 | <.0001 |
| **Wilcoxon** | 17.8661 | 2 | 0.0001 |
| **−2Log(LR)** | 10.6663 | 2 | 0.0048 |

The Kaplan-Meier estimates of the survival curves of the three groups are given in Figure 9.1.

The SAS output results in Table 9.5 are explained in the next section.

Notes:

1. Note that the survival function is only estimated at death times that are not censored. The SAS output also includes the summary

statistics concerning sample size and percent of censored observations, and so on not displayed here.

2. Adding a "strata" statement to the SAS program results in the calculations being performed within each level of the strata variable.

3. The last observation was censored so the estimate of the mean survival time is biased.

4. The Kaplan-Meier survival curves indicate that, compared to the placebo group, the time to cyst disappearance is shorter with treatment 1 and longer with treatment 2 (Figure 9.1).

The following R code sets up the data and performs the analysis:
**library(survival)**

```
cow <- c(1:20)
treat <- c(rep(1,7),rep(2,6),rep(3,7))
time <- c(4,6,8,8,9,10,12,7,12,15,16,18,22,19,24,18,20,22,27,30)
censor <- c(rep(1,12),0,1,1,rep(0,5))
cyst <- data.frame(cow,treat,time,censor)

fit <- survfit(Surv(time, censor) ~ treat, data=cyst)
summary(fit)
plot(fit)
```

The R output (not shown) produces the estimated survival probabilities at each time point, the standard errors, and 95% confidence interval of the survival probabilities.

In order to determine whether there is a significant difference between groups in a data set, a statistical test must be used for comparison of survival curves. Without any prior knowledge of the distribution that may be appropriate for the data, a nonparametric (distribution-free) test is preferable. A test designed for this purpose is the log-rank test by Peto and Peto (1972).

---

## 9.6 Nonparametric Techniques for Group Comparisons

### 9.6.1 The Log-Rank Test

Suppose that we have two treatment groups A and B, and it is of interest to compare the survival in these two groups. For a hypothesis test of

$$H_0: \text{No difference in survival between the two groups}$$

$$\text{versus}$$

$$H_A: \text{Difference in survival between the two groups}$$

The test statistic is given by

$$\chi^2 = \frac{U_L^2}{V_L} \tag{9.2}$$

where

$$U_L = \sum_{j=1}^{r} \left( d_{1j} - e_{1j} \right)$$

$$V_L = \sum_{j=1}^{r} \frac{n_{1j} n_{2j} d_j \left( n_j - d_j \right)}{n_j^2 \left( n_j - 1 \right)} = \sum_{j=1}^{r} v_j$$

for $n_{1j}, n_{2j}, d_{1j}, d_{2j}, e_{1j}, e_{2j}$ defined as follows:

$t_{(j)} = j^{\text{th}}$ death time(regardless of group)
$n_{1j}$ = number at risk in group A just before time $t_{(j)}$
$n_{2j}$ = number at risk in group B just before time $t_{(j)}$.
$d_{1j}$ = number of deaths in group A at time $t_{(j)}$
$d_{2j}$ = number of deaths in group B at time $t_{(j)}$
$e_{1j}$ = expected number of individuals dying in group A at time $t_{(j)}$
$e_{2j}$ = expected number of individuals dying in group B at time $t_{(j)}$

where

$$e_{kj} = \frac{n_{kj} d_j}{n_j}$$

for $k = 1, 2$ so that $n_j$ is the total number at risk in both groups just before $t_{(j)}$ and $d_j$ is the total number of deaths in both groups at $t_{(j)}$.

The test statistic is calculated by constructing a table of the nature of Table 9.6.

**TABLE 9.6**

Calculations for Log-Rank Test Statistic

| Death Time | $d_{1j}$ | $n_{1j}$ | $d_{2j}$ | $n_{2j}$ | $d_j$ | $n_j$ | $e_{1j} = n_{1j}d_j/n_j$ | $d_{1j}-e_{1j}$ (1) | $v_{1j}$ (2) |
|---|---|---|---|---|---|---|---|---|---|
| $t_{(1)}$ | | | | | | | | | |
| $t_{(2)}$ | | | | | | | | | |
| $\vdots$ | | | | | | | | | |
| $t_{(n)}$ | | | | | | | | | |

Under the null hypothesis of no differences between groups A and B, the test statistic is distributed as $\chi^2_{(1)}$.

**Example 9.2: Optimal Cut-Off Point in the MELD Score Using the ROC Curve**

**R Code**

```
library("ggplot2")
library("ROCR")
library("verification")
library(OptimalCutpoints)
liver <-read.csv(file.choose())
pred <- with(liver,prediction(meld,status))
perf <- performance(pred,"tpr","fpr")
auc <-performance(pred, measure = "auc")@y.values[[1]]
rd <- data.frame(x=perf@x.values[[1]],y=perf@y.values[[1]])
p <- ggplot(rd,aes(x=x,y=y)) + geom_path(size=1)
p    <-   p    +    geom_segment(aes(x=0,y=0,xend=1,yend=1),
colour="black",linetype= 2)
p <- p + geom_text(aes(x=1, y= 0, hjust=1, vjust=0, label=paste
(sep = "", "AUC = ",round(auc,3) )),colour="black",size=4)
p <- p + scale_x_continuous(name= "False positive rate")
p <- p + scale_y_continuous(name= "True positive rate")

#geting the cutoff point

optimal.cutpoint.Youden <- optimal.cutpoints(X = meld ~ status,
tag.healthy = 0,
methods = "Youden", data = liver, pop.prev = NULL, categorical.cov
= "doner.type",
control = control.cutpoints(), ci.fit = FALSE, conf.level = 0.95,
trace = FALSE)
summary(optimal.cutpoint.Youden)
plot(optimal.cutpoint.Youden)
x=23
liver$category[liver$meld< x] = "low"
liver$category[liver$meld>=x] = "high"
new=data.frame(liver)
```

Based on the ROC curve in Figure 9.2, the optimal cut-off point for the MELD in the "liver" data set is x = 23. Thereafter, we categorize the MELD into "low" if MELD is lower than 23 and "high" if MELD is above 23.

The next step is to create a new data set that includes the "liver" data and the new category variable.

```
liver$category[liver$meld< x] = "low"
liver$category[liver$meld>=x] = "high"
new=data.frame(liver)
# Using the Kaplan-Meier test to compare the survival curves of the
low and high MELD cores.
test=survdiff(Surv(year,status)~category,data=new)
```

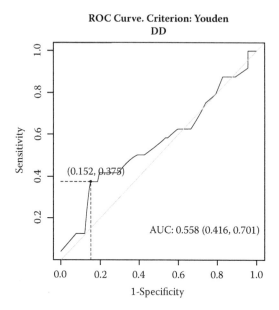

**FIGURE 9.2**
The ROC for the MELD score.

```
R_Output:
Call:
survdiff(formula = Surv(years, status) ~ category, data = new)

n=176, 2 observations deleted due to missingness.

                   N Observed  Expected (O-E)^2/E (O-E)^2/V
category=high 69      23         12.6      8.57      13.6
category=low 107      14         24.4      4.43      13.6

Chisq= 13.6 on 1 degrees of freedom, p= 0.000232.
```

The –value of the log-rank chi-square test is 0.000232.
The following code produces the survival curves (see Figure 9.3):

```
plot(survfit(Surv(years)~category,data=new),
xlab="Time in years",
ylab="Survival probablility",
col=c("blue","red"),lwd=2)
legend("topright", legend=c("HIGH", "LOW"),
col=c("blue","red"),lwd=2)
```

The survival curves produced by R for the two MELD groups are
shown in Figure 9.3.

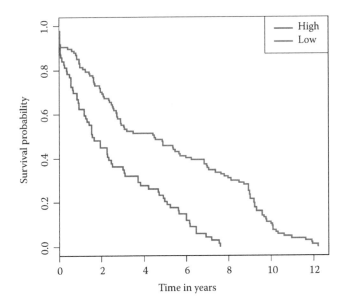

**FIGURE 9.3**
Survival curves for the high MELD Score (blue) and the low MELD score (red).

## 9.6.2 Log-Rank Test for More Than Two Groups

Often it is desirable to make comparisons of the survival between three or more groups; in this case, an extension of the log-rank test is used. If there are $q$ groups we wish to make comparisons between, then the following are calculated for each group $i = 1,...,q$:

$$U_{Li} = \sum_{j=1}^{r} \left( d_{ij} - \frac{n_{ij}d_j}{n_j} \right)$$

$$U_{Wi} = \sum_{j=1}^{r} n_j \left( d_{ij} - \frac{n_{ij}d_j}{n_j} \right)$$

The vectors

$$U_L = \begin{bmatrix} U_{L1} \\ U_{L2} \\ \vdots \\ U_{Lq} \end{bmatrix}$$

and

$$U_W = \begin{bmatrix} U_{W1} \\ U_{W2} \\ \vdots \\ U_{Wq} \end{bmatrix}$$

are then formed. In addition, the variances and covariances are needed and are given by the formula

$$V_{Lii'} = \sum_{j=1}^{r} \frac{n_{ij}d_j\left(n_j - d_j\right)}{n_j\left(n_j - 1\right)}\left(\delta_{ii} - \frac{n_{ij}}{n_j}\right)$$

where $\delta_{ii'}$ is such that

$$\delta_{ii'} = \begin{array}{l} 1 \text{ if } i = i' \\ 0 \text{ otherwise} \end{array}$$

The variance–covariance matrix is then given by

$$V = \begin{bmatrix} V_{L11} & V_{L12} & \cdots & V_{L1i} \\ V_{L21} & V_{L22} & \cdots & V_{L2i} \\ \vdots & \vdots & \ddots & \vdots \\ V_{Li1} & V_{Li2} & \cdots & V_{Lii} \end{bmatrix}$$

where $V_{Lij} = \mathrm{cov}(U_{Li}, U_{Lj})$ and $V_{Lii} = \mathrm{var}(U_{Li})$.

To test a null hypothesis of

$H_0$: no difference in survival between all groups

versus

$H_A$: difference in survival between at least two groups

the test statistic given by

$$\chi^2 = U_L' V_L^{-1} U_L \tag{9.3}$$

has a $\chi^2_{(q-1)}$ distribution under the null hypothesis, where $q$ is the number of strata.

For calculation of the stratified log-rank test in SAS, the same strata statement is used for calculation of the log-rank test between two groups.

A stratified test will be performed automatically for variables having more than two levels.

---

## 9.7 Parametric Methods

All parametric methods involve specification of a distributional form for the probability density function $f(t)$. This in turn specifies the survival function $S(t)$ and the hazard function $h(t)$ using the relationships previously defined. The two parametric models to be discussed are the exponential and the Weibull. Their survival and hazard functions are given and their properties reviewed. Much of the work in this section can be found in Lawless (1982).

### 9.7.1 Exponential Model

An exponential density function is given by

$$f(t) = \lambda \exp(-\lambda t), \; t \geq 0, \; \lambda > 0$$

and zero elsewhere.
   Therefore

$$S(t) = \exp(-\lambda t), \; t \geq 0, \; \lambda > 0 \tag{9.4}$$

and

$$h(t) = \lambda$$

Notice that the hazard function is independent of time, implying the instantaneous conditional failure rate does not change within a lifetime. This is also referred to as the memory-less property of the exponential distribution, since the age of an individual does not affect their probability of future survival. When $\lambda = 1$, the distribution is referred to as the unit exponential.
   In practice, most failure times do not have a constant hazard of failure and thus the application of the exponential model for survival analysis is limited. The exponential model is in fact a special case of a more general model that is widely applicable: the Weibull model.

### 9.7.2 Weibull Model

A continuously distributed random variable is said to have a Weibull distribution if its probability density function is given by

$$f(t) = \alpha \gamma (\alpha t)^{\gamma - 1} \exp[-(\alpha t)^{\gamma}], t \geq 0 \text{ and } \alpha > 0, \gamma > 0.$$

The survival function is given by

$$S(t) = \exp[-(\alpha t)^{\gamma}].$$                                      (9.5)

Moreover, the hazard function is $h(t) = \alpha\gamma(\alpha t)^{\gamma-1}$, and

$$\ln[-\ln S(t)] = a + bt^{*}.$$                                      (9.6)

In the linear equation (Equation 9.6) $a = \gamma \ln \alpha$, $b = \gamma$, and $t^{*} = \ln(t)$.

For the Weibull distribution, $\alpha$ and $\gamma$ are the scale and shape parameters, respectively. The specific case of $\gamma = 1$ defines the exponential model with constant hazard previously discussed, $\gamma > 1$ implies a hazard that increases with time, and $\gamma < 1$ yields a hazard decreasing with time. It is evident that the Weibull distribution is widely applicable since it allows modeling of populations with various types of failure risk.

**Example: 9.3: Veteran Administration Lung Cancer Survival Data**

- Treatment: 1 = standard, 2 = test
- Cell type: 1 = squamous, 2 = small cell, 3 =a deno, 4 = large
- Survival in days
- Status: 1 = dead, 0 = censored
- Karnofsky score
- Months from diagnosis
- Age in years
- Prior therapy: 0 = no, 10 = yes

We shall fit several parametric regression models including exponential, gamma, Weibull, and lognormal.

```
proc lifereg data= lungcancer;
   class celltype;
    model survtime*status(0)= rx celltype karno months_fromdiag
age prior_rx/ D=Exponential COVB;
     PROBPLOT;
    run;
   proc lifereg data= lungcancer;
   class celltype;
    model survtime*status(0)= rx celltype karno months_fromdiag
age prior_rx/ D=Gamma COVB;
     PROBPLOT;
    run;
   proc lifereg data= lungcancer;
   class celltype;
    model survtime*status(0)= rx celltype karno months_fromdiag
age prior_rx/ D=LNORMAL COVB;
     PROBPLOT;
    run;
   proc lifereg data= lungcancer;
   class celltype;
```

```
model survtime*status(0)= rx celltype karno months_fromdiag
age prior_rx/ D=Weibull COVB;
    PROBPLOT;
    run;
```

Examining the output of the regression models shows no tangible differences in the parameter estimates and their precisions. However, there are some minor differences in the "fitness statistics" summarized by the AIC: Weibull (AIC) = 1451.103, exponential (AIC) = 1450.318, lognormal (AIC) = 1449.270, and the gamma (AIC) = 1447.404. Since a smaller AIC indicates a better fit, it seems that the gamma regression model gives the best fit for this data. The results of the gamma regression model are presented in Table 9.7.

## 9.8 Semiparametric Models

### 9.8.1 Cox Proportional Hazards Model

One positive feature of the parametric methods previously discussed is that specification of a form for the probability density function allows the likelihood to be constructed. Maximum likelihood estimates and standard errors may then be obtained for all parameters in the model. However, the drawback in parametric modeling lies in the fact that it may not be desirable to specify a probability distribution function for a particular set of data, making nonparametric calculations more attractive. The ideal situation would involve no distributional restrictions on the density, yet maximum likelihood estimates of regression parameters (and thus treatment effects) would be readily available. An analysis with such properties may be performed using the Cox proportional hazards model (Cox, 1972).

As its name implies, the proportional hazards model is constructed by assuming that the hazard function of the $i$th individual is the product of a baseline hazard common to all individuals, denoted by $h_0(t)$ and a function of the covariate vector, $x_i = (x_{i1}, x_{i2}, \ldots x_{ip})$ for that individual, $\varphi(x)$:

$$h_i(t) = h_0(t)\varphi(x_i) \qquad (9.7)$$

Rearrangement yields the hazard ratio or relative hazard to be the non-negative function

$$\varphi(x_i) = \frac{h_i(t)}{h_0(t)} \qquad (9.8)$$

The covariate function is specified to be $\varphi(x_i) = exp(x_i\beta)$.

**TABLE 9.7**

Parameter Estimates of the Gamma Regression Model

| Parameter | | DF | Analysis of Maximum Likelihood Parameter Estimates | | | | | |
|---|---|---|---|---|---|---|---|---|
| | | | Estimate | Standard Error | 95% Confidence Limits | | Chi-Square | Pr > ChiSq |
| Intercept | | 1 | 2.6622 | 0.7894 | 1.1149 | 4.2094 | 11.37 | 0.0007 |
| Rx | | 1 | -0.1682 | 0.1855 | -0.5318 | 0.1953 | 0.82 | 0.3644 |
| celltype | adeno | 1 | -0.9092 | 0.2896 | -1.4769 | -0.3415 | 9.85 | 0.0017 |
| celltype | large | 1 | -0.1501 | 0.2910 | -0.7205 | 0.4202 | 0.27 | 0.6059 |
| celltype | smallcel | 1 | -0.7440 | 0.2486 | -1.2314 | -0.2567 | 8.95 | 0.0028 |
| celltype | squamous | 0 | 0.0000 | . | . | . | . | . |
| karno | | 1 | 0.0341 | 0.0050 | 0.0243 | 0.0438 | 46.44 | <.0001 |
| months_fromdiag | | 1 | 0.0002 | 0.0088 | -0.0171 | 0.0175 | 0.00 | 0.9798 |
| age | | 1 | 0.0082 | 0.0089 | -0.0093 | 0.0256 | 0.84 | 0.3580 |
| prior_rx | | 1 | -0.0892 | 0.2159 | -0.5124 | 0.3339 | 0.17 | 0.6794 |
| Scale | | 1 | 1.0005 | 0.0707 | 0.8711 | 1.1491 | | |
| Shape | | 1 | 0.4502 | 0.2227 | 0.0137 | 0.8868 | | |

Further rearrangement shows that

$$\log\left(\frac{h_i(t)}{h_0(t)}\right) = \beta_1 x_{i1} + \ldots \beta_p x_{ip}$$

Therefore, the proportional hazards model is a linear model for the log hazard ratio.

### 9.8.2 Estimation of Regression Parameters

Cox (1972) introduced the method of partial likelihood to estimate the regression parameters and avoid the specification of the baseline hazard function $h_0(t)$.

The partial likelihood for the proportional hazards model is given by

$$L(\beta) = \prod_{j=1}^{r} \frac{\exp\left(\beta' x_{(j)}\right)}{\sum_{l \in R(t_i)} \exp\left(\beta' x_{(l)}\right)}$$

where $R(t_{(j)})$ is the risk set (individuals alive) at the $j$th ordered death time, $t_{(j)}$. Thus the likelihood takes the product over the $j$th ordered death times of all terms and is given by

$$L_j(\beta) = \frac{\exp\left(\beta' x_{(j)}\right)}{\sum_{l \in R(t_i)} \exp\left(\beta' x_{(l)}\right)}$$

$$= \Pr\left(\begin{array}{c} \text{failure occurs to observed} \\ \text{individual given the risk set} \\ R(t_i) \text{ and a failure at } t_{(i)} \end{array}\right)$$

(9.9)

As can be seen, $L_i(\beta)$ is the ratio of the hazard for the individual who died at the $i$th ordered death time divided by the sum of hazards for the individuals who were at risk when the $i$th ordered death occurred. Notice that individuals with censored failure times do not contribute a term to the likelihood, however, they are included in the risk sets. Hence, $L(\beta)$ may be expressed as

$$L(\beta) = \prod_{i=1}^{n} \left(\frac{\exp\left(\beta' x_{(i)}\right)}{\sum_{l \in R(t_{(j)})} \exp\left(\beta' x_{l}\right)}\right)^{\delta_i}$$

(9.10)

where

$\delta_i = 1$ if the $i^{th}$ individual fails during the study

$\quad = 0$ if the failure time of the $i^{th}$ individual is censored

Taking the product over the $j$ uncensored failure times in Equation 9.9 is equivalent to taking the product over the $n$ censored and uncensored failure times in Equation 9.10 due to the indicator variable $\delta_i$.

Notice that the likelihood 9.10 is referred to as a partial likelihood due to the fact that it is based on the product of probabilities of failure rather than official density functions. Maximum likelihood estimation of the regression parameters $\beta$ occurs by treating the partial likelihood given in Equation 9.10 as a true likelihood, so that differentiation of the log and subsequent maximization is possible. Variance estimates are found using the matrix of partial second derivatives. Newton-Raphson techniques are required for the maximization.

### 9.8.3 Treatment of Ties in the Proportional Hazards Model

The proportional hazards model implicitly assumes (due to its continuous nature) that the exact time of failure or censoring is known for each individual under study. For this situation, there are no ties in failure times. Such accuracy is not usually encountered in practice, where frequently survival times are only available for the nearest day, week, or month. This may result in tied failure times in the data. Several methods exist for dealing with tied failure times in the Cox proportional hazards model. Two will be discussed here.

Note that in the treatment of ties, it is assumed that

1. There are $r$ distinct deaths at $t_{(j)}$, $j = 1,\ldots,r$.
2. There may be $d_j$ deaths at $t_{(j)}$.
3. $s_j = x_1 + x_2 + \ldots + x_{d_j}$, $j = 1,\ldots,r$, where $x_1, x_2, \ldots, x_{d_j}$ are the covariate vectors of the individuals dying at $t_{(j)}$.

In addition, it is assumed that if an uncensored failure and a censored failure (i.e., a death and a drop out) occur at the same time, the uncensored failure occurs first, so that the following discussion focuses on ties in uncensored failure times.

Breslow (1972) suggested the following approximation to the partial likelihood function to account for tied failure times in the data:

$$\prod_{j=1}^{r} \frac{\exp\left(\beta' s_j\right)}{\left[\sum_{l \in R\left(t_{(j)}\right)} \exp\left(\beta' x_l\right)\right]^{d_j}}$$

where all possible sequences of deaths are summed for the respective component of the likelihood function. Breslow's approximation is simple to compute and has the advantage of being quite accurate when the number of ties at a given death time is small.

**Example 9.3: Cox Regression for the Veteran Administration Data**

```
proc phreg data= lungcancer;
class celltype;
model survtime*status(0)= rx celltype karno months_fromdiag age
prior_rx /TIES=BRESLOW;
run;
```

The results of fitting Cox regression model for the data in this example are summarized in Table 9.8.

From the earlier results and similar to the gamma regression model (see Table 9.7) the cell type and the Karnofsky score are the only predictors of survivorship.

## 9.9 Survival Analysis of Competing Risk

The notations and terminology used in this section are the same as those used throughout this chapter. Changes are made whenever needed.

The term *competing risk* is used to describe an event or events that either preclude the observation of the event of interest or changes the risk that this event (our event of interest) occurs.

In the standard survival analysis we know that the hazard function $h(t)$, the survival function $S(t)$, and the cumulative incidence function $F(t) = 1 - S(t)$ are fundamental in reporting the results of the analysis.

The statistical analysis tackles the competing risks using two quantities. The first is called cause specific hazard (csh), and the other is called subdistributional hazard (sdh). Applications of (csh) is when the main interest is studying disease etiology, whereas (sdh) is used in predicting disease burden.

An example of competing events is encountered in cancer trials. After treatment, patients are observed and the event of interest is treatment failure or local recurrence. Clearly, death is an event that precludes these events, that is, observing treatment failure becomes impossible if death precedes treatment failure or local recurrence.

**TABLE 9.8**

Fitting Cox Regression Model for the Veteran's Administration Lung Cancer Data

| Parameter | | DF | Parameter Estimate | Standard Error | Chi-Square | Pr > ChiSq | Hazard Ratio | Label |
|---|---|---|---|---|---|---|---|---|
| | | | | | Analysis of Maximum Likelihood Estimates | | | |
| rx | | 1 | 0.28994 | 0.20721 | 1.9579 | 0.1617 | 1.336 | |
| Cell type | adeno | 1 | 1.18830 | 0.30076 | 15.6101 | <.0001 | 3.281 | celltype adeno |
| Cell type | large | 1 | 0.39963 | 0.28266 | 1.9988 | 0.1574 | 1.491 | celltype large |
| Cell type | Small cel | 1 | 0.85649 | 0.27519 | 9.6866 | 0.0019 | 2.355 | celltype smallcel |
| karno | | 1 | -0.03262 | 0.00551 | 35.1124 | <.0001 | 0.968 | |
| months_fromdiag | | 1 | -0.0000916 | 0.00913 | 0.0001 | 0.9920 | 1.000 | |
| age | | 1 | -0.00855 | 0.00930 | 0.8443 | 0.3582 | 0.991 | |
| prior_rx | | 1 | 0.07232 | 0.23213 | 0.0971 | 0.7554 | 1.075 | |

Table 9.9 gives an illustration of (csh).

For example, at time 4 the risk of failure for event 1 is 2/30 = 0.067, which estimates the cause specific hazard.

When we have more than one type of event and if these events are related, the KM estimates are biased. This bias arises because the survival function, and cumulative incidence function is a complementary of survival function. This bias arises because the KM method assumes that all events are independent, and thus, censors events other than the event of interest. For example, disease relapse is an event of interest in studies of bone marrow transplantation (BMT), as is mortality related to complications of transplantation (transplant-related mortality or TRM). Relapse and TRM are not independent in this setting because these two events are likely related to immunologic effector mechanisms following BMT, whereby efforts to reduce TRM may adversely affect the risk of relapse; moreover, patients who die from TRM cannot be at further risk of relapse. Therefore, the KM method is inappropriate for estimating the cumulative incidence rate of relapse in the presence of TRM because it censors TRM. More examples of competing risk in survival analysis data are given in Klein and Andersen (2005), Agarwal et.al. (2008), and Lim et al. (2010). A self-learning material is given in Kleinbaum and Klein (2005).

In standard survival analysis, we used the log-rank test to compare curves using the KM method. In the presence of competing risks, however, this is inappropriate, for the same reason given earlier. Instead, Gray (1988) investigated this issue and proposed a class of tests for comparing the cumulative incidence curves of a particular type of failure among different groups in the presence of competing risks.

Moreover, when there is a difference in the cumulative incidence curves among different treatment groups, it is also important to determine whether this difference is solely due to the treatment or to the confounding factors, such as age or baseline comorbid conditions. In standard survival analyses, we have addressed this question by fitting a Cox proportional hazards model.

**TABLE 9.9**

Cause-Specific Hazard

| $t$ | $d_1$ | $d_2$ | $n_t$ | $h_{1(t)}$ | $h_{2(t)}$ | $SDH(1)$ |
|---|---|---|---|---|---|---|
| 0 | 0 | 0 | 40 | 0/40 | 0/40 | 0/40 |
| 1 | 2 | 2 | 40 | 2/40 | 2/40 | 2/40 |
| 2 | 2 | 0 | 36 | 2/36 | 0/36 | 2/38 |
| 3 | 3 | 1 | 34 | 3/34 | 1/34 | 3/36 |
| 4 | 2 | 0 | 30 | 2/30 | 0/30 | 2/33 |
| 5 | 1 | 1 | 28 | 1/28 | 1/28 | 1/31 |
| 6 | 0 | 3 | 26 | 0/26 | 3/26 | 0/30 |

*Note:* $t$, time to event; $d_j$, number of failures for event ($j = 1,2$); $h_{j(t)}$, hazard of failure at time $t$ for event $j$; $n_t$, number of risks at time $t$; $SDH(1)$, subdistribution hazard for event (I).

In fact one may attempt to construct a cause-specific standard Cox model for a particular failure treating other competing risks censored. However, the effect of a covariate on an event from either a cause-specific (e.g., relapse) model or cause nonspecific (e.g., relapse and TRM combined) model may be very different from the effect of the covariate of the event (e.g., relapse) in the presence of competing risks (e.g., TRM). Fine and Gray (1999) proposed a method for direct regression modeling of the effect of covariates on the cumulative incidence function for competing risks data. As in any other regression analysis, modeling cumulative incidence functions for competing risks can be used to identify potential prognostic factors for a particular failure in the presence of competing risks or to assess a prognostic factor of interest after adjusting for other potential risk factors in the model.

### 9.9.1 Cause-Specific Hazard

The manner in which risk sets are defined in standard survival analyses may be modified to allow for competing events. Recall that the *risk set* is defined as the group of individuals that have not experienced the outcome and therefore are at risk for the event of interest at time $t$. Individuals who have a competing event can be removed from all later risk sets for the event of interest. Table 9.9 illustrates this approach in discrete time. At time 0, there are 40 individuals at risk. At time 1, two individuals have event 1, and another two individuals have event 2, such that the risk set for time 2 is now $36 = 40 - 2_{\text{event 1}} - 2_{\text{event 2}}$. Thus, individuals with time 2 event 1 or time 2 event 2 prior to time $t$ are excluded from the risk set at time $t$.

An estimate of the hazard for event 2 can be described in the discrete time setting as the number of individuals who experience the event divided by the number at risk at time $t$. For example, at time 3, this would be $1/34 = 0.03$, which estimates the *cause-specific hazard*, which is formally defined as $h_j(t) = P$ $(T = t, J = j \mid T \geq t)$, where $J = j$ indicates whether event 1 ($j = 1$) or event 2 ($j = 2$) is being estimated.

The cause-specific hazard can be extended to continuous time (Pintilie, 2006):

$$h_j(t) = \lim_{\Delta t \to 0} \left\{ \frac{P(t < T \leq t + \Delta t, J = j \mid T > 1)}{\Delta t} \right\} = \frac{f_j^*(t)}{S(t)}$$

where $f_j^*(t) = P(T = t, 1 = j)$ is a "sub"-density function (an asterisk indicates an improper, i.e., "sub"-density function that integrates to <1), and reflects the net survival function of both events 1 and 2, that is, $S(t) = P(T > t) =$

$$\exp\left[-\int_0^t \sum_{j=1}^2 h_j(u)du\right] = \exp\left[-\int_0^t h(u)du\right], \text{ where } \lambda(t) \text{ is the net hazard for having}$$

either event 1 or event 2.

Consequently, a proportional hazards model can be constructed for the cause-specific hazard:

$$\lambda_j(t|z) = \lambda_{0j}(t) \exp\left(z^T \beta_j\right) j = 1, 2 \tag{9.11}$$

where $\lambda_{0j}$ is the arbitrary baseline cause-specific hazard, and $\beta_j, j = 1, 2$ are the corresponding regression coefficients, where $\exp(\beta_j) = {}_{cs}RHj$ is interpretable as the relative change in the cause-specific hazard for the $j$th event corresponding to a one-unit increase in the corresponding covariate.

### 9.9.2 Subdistribution Hazard

In light of the strong assumption of independence between events to allow interpretation of the cause-specific cumulative incidence function (${}_{cs}CIF$), the competing risk literature has focused on an alternative measure of risk: the subdistribution cumulative incidence function (${}_{sd}CIF$). This function is defined as the joint probability of an event prior to time $t$ and that the event is of type $j$: $F_j^*(t) = P(T < t, J = j)$..

For individuals who have the competing cause, risk sets were constructed so that they include both individuals without any event and those who have had the competing event.

With this structure, a different hazard function is defined as the probability of the event given that an individual has survived up to time $t$ without any event or has had the competing event prior to time $t$. This is the subdistribution hazard. For example at $t = 3$, the subdistribution hazard is $3/36 = .083$, which is smaller than the cause-specific hazard of $3/34$ because of the larger risk set.

For the discrete time setting, the subdistribution hazard is $\lambda_j(t) = P(T = t, J = j \mid T \geq t$ or $(T < t$ and $J \neq j))$. In continuous time, the subdistribution hazard is the following (Fine and Gray, 1999):

$$\lambda_j(t) = \lim_{\Delta t \to 0} \left\{ \frac{P[t < T \leq t + \Delta t, J = j | T > t \cup (T < t \cap J \neq j)]}{\Delta t} \right\}$$

$$= \frac{f_j^*(t)}{1 - F_j^*(t)} = \frac{f_j^*(t)}{P(J \neq j) + S_j^*(t)} \tag{9.12}$$

where $F_j^*(t) = P(T < t, J = j), S_j^*(t) = P(T > t, J = j)$, and $f_t^*(t) = \dfrac{\partial F_j^*(t)}{\partial t}$ are the subdistribution cumulative incidence, subsurvivor, and subdensity functions (note that $P(J = j) = S_j^*(t) + F_j^*(t)$).

In summary:

- We use competing risk when observing an event of interest that is precluded by a preceding competing event(s).
- Standard survival analysis using Kaplan-Meier is not appropriate in the presence of competing events as it produces biased estimates of cumulative incidence because it produces separate probabilities of event for each cause.
- The fundamental functions of the competing risks analyses are:
  1. Cumulative incidence—Since the KM estimator produces estimates of the survival function $S(t)$, we obtain the cumulative incidence as $h(t) = 1 - S(t)$.
  2. Cause-specific hazard—It is the conditional probability that a subject dies in the interval $[t, t + \Delta t]$ and the event of interest is the $i$th cause, given that the subject was alive just before time $t$.
  3. The aforementioned methods do not permit the inclusion of covariates of the marginal probability function. In fact many authors noted that the effect of a single covariate on the cause-specific hazard function of a particular event may be substantially different from the effect of the covariate on the corresponding cumulative incidence function. Fine and Gray (1999) suggested an approach that combines estimates of the cause-specific hazard function under proportional hazard formulation using the partial likelihood principle and weighting technique.

**Example 9.5: Competing Risk in SAS—The Bone Marrow Transplant (BMT) Data**

Bone marrow transplant (BMT) is a standard treatment for acute leukemia. Klein and Moeschberger (2003) present a set of BMT data for 137 patients, grouped into three risk categories based on their status at the time of transplantation: acute lymphoblastic leukemia (ALL), acute myelocytic leukemia (AML) low-risk, and AML high-risk. During the follow-up period, some patients might relapse or some patients might die while in remission. Consider relapse to be the event of interest. Death is a competing risk because death impedes the occurrence of leukemia relapse. The Fine and Gray (1999) model is used to compare the risk categories on the disease-free survival.

   The following DATA step creates the data set Bmt. The variable Disease represents the risk group of a patient, which is either ALL, AML-Low Risk, or AML-High Risk. The variable T represents the disease-free survival in days, which is the time to relapse, time to death, or censored. The variable Status has three values: 0 for censored observations, 1 for relapsed patients, and 2 for patients who die before experiencing a relapse.

Competing Risk in SAS (code):

```
data Risk;
  Disease=1; output;
  Disease=2; output;
  Disease=3; output;
  format Disease DiseaseGroup.;
  run;
ods graphics on;
ODS RTF;
proc phreg data=Bmt plots(overlay=stratum)=cif;
  class Disease (order=internal ref=first);
  model T*Status(0)=Disease / eventcode=1;
  Hazardratio 'Pairwise" Disease / diff=pairwise;
  baseline covariates=Risk out=out1 cif=_all_ / seed=191;
run;
ODS GRAPHICS OFF;
ODS RTF CLOSE;
```

The preceding SAS statements use the PHREG procedure to fit the proportional subdistribution hazards model. To designate relapse (Status=1) as the event of interest, you specify EVENTCODE=1 in the MODEL statement. The HAZARDRATIO statement provides the hazard ratios for all pairs of disease groups. The COVARIATES= option in the BASELINE statement specifies the data set that contains the covariate settings for predicting cumulative incidence functions; and the OUT= option saves the prediction results in a SAS data set. The PLOTS= option in the PROC PHREG statement displays the cumulative incidence curves. These curves are given in Figure 9.4.

## SAS Output

Analysis of Maximum Likelihood Estimates

| Parameter | | DF | Parameter Estimate | Standard Error | Chi-Square | Pr > ChiSq | Hazard Ratio | Label |
|---|---|---|---|---|---|---|---|---|
| Disease | AML-Low Risk | 1 | –0.80340 | 0.42846 | 3.5160 | 0.0608 | 0.448 | Disease AML-Low Risk |
| Disease | AML-High Risk | 1 | 0.50849 | 0.36618 | 1.9283 | 0.1649 | 1.663 | Disease AML-High Risk |

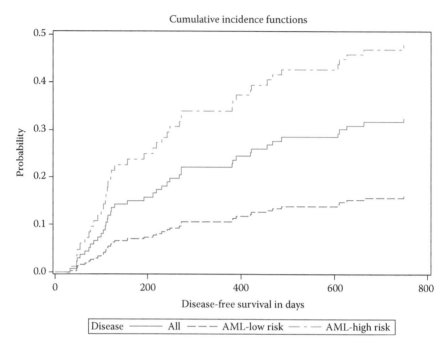

**FIGURE 9.4**
Plots of cumulative cause-specific hazards.


Pairwise: Hazard Ratios for Disease

| Description | Point Estimate | 95% Wald Confidence Limits | |
|---|---|---|---|
| Disease ALL vs AML-Low Risk | 2.233 | 0.964 | 5.171 |
| Disease AML-Low Risk vs ALL | 0.448 | 0.193 | 1.037 |
| Disease ALL vs AML-High Risk | 0.601 | 0.293 | 1.233 |
| Disease AML-High Risk vs ALL | 1.663 | 0.811 | 3.408 |
| Disease AML-Low Risk vs AML-High Risk | 0.269 | 0.127 | 0.573 |
| Disease AML-High Risk vs AML-Low Risk | 3.713 | 1.745 | 7.900 |


## Competing Risk in R

```
library(cmprsk)
bmtcrr <- read.csv(file.choose())
attach(bmtcrr)
data=data.frame(bmtcrr)
attach(data)
names(data)
nrow(data)
```

```
table(data$Status)
table(Disease)
CI.overall = cuminc(ftime = data$ftime, fstatus = data$Status)
CI.overall
plot(CI.overall, curvlab = c("Alive", "Dead"), xlab = "weeks")
CI.estimates = cuminc(ftime = data$ftime, fstatus = data$Status,
  group = data$Disease)
CI.estimates
#The result can again be plotted, this time with separate curves for
  each
#risk class
plot(CI.estimates, lty = c(1, 1, 1, 2, 2, 2, 3, 3,3), col = c("black",
  "blue","green",
"black", "blue","green"), curvlab = c("Alive, Disease 1", "Alive,
  Disease 2","Alive, Disease 3",
"Dead, Disease 1", "Dead, Disease 2", "Dead, Disease 3"), xlab =
  "weeks")
CI.estimates$Tests
```

We shall leave it as an exercise to reproduce the R output, if you closely follow the preceding R-code.

## 9.10 Time-Dependent Variables

In many survival data, individuals are monitored for the duration of the study. During this period, the values of certain explanatory variables may be recorded on a regular basis. For example, the size of the tumor, and other variables, may be recorded at frequent intervals. If account can be taken of the values of explanatory variables as they evolve, a more satisfactory model for the hazard of death at any given time would be obtained.

Variables whose values change over time are known as *time-dependent variables*, and in this section, we see how such variables can be incorporated in models used in the analysis of survival data.

### 9.10.1 Types of Time-Dependent Variables

It is useful to consider two types of variables that change over time, which may be referred to as *internal variables* and *external variables*.

Internal variables relate to a particular individual in a study and can only be measured while a patient is alive. Such data arises when repeated measurements of certain characteristics are made on a patient over time, and examples include white blood cell count, systolic blood pressure, and serum cholesterol level. Variables that describe changes in the status of a patient are also of this type. In each case, such variables reflect the condition of the patient and their values may well be associated with the survival time of the patient.

On the other hand, external variables are time-dependent variables that do not necessarily require the survival of a patient for their existence. One type of external variable is a variable that changes in such a way that its value will be known in advance at any future time. The most obvious example is the age of a patient, in that once the age at the time origin is known, that patient's age at any future time will be known exactly.

Time-dependent variables also arise in situations where the coefficient of a time-constant explanatory variable is a function of time. Under the Cox proportional hazards model, the coefficient is a log-hazard ratio, and so under this model, the hazard ration is constant over time. If this ration were in fact a function of time, then the coefficient of the explanatory variable that varies with time is referred to as a *time-varying coefficient*.

In other words, a term that involves a time-varying coefficient can be expressed as a time-dependent variable with a constant coefficient. However, if $\beta(t)$ is a nonlinear function of one or more unknown parameters, for example $\beta_0 exp(\beta_1 t)$, the term is not easily fitted in a model.

These different types of time-dependent variables can be introduced into the Cox proportional hazards model. The resulting model will simply be referred to as the Cox regression model and is described in the following section.

### 9.10.2 Model with Time-Dependent Variables

Extending the Cox model to the situation in which some of the explanatory variables are time-dependent, we write $x_{ji}(t)$ for the value of the $j$th explanatory variable at time $t$ in the $i$th individual. The Cox regression model then becomes

$$h_i(t) = \exp\left\{ \sum_{j=1}^{p} \beta_j x_{ji}(t) \right\} h_0(t) \tag{9.13}$$

In this model, the baseline hazard function $h_0(t)$ is interpreted as the hazard function for an individual for whom all the variables are zero at the time origin, and remain at this same value through time.

It is important to note that in the model given in Equation 9.13, the values of the variables $x_{ji}(t)$ depend on the time $t$, and so the relative hazard $h_i(t)/h_0(t)$ is also time-dependent. This means that the hazard of death at time $t$ is no longer proportional to the baseline hazard, and the model is no longer a proportional hazards model.

To provide an interpretation of the $\beta$-parameters in this model, consider the ratio of the hazard functions at time $t$ for two individuals: the $r$th and $s$th. This is given by

$$\frac{h_r(t)}{h_s(t)} = \exp\left[ \beta_1\{x_{r1}(t) - x_{s1}(t)\} + \cdots + \beta_p\{x_{rp}(t) - x_{sp}(t)\} \right] \tag{9.14}$$

The coefficients $\beta_j$, $j = 1, 2, ..., p$ can therefore be interpreted as the log-hazard ratio for two individuals whose value of the $j$th explanatory variable at any time $t$ differs by one unit, with the two individuals having the same values of all the other $p-1$ variables at that time.

**Example 9.5: Stint Redo with Time-Varying Covariate (Hypertension Status)**

The data represents the follow-up time for heart patients who received a stint. Quite often some patients require restinting. We would like to identify the potential risk factors that are associated with the event of restinting. The time variable here is the time between stints and restints. The time-dependent covariate "hypertension" is measured at five different occasions. Specifically for each of the five follow-up periods, we specified a dummy variable coded 1 if the patient had elevated blood levels during the follow-up visit; otherwise the variable was coded 0. A very efficient SAS code following Allison (2010) is given below. Note that from Table 9.10 the only predictor for restinting is patient's age.

**SAS Code**

```
data stint;
input redo treat age gender smoking diabetes educ ht1 ht2 ht3 ht4
ht5 time;
fup=time+5;
cards;
proc PHREG data=stint;
class gender;
model fup*redo(0)= treat    age smoking    diabetes educ hyper
/TIES=EFRON;
ARRAY ht(*) ht1-ht5;
hyper=ht[time];
RUN;
```

**TABLE 9.10**

SAS Output for the PHREG Model Accounting for the Time-Varying Covariate

| | | | Analysis of Maximum Likelihood Estimates | | | |
| Parameter | DF | Parameter Estimate | Standard Error | Chi-Square | Pr > ChiSq | Hazard Ratio |
|---|---|---|---|---|---|---|
| treat | 1 | −0.45171 | 0.28340 | 2.5405 | 0.1110 | 0.637 |
| age | 1 | 0.04293 | 0.02105 | 4.1602 | 0.0414 | 1.044 |
| smoking | 1 | 0.13529 | 0.31833 | 0.1806 | 0.6708 | 1.145 |
| diabetes | 1 | −0.26865 | 0.27948 | 0.9240 | 0.3364 | 0.764 |
| educ | 1 | 0.01631 | 0.15146 | 0.0116 | 0.9143 | 1.016 |
| hyper | 1 | −0.38866 | 0.32084 | 1.4675 | 0.2257 | 0.678 |

## 9.11 Joint Modeling of Longitudinal and Time to Event Data

The main objective of this section is to briefly introduce the research in an area that aims at characterizing the relationship between a longitudinal response process and a time-to-event (see Tsiatis et al., 1995; Wulfsohn and Tsiatis, 1997). Considerable recent interest has focused on so-called joint models, where models for the event time distribution and longitudinal data are taken to depend on a common set of latent random effects.

In many biomedical studies it is common that both longitudinal measurements of a response variable and the time to some event of interest are recorded during follow-up. A typical example is the AIDS study where CD4 count (De Gruttola and Tu, 1994; Faucett and Thomas, 1996) and viral load are collected longitudinally and the time to AIDS or death is also monitored. Another example is the Scleroderma Lung Study (Tashkin et al., 1987), a double-blind, randomized clinical trial to evaluate effectiveness of oral cyclophosphamide (CYC) versus placebo in the treatment of lung disease due to scleroderma.

In this study the primary outcome is forced vital capacity (FVC, as % predicted), determined at three-month intervals from the baseline. The event of interest is the time-to-treatment failure or death. A treatment failure occurs when %FVC of a patient in either group falls by ≥15% after three months into the treatment. In both examples the two endpoints are known to be correlated, which may introduce unignorable nonresponse missing values for the longitudinal outcome after event times. This type of missing data cannot be handled correctly by standard methods such as mixed effects models (Laird and Ware, 1982) and generalized estimating equations (Liang and Zeger, 1986; Zeger and Liang, 1986; Zeger, Liang, and Albert, 1988).

A potential difficulty for making inference on the longitudinal process is that occurrence of the time-to-event may induce an informative censoring, as discussed by Wu and Carroll (1988), Hogan and Laird (1997), and many other authors. For example, subjects with more serious HIV disease may be more likely to experience the event or withdraw from the study earlier than healthier individuals, leading to fewer CD4 measurements, and to have sharper rates of CD4 decline. Failure to take appropriate account of this phenomenon, for example, by using ordinary longitudinal data techniques, can lead to biased estimation of average quantities of interest. Valid inference requires a framework in which potential underlying relationships between the event and longitudinal process are explicitly acknowledged. Current research shows an increasing popularity of the joint likelihood approaches, due to their efficiency and their advantages compared with the methodologies mentioned earlier (Zeger and Cai, 2005). Joint models take into account the association between the longitudinal and the survival process by simultaneously determining the parameter estimates for both processes.

Rizopoulos (2010) has proposed a joint model where the time-to-event process is of main interest and by a longitudinal time-dependent covariate assumed to be measured with error.

---

## 9.12 Submodel Specification

The joint model consists of two linked submodels, known as the longitudinal submodel and the survival submodel. To introduce this methodology we will use the same notation as outlined in the present chapter. To understand the model development we shall modify the notations used in this chapter.

Let $T_i$ be the event time, $C_i$ the censoring time, and $\delta_i = 1(T_i \leq C_i)$ the event indicator for the $i$th subject.

Let $y_i(t)$ be the observed value of the time-dependent covariate at time point $t$, equivalently, $y_{ij} = \{y_i(t_{ij}), \ j = 1,...,n_i\}$. Thus, $m_i(t)$ denotes the *true* and unobserved value of the respective longitudinal outcome at time $t$, uncontaminated with the measurement error value of the longitudinal outcome, so it is different from $y_i(t)$.

### 9.12.1 The Survival Submodel

Our aim is to associate the *true* and unobserved value of the longitudinal outcome at time $t$, $m_i(t)$, with the risk for an event $T_i$. The relative risk model can be written as

$$\lambda_i(t|M_i(t), w_i) = \lambda_0(t) \exp\{\gamma^T w_i + \alpha m_i(t)\}, t > 0 \tag{9.15}$$

where $M_i(t) = \{m_i(s), 0 \leq s < t\}$ denotes the history of the true (unobserved) longitudinal process up to time $t$. Let $\lambda_0(\cdot)$ denote the baseline risk function and $w_i$ the vector of baseline covariates. The interpretation of the regression coefficients is exactly the same:

- $\exp(\gamma_j)$ denotes the ratio of hazards for one unit change in the $j$th covariate at any time $t$.
- $\exp(\alpha)$ denotes the relative increase in the risk for an event at time $t$ that results from one unit increase in $m_i(t)$ at this point.

In the expression 9.15 we can note that it depends only on a single value of the time-dependent marker $m_i(t)$. To take into account the whole covariate history $M_i(t)$ to determine the survival function, we obtain

$$S_i(t|M_i(t), w_i) = P(T_i > t|M_i(t), w_i)$$

$$= \exp\left(-\int_0^t \lambda_0(s) \exp\{\gamma^T w_i + \alpha m_i(s)\} ds\right) \tag{9.16}$$

Remember again that both are written as a function of a baseline hazard $\lambda_0(t)$. Regardless of the fact that the literature recommends to leave $\lambda(t_0)$ completely unspecified in order to avoid the impact of misspecifying the distribution of survival times. In the joint modeling framework it can lead to an underestimation of the standard error of the parameter estimates (Hsieh et al., 2006). There are several options to use a risk function corresponding to a known parametric distribution, such as

- The *Weibull model*. Let $Y$ follow a Weibull distribution with parameters $t$ and $p$, $Y \sim W(\lambda, p)$ and the hazard is obtained as

$$\lambda(t) = \lambda p(\lambda t)^{p-1}$$

  where if $p > 1$ indicates that the failure rate increases with time, decreasing if $p < 1$, and constant over time if $p = 1$, called also *exponential model*.

But it is more desirable to flexibly model the baseline risk function. Among the proposals encountered, we would like to highlight those that follow.

- The *piecewise-constant model*, where the baseline risk function takes the form

$$\lambda_0(t) = \sum_{q=1}^{Q} \xi_q I\left(v_{q-1} < t \le v_q\right)$$

  where $0 = v_0 < v_1 < \ldots < v_Q$ denotes a partition of the time scale, with $v_Q$ being larger than the largest observed time, and $v_q$ denotes the value of the hazard in the interval $(v_{q-1}, v_q]$.

- The *regression splines model*, where the log baseline risk function $\log \lambda_0(t)$ is given by

$$\log \lambda_0(t) = \kappa_0 + \sum_{d=1}^{m} \kappa_d B_d(t, q)$$

  where $\kappa^T = (\kappa_0, \kappa_1, \ldots, \kappa_m)$ are the spline coefficients, $q$ denotes the degree of the B-splines basis functions $B(\cdot)$, and $m = \tilde{m} + q - 1$, with $\tilde{m}$ denoting the number of interior knots.

In both models, the specification of the baseline hazard becomes more flexible as the number of knots increases.

### 9.12.2 Estimation: JM Package

In the previous chapter the estimation of the parameters was based on the maximum likelihood approach for both the survival and the longitudinal processes. Rizopoulos (2010) has also used the likelihood method for joint models, as perhaps the most commonly used approach in the joint literature. Further readings are available in Henderson et al. (2000) and Elashoff (2008).

Although software capable of fitting joint models has recently been developed, we find different approaches to model specification across software packages. We shall use one available package in R, which is a random effect jointly modeled with survival analysis set up. The package that we use is named "JM" (Rizopoulos, 2010).

The R package JM constitutes a useful tool for the joint modeling of longitudinal and time-to-event data. In addition it contains all the methodology explained earlier. In order to adjust the submodels, two additional packages are necessary. The linear mixed effects modeling is based on the output of the function **lme()** from package **nlme()**, and the survival fit is implemented by either function **coxph()** or function **survreg()** of package survival. Then, the joint model is fitted by **jointModel()**, which includes as main arguments the two separately fitted models to extract all the required information.

#### Example 9.6: AIDS Data

The AIDS data set has 467 patients with advanced human immunodeficiency virus during antiretroviral treatment who had failed or were intolerant to zidovudine therapy. The objective of the study was to compare the efficacy and safety of two alternative antiretroviral drugs, namely, didanosine (ddI) and zalcitabine (ddC) in the time-to-death. Patients were randomized to receive either ddI or ddC, and CD4 cell counts were recorded at study entry, as well as at 1, 6, 12, and 18 months. Study details can be found in Abrams et al. (1994).

#### R Code: Joint modeling of survival and longitudinal AIDS data

```
library(survival)
library(nlme)
library(JM)
library(asaur)
aids=read.csv(file.choose())
td.Cox <- coxph(Surv(start, stop, event) ~ drug + CD4, data = aids)

td.Cox
# Joint model fit for the AIDS dataset
lmeFit.aids <- lme(CD4 ~ obstime + obstime:drug,
    random = ~ obstime | patient, data = aids)
```

```
coxFit.aids <- coxph(Surv(Time, death) ~ drug, data = aids.id, x =
TRUE)
jointFit.aids <- jointModel(lmeFit.aids, coxFit.aids,
  timeVar = "obstime", method = "piecewise-PH-aGH")
summary(jointFit.aids)
Call:
jointModel(lmeObject = lmeFit.aids, survObject = coxFit.aids,
  timeVar = "obstime", method = "piecewise-PH-aGH")
```

**Data Descriptives:**

Longitudinal Process          Event Process
Number of Observations: 1405  Number of Events: 188 (40.3%)
Number of Groups: 467

**Joint Model Summary:**

Longitudinal Process: Linear mixed-effects model
Event Process: Relative risk model with piecewise-constant
                baseline risk function
Parameterization: Time-dependent

```
 log.Lik    AIC      BIC
-4328.261 8688.523 8754.864
```

**Variance Components:**

```
             StdDev    Corr
(Intercept)  4.5839    (Intr)
obstime      0.1822   -0.0468
Residual     1.7377
```

Coefficients:

**Longitudinal Process**

```
                  Value   Std.Err  z-value  p-value
(Intercept)      7.2203   0.2218   32.5537  <0.0001
obstime         -0.1917   0.0217   -8.8374  <0.0001
obstime:drugddI  0.0116   0.0302    0.3834   0.7014
```

**Event Process**

```
            Value   Std.Err   z-value   p-value
drugddI     0.3348   0.1565    2.1397    0.0324
Assoct     -0.2875   0.0359   -8.0141   <0.0001
log(xi.1)  -2.5438   0.1913  -13.2953
log(xi.2)  -2.2722   0.1784  -12.7328
log(xi.3)  -1.9554   0.2403   -8.1357
log(xi.4)  -2.5011   0.3412   -7.3297
log(xi.5)  -2.4152   0.3156   -7.6531
log(xi.6)  -2.4018   0.4007   -5.9941
log(xi.7)  -2.4239   0.5301   -4.5725
```

## 9.13 Modeling Clustered Survival Data

As with all types of response variables, techniques must be available
for analyses performed on correlated time-to-event data. The complexity
of studies involving multiple treatment centers, family members, and

measurements repeatedly made on the same individual require methods to account for correlation in the data. Such is the case for any type of response, be it continuous, binary, or a time to event. Use of the multivariate normal distribution allows correlation to be accounted for in continuous data where techniques are well established. For the situation of binary responses, work over the past few decades has resulted in tests adjusted for correlation in the data. However, for the time-to-event case, methods of accounting for correlation in the data have only recently been developed. Two methods with existing programs in computer software packages are currently available: the marginal approach and frailty models. They will be discussed subsequently and their properties contrasted. Situations in which each method is desirable are discussed.

### 9.13.1 Marginal Models (GJE Approach)

The Generalized Jackknife Estimator (GJE) (Therneau, 1993) produces robust parameter estimates and variances.

### 9.13.2 Random Effects Models (Frailty Models)

The notion of frailty provides a convenient way to introduce random effects, association, and unobserved heterogeneity into models for survival data. In its simplest form, a frailty is an unobserved random proportionality factor that modifies the hazard function of an individual or of related individuals. In essence, the frailty concept goes back to work of Greenwood and Yule (1920) on "accident proneness." The term *frailty* itself was introduced by Vaupel et al. (1979) in univariate survival models and the model was substantially promoted by its application to multivariate survival data in a seminal paper by Clayton (1978) (without using the notion "frailty") on chronic disease incidence in families.

Frailty models are extensions of the proportional hazards model, which is best known as the Cox model (Cox, 1972), the most popular model in survival analysis. Normally, in most clinical applications, survival analysis implicitly assumes a homogenous population to be studied. This means that all individuals sampled into that study are subject, in principle, to the same risk (e.g., risk of death, risk of disease recurrence). In many applications, the study population cannot be assumed to be homogeneous but must be considered as a heterogeneous sample, that is, a mixture of individuals with different hazards. For example, in many cases it is impossible to measure all relevant covariates related to the disease of interest, sometimes because of economical reasons, sometimes because the importance of some covariates is still unknown. The frailty approach is a statistical modeling concept that aims to account for heterogeneity, caused by unmeasured covariates. In statistical terms, a frailty model is a random effect model for time-to-event data, where the random effect (the frailty) has a multiplicative effect on the baseline hazard function. One can distinguish two broad classes of frailty models:

1. Models with a univariate survival time as endpoint
2. Models that describe multivariate survival endpoints (e.g., competing risks, recurrence of events in the same individual, occurrence of a disease in relatives)

In the first case, a univariate (independent) lifetime is used to describe the influence of unobserved covariates in a proportional hazards model (heterogeneity). The variability of survival data is split into a part that depends on risk factors, and is therefore theoretically predictable, and a part that is initially unpredictable, even when all relevant information is known. A separation of these two sources of variability has the advantage that heterogeneity can explain some unexpected results or give an alternative interpretation of some results, for example, crossing-over effects or convergence of hazard functions of two different treatment arms (see Manton and Stallard, 1981) or leveling-off effects (i.e., the decline in the increase of mortality rates), which could result in a hazard function at old ages parallel to the x-axis (Aalen and Tretli, 1991). More interesting, however, is the second case when multivariate survival times are considered. There one aims to account for the dependence in clustered event times, for example, in the lifetimes of patients in study centers in a multicenter clinical trial, caused by center-specific conditions (see Andersen et al., 1999). A natural way to model dependence of clustered event times is through the introduction of a cluster-specific random effect—the frailty. This random effect explains the dependence in the sense that had we known the frailty, the events would be independent. In other words, the lifetimes are conditional independent, given the frailty. This approach can be used for survival times of related individuals like family members or recurrent observations on the same person. The extension of the Cox regression model that incorporates the frailty component is

$$h(t|v_i) = v_i h_0(t) \exp\left(\beta x_{ij}\right) = h_0(t) \exp\left(\beta x_{ij} + z_i\right) \tag{9.17}$$

showing that $v_i = \exp(z_i)$ actually behaves as an unknown covariate for the $i$th cluster in the model.

Using previous relationships between the survival and hazard function, we have the conditional survival function as

$$S(t|v_i) = \exp\left[-v_i H_0(t) \exp\left(\beta x_{ij}\right)\right]$$

and the conditional likelihood as

$$L(\gamma, \beta | v_i) = \prod_{i=1}^{k} \prod_{j=1}^{n_j} \left( h\left(t_{ij}|v_i\right)^{\delta_{ij}} S\left(t_{ij}|v_i\right) \right)$$

Note that for the Weibull failure time distribution with $\alpha = 1$, the cumulative base line hazard is

$$H_0(t|z) = \exp\left(\gamma \log(t) + x\beta + z\right)$$

where there are $k$ clusters, the $i$th one being of size $n_i$. Substitution gives

$$L(\gamma, \beta | v_i) = \prod_{i=1}^{k} \prod_{j=1}^{n_j} \left( \left[ h_0(t) v_i \exp\left(\beta X_{ij}\right) \right]^{\delta_{ij}} \exp\left[ -v_i H_0(t) \exp\left(\beta X_{ij}\right) \right] \right) \quad (9.18)$$

The marginal (i.e., independent of $v_i$) likelihood, $L(\gamma, \beta)$, is obtained through integration of the random effect distribution. The following sections illustrate the methodology when the frailty distributions are conveniently chosen.

### 9.13.2.1 Weibull Model with Gamma Frailty

A common assumption is for the random effect to follow a gamma distribution with mean one and variance $\tau$, that is,

$$g(v_i) = \frac{v_i^{1/\tau - 1} \exp(-v_i/\tau)}{\Gamma(1/\tau)\tau^{1/\tau}}$$

Therefore, $E(v_i) = 1$ and $\text{var}(v_i) = \tau$. The failure times are assumed to follow Weibull distribution with baseline hazard function $h_0(t) = \dfrac{f(t)}{S(t)} = \alpha\gamma(\alpha t)^{\gamma-1}$. The marginal likelihood is then obtained as

$$L(\gamma, \beta, \tau) = \prod_{i=1}^{k} \prod_{j=1}^{n_j} \int_0^{\infty} L(\gamma, \beta | v_i) g(v_i) dv_i \quad (9.19)$$

Inference on the regression parameters, baseline hazard parameter, and dispersion parameter is then possible using maximum likelihood procedures. The Newton-Rahpson method is used for the optimization of the likelihood function and obtaining the parameters estimates.

#### Example 9.7: Ventilating Tube Data

It was desired to examine the significance of treatment in delaying time-to-tube failure after accounting for correlation within ears. Notice that maximum likelihood estimates are obtained for the cluster level treatment effect $\beta$, the Weibull baseline hazard parameter $\gamma$, and the dispersion parameter $\tau$, but not for within-cluster covariates due to the fact that each child received the same medicine in each ear.

The SAS code for the data step and the PHREG procedure is

```
data ear;
input subject treat ear time censor;
```

```
cards;
1001 1 1 3.10 1
1001 1 2 4.80 0
  .  .  .  .  .
  .  .  .  .  .
2017 2 2 8.70 1
;

proc phreg data=ear covs(aggregate) covm ;
model time*censor(0)= treat ;
id subject;
run;
```

The partial output is

Testing Global Null Hypothesis: BETA=0

| Test | Chi-Square | DF | Pr > ChiSq |
|---|---|---|---|
| Likelihood Ratio | 4.3407 | 1 | 0.0372 |
| Score (Model-Based) | 4.4225 | 1 | 0.0355 |
| Score (Sandwich) | 3.8127 | 1 | 0.0509 |
| Wald (Model-Based) | 4.3788 | 1 | 0.0364 |
| Wald (Sandwich) | 3.5996 | 1 | 0.0578 |

Analysis of Maximum Likelihood Estimates With Model-Based Variance Estimate

| Variable | DF | Parameter Estimate | Standard Error | Chi-Square | Pr > ChiSq | Hazard Ratio |
|---|---|---|---|---|---|---|
| Treat | 1 | −0.35292 | 0.16865 | 4.3788 | 0.0364 | 0.703 |

Analysis of Maximum Likelihood Estimates With Sandwich Variance Estimate

| Variable | DF | Parameter Estimate | Standard Error | StdErr Ratio | Chi-Square | Pr > ChiSq | Hazard Ratio |
|---|---|---|---|---|---|---|---|
| treat | 1 | −0.35292 | 0.18601 | 1.103 | 3.5996 | 0.0578 | 0.703 |

It is well known that parametric survival regression models have more power than semiparametric models such as Cox regression. From the output we find that both models provide the same interpretation for the treatment effect, which is the reduction of the hazard of failure. The $p$-value provided by the parametric gamma frailty model is much smaller than the one obtained from the Cox regression model after accounting for the possible correlation between ears.

Frailty models have a great deal of potential in accounting for correlation arising in clustered survival data (Hougaard, 1995). Although the gamma frailty distribution has been examined here, other possibilities include the inverse-Gaussian and lognormal. The inverse-Gaussian appears to be particularly well-suited to the situation in which survival times are positively skewed as well as correlated. However, these types of models have the downfall of being difficult to fit due to a complex distributional structure, and divergence is not uncommon when attempting to maximize the likelihood.

### SAS Code for Weibull Survival Distribution with Lognormal Frailty

```
proc nlmixed data=ear;
bounds gamma > 0;
linp = b0+b1*(treat-2) +z;
alpha = exp(-linp);
G_t = exp(-(alpha*time)**gamma);
g = gamma*alpha*((alpha*time)**(gamma-1))*G_t;
LL = (censor=0)*log(g)+(censor=1)*log(G_t);
model time ~ general(LL);
random z ~ normal(1,exp(2*logsig)) subject=subject out=EB;
predict 1-G_t out=cdf;
run;
```

### Results of Fitting Frailty Model to the Ears Data
The R code to read the data and fit the gamma frailty model is

```
ear <- read.csv(file.choose())
fit <- coxph(Surv(time, censor)~ treat + frailty(subject,
dist="gamma"), ear)
summary(fit)
```

The results of fitting the model are

```
Call:
coxph(formula = Surv(time, censor) ~ treat + frailty(subject,
  dist = "gamma"), data = ear)

 n= 156
                          coef  se(coef) se2    Chisq  DF    p
treat                     -0.506 0.228    0.177  4.93   1.0   0.026
frailty(subject, dist = "                      42.24  24.7  0.015

   exp(coef) exp(-coef) lower .95 upper .95
treat   0.603    1.66    0.386    0.943

Iterations: 7 outer, 40 Newton-Raphson
  Variance of random effect= 0.267  I-likelihood = -584.8
Degrees of freedom for terms= 0.6 24.7
Rsquare= 0.348  (max possible= 0.999 )
Likelihood ratio test= 66.7 on 25.3 df,  p=1.36e-05
Wald test       = 4.93 on 25.3 df,  p=1
```

Another illustration of the Weibull-lognormal frailty model is given in Example 9.8.

**TABLE 9.11**

Maximum Likelihood Estimates of the Weibull Lognormal Frailty Model
Parameters

| | | | | | | Parameter Estimates | | | | |
|---|---|---|---|---|---|---|---|---|---|---|
| Parameter | Estimate | Standard Error | DF | t Value | Pr > \|t\| | Alpha | Lower | Upper | Gradient |
| b0 | −4.2856 | 0.1277 | 71 | −33.56 | <.0001 | 0.05 | −4.5403 | −4.0310 | −3.07E−6 |
| b1 | −1.8937 | 0.05916 | 71 | −32.01 | <.0001 | 0.05 | −2.0116 | −1.7757 | 0.000013 |
| b2 | −0.2216 | 0.03862 | 71 | −5.74 | <.0001 | 0.05 | −0.2986 | −0.1446 | −0.00007 |
| b3 | 0.03927 | 0.07457 | 71 | 0.53 | 0.6001 | 0.05 | −0.1094 | 0.1880 | 0.000016 |
| b4 | 0.1155 | 0.07207 | 71 | 1.60 | 0.1135 | 0.05 | −0.02822 | 0.2592 | −0.00002 |
| s2 | 0.3247 | 0.07595 | 71 | 4.28 | <.0001 | 0.05 | 0.1733 | 0.4761 | −0.00006 |
| alpha | 7.3119 | 0.1868 | 71 | 39.14 | <.0001 | 0.05 | 6.9395 | 7.6844 | −5.23E−6 |

**Example 9.8: Weibull Model with Lognormal Frailty (Data: Culling)**

Why are cows being culled? The data has an outcome variable (time-to-event): age at culling in 72 herds. The risk factors are parity, presence or absence of diseases, herd size, and amount of milk. Note that all continuous variables are centralized. The SAS code is given as

```
Input cluster cowid time censor parity herdsiz milk disease
  proc nlmixed data=herd;
  linp = exp(b0+b1*parity+b2*milk+b3*herdsiz+b4*disease+z);
  ll=-linp*time**(alpha+1)+censor*(LOG(alpha+1)+
    alpha*LOG(time)+LOG(linp));
  model time ~ general(LL);
  random z ~ normal(0,s2) subject=cluster ;
  PARMS b0=1 b1=0 b2=0 b3=0 b4=0 s2=1 alpha=0;
  run;
```

**SAS Output**

The maximum likelihood estimation of the fully parametric frailty Weibull model shows that parity and herd size are the most significant predictors of age at culling. The results are summarized in Table 9.11.

## 9.14 Sample Size Requirements for Survival Data

Suppose that we would like to compare between two survival curves of two independent groups. To compute sample size, the following is needed: the estimated proportions of subjects in each group who are "event-free" at a fixed time; or the estimated hazard ratio ($h = e^{\beta}$, where $\beta$ is the Cox model coefficient corresponding to the treatment effect) and the estimated control group probability of survival at a fixed time; or the estimated median survival times or exponential hazard rates in each group.

### 9.14.1 Sample Size Based on Log-Rank Test

This calculation assumes patients are followed for a fixed length of time ($t$), and the hazard ratio ($h$) is constant over time. Suppose that we have two treatment groups, if $p_i$ denotes the proportion of subjects who are event-free at time $t$ for group $i (i = 1, 2)$, then

$$h = \frac{\ln(p_1)}{\ln(p_2)}$$

and the sample size per group is

$$n = \frac{(z_{\alpha/2} + z_\beta)^2 (h + 1)^2}{(2 - p_1 - p_2)(h - 1)^2}$$

**Example 9.9**

We need to determine if a new drug for treatment of lung cancer lengthens survival time. All patients in the trial will be followed for two years. Find the sample size needed for an $\alpha = 0.05$ level test to have 80% power to detect $h = 2$ if the two-year survival rate under standard therapy is $p_1 = 0.30$.

$$h = 2 = \frac{\ln(.30)}{\ln(p_2)} \quad \Rightarrow \quad p_2 = 0.55$$

$$n = \frac{(1.96 + 0.84)^2 (2 + 1)^2}{(2 - .30 - .55)(2 - 1)^2} = 62$$

### 9.14.2 Exponential Survival and Accrual

This calculation assumes patients enter the study at a uniform rate until the trial ends in $t$ years, and the survival curves are exponential with parameter $\lambda_i$, the hazard for group $i (i = 1, 2)$.

The sample size per group is

$$n = \frac{(z_{\alpha/2} + z_\beta)^2 [\phi(\lambda_1) + \phi(\lambda_2)]}{(\lambda_1 - \lambda_2)^2} \qquad (9.20)$$

where

$$\phi(\lambda) = \frac{\lambda^3 t}{\lambda t + e^{-\lambda t} - 1}$$

If the survival curves are exponential, the hazard for group $i$ can be estimated using

$$\lambda_i = -\frac{1}{t_{50}} \ln(0.5)$$

where $t_{50}$ is the median survival time for group $i$.

Or, if you have $\lambda$ for the control group and $h$ use

$$h = \frac{\lambda_1}{\lambda_2}$$

to determine

$$\lambda_2 = \frac{\lambda_1}{h}$$

**Example 9.10**

In the previous example, suppose the trial lasts two years and the survival curves are exponential. The median survival with the standard drug is one year, corresponding to $\lambda_1 = -\ln(0.5) = 0.693$.

$$h = 1.5 = .693/\lambda_2 \Rightarrow \lambda_2 = 0.462$$

$$\phi(\lambda_1) = \frac{(0.693)^3 \times 2}{(0.693)2 + e^{-.693 \times 2} - 1} = 1.046; \ \phi(\lambda_2) = 0.615$$

$$n = \frac{(1.96 + 1.28)^2 [1.046 + 0.615]}{(0.693 - 0.462)^2} = 324$$

## 9.14.3 Sample Size Requirements for Clustered Survival

This section presents a simple sample size formula for log-rank statistics applied to clustered survival data with variable cluster sizes and arbitrary treatment assignments within clusters. This formula is based on the asymptotic normality of log-rank statistics under certain local alternatives. The derived sample size expression reduces to the formula given in Section 9.14.1, cases of no clustering or within-clustering independence. The results presented here are based on the work by Schoenfeld (1983) and Gangnon and Kosorok (2004).

Assume that

1. We have $n$ clusters, and the average cluster size is $m$.
2. The entire cluster is randomized as a whole (i.e., no within-cluster randomization).
3. The log hazard ratio is denoted by $\vartheta$.
4. $P_1 = n_1/n$ is the proportion of clusters receiving the first treatment, and $P_2 = 1 - P_1$ is the proportion of clusters receiving the second treatment.
5. $D \equiv$ probability of observing an event, hence the required number of events, $k = mnd$, is estimated as

$$k = \frac{(Z_{\alpha/Z} + Z\beta)^2}{P_1 P_2 \theta^2} [1 + (m-1)\rho]$$

and

$$n = \frac{(Z_{\alpha/Z} + Z_\beta)[1 + (m-1)\rho]}{mDP_1 P_2 \theta^2}$$

Here $\rho$ is the intraclass correlation.

Note that in most settings, it is difficult to specify $\rho$ in advance, because $\rho$ depends on the censoring distribution and the dependence between event times within a cluster. The sample size should be based on a conservative choice of $\rho$, such as the largest $\rho$ value for clustered data.

Remark: If the cluster size $m = 2$, and one unit receives treatment 1, and the other receives treatment 2, the sample size formula is

$$n = \frac{2(Z_{\alpha/2} + Z_\beta)^2}{D\theta^2} (1 - \rho)$$

In this case one should choose the smallest possible value for $\rho$ to get a conservative estimate of $n$ (the number of pairs).

**Example 9.11**

In a cluster randomized trial, we want to determine if a new drug for treating a familial disease lengthens survival time. The follow-up period is three years. We need to estimate the number of clusters for an $\alpha = 0.05$ level test to have 90% power to detect hazard ratio of 2. Assume an average cluster size $m = 5$, and intracluster correlation $\rho = 0.5$, and equal number of clusters in both arms ($P_1 = P_2 = 1/2$). We may also expect 10% of the subjects to be censored.

Here, $D = 90\%$, $\vartheta = \ln(2) = 0.69$.

$$n = \frac{(1.96 + 1.28)^2 (1 + 4(.5))}{(5)(.25)(.9)(.69)^2} = 27$$

That is we need 27 clusters to be randomized in each arm.

---

## Exercises

9.1. The following data shows survival times (in months) of patients with Hodgkin's disease who were treated with nitrogen mustards. Group A patients received little or no prior therapy, whereas Group B patients receive heavy prior therapy. Starred observations are censoring times.

Group A         1.20,1.11,4.96, 5.25,5.40, 5.92,8.89, 10.98, 11.18, 13.11, 14.21,
                16.33, 19.77, 21.08, 21.84*, 22.07, 31.38*, 32.62*, 36.18*, 44.99

Group B         1.01, 2.82, 3.61, 5.20, 5.49, 6.72, 7.31, 8.08, 9.11, 14.49*,
                16.85,18.92*, 26.59*, 30.26*, 46.74*

a. Calculate and compare product-limit estimates for two groups. Does there appear to be a difference in the one-year survival probability for the two types of patients?

b. Use any two parametric models to detect group differences in the survival probabilities. Plot the Kaplan-Meier curves. Which model has the smallest AIC?

c. Use (hazard) plots of the empirical hazard function $\hat{H}(t)$ to examine and compare the two life distributions.

9.2. The following data are remission times, in weeks, for a group of 30 leukemia patients in a certain type of therapy; starred observations are censoring times: 1, 1, 2, 5, 5, 6, 6, 6, 7, 8, 9, 9, 10, 12, 13, 14, 18, 20, 24, 26, 29, 32*, 42, 45*, 55*, 59, 66, 73*, 86*, 100.

a. Estimate the median remission time (1) using the nonparametric method described in this chapter and (2) assuming that the underlying distribution of remission times is exponential. Obtain and compare confidence intervals for the median using the two methods.

b. Similarly compare estimates of $S(26)$, the probability of a remission lasting more than 26 weeks, using the nonparametric product-limit estimate and the exponential model, respectively.

9.3. Fit the EARS DATA using the Weibull survival model with gamma frailty in SAS. Compare the results to the fitting obtained from the "PHREG" with "aggregate" option to account for the correlation. Interpret the results obtained from these models

9.4. For the example on sample size for clustered survival data, what is the required number of events needed to achieve the study objectives?

9.5. The data file "transplant_donor_type" shows the survival time in years for liver transplanted patients who received DD (deceased donors) and LD (living donors) organ transplants. The MELD scores are also given.

a. Use the ROC function in R to dichotomize the MELD score to discriminate between the living (status=0) and the dead (status=1). Use the optimal score to create two groups high (above the cut-off point) and low (below the cut-off point), creating "meldgroup".

b. Compare between the DD and LD survival curves, stratifying by the MELD category, using the R function:

```
survdiff(surv(time,status)~donor_type+strata(meldgroup)
```

# 10

## *Introduction to Propensity Score Analysis*

### 10.1 Introduction

Propensity score analysis methods aim to reduce bias in treatment effect estimates obtained from observational studies, which are studies estimating treatment effects with research designs that do not have random assignment of participants to conditions. Research designs (see Chapter 1) to estimate treatment effects that do not have random assignments to conditions are also referred to as quasi-experimental or nonexperimental designs.

Propensity score analysis methods have become a common choice for estimating treatment effects with nonexperimental data in the social sciences (Thoemmes and Kim, 2011). The use of propensity scores to reduce selection bias in nonexperimental studies was proposed by Rosenbaum and Rubin (1983, 1985) and was connected to earlier work by Rubin (1974) on matching methods for selecting an untreated group that was similar to the treated group with respect to covariates. Propensity scores solve a difficult problem with multivariate matching: If there are many covariates, it is difficult to find an appropriate match for each treatment participant with respect to all covariates. With propensity scores, each individual has a unique score that summarizes the relationship between covariates and the treatment assignment. Rosenbaum and Rubin (1983, 1985) have shown that adjustment for the propensity score is sufficient to remove all bias-related covariates.

One of the most serious sources of bias in clinical and epidemiological research is *confounding*. The next section of this chapter details the effects of confounding on the effect of intervention. Thereafter, the issue of propensity score matching will be outlined as an effective way to deal with confounding.

### 10.2 Confounding

#### 10.2.1 Definition of Confounding

Compounding is a distortion (inaccuracy) in the estimated measure of association that occurs when the primary exposure of interest is mixed up with some other factor that is associated with the outcome. In the following diagram, the primary goal is to ascertain the strength of association between fat intake and prostate cancer. Age is a confounding factor because it is associated with the exposure (meaning that older people are more likely to be eating unhealthy food), and it is also associated with the outcome (because older people are at greater risk of developing prostate cancer).

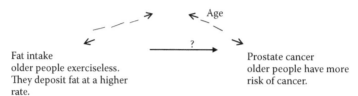

Fat intake
older people exerciseless.
They deposit fat at a higher
rate.

Prostate cancer
older people have more
risk of cancer.

### 10.2.2 Identification of Confounding

There are three conditions that must be satisfied for confounding to exist:

1. The confounding factor must be associated with both the risk factor of interest and the outcome.
2. The confounding factor must be distributed unequally among the groups being compared.
3. A confounder cannot be an intermediary step in the causal pathway from the exposure of interest to the outcome of interest.

Since most diseases are multifactorial, that is multiple risk factors may be associated with the disease, potentially there are several possible confounders.

- A confounder can be another risk factor for the disease. For example, in a hypothetical cohort study testing the association between exercise and heart disease, age is a confounder because it is a risk factor for heart disease.

- Similarly a confounder can also be a preventive factor for the disease. If those people who exercised regularly were more likely to take aspirin, and aspirin reduces the risk of heart disease, then aspirin use would be a confounding factor that would tend to exaggerate the benefit of exercise.

- A confounder can also be a surrogate or a marker for some other cause of disease. For example, socioeconomic status may be a confounder in this example because lower socioeconomic status is a marker for a complex set of poorly understood factors that seem to carry a higher risk of heart disease.

A simple, direct way to determine whether a given risk factor caused confounding is to compare the estimated measure of association before and after adjusting for confounding. In other words, compute the measure of association both before and after adjusting for a potential confounding factor. If the difference between the two measures of association is 10% or more, then confounding was present. If it is less than 10%, then there was little, if any, confounding. Other investigations will determine whether a potential confounding variable is associated with the exposure of interest and whether it is associated with the outcome of interest. If there is a clinically meaningful relationship between the variable and the risk factor and between the variable and the outcome (regardless of whether that relationship reaches statistical significance), the variable is regarded as a confounder. The confounder if not accounted for its effect may account for all or part of an apparent association. Moreover it may cause an overestimate of the true association (positive confounding) or an underestimate of the association (negative confounding).

### 10.2.3 Control of Confounding in Study Design

It is possible to minimize confounding by utilizing certain strategies in the design of a study:

- Restriction
- Matching
- Randomization (in intervention studies only)

There are also analytical techniques that provide a way of adjusting for confounding in the analysis, provided one has information on the status of the confounding factors in the study subjects. These techniques are:

- Stratification
- Multiple variable regression analysis

#### 10.2.3.1 Restriction

One of the conditions necessary for confounding to occur is that the confounding factor must be distributed unequally among the groups being compared. Consequently, one of the strategies employed for avoiding confounding is to restrict admission into the study to a group of subjects who have the same levels of the confounding factors. For example, in a hypothetical study looking at the association between physical activity and heart disease, suppose that age and gender were the only two confounders of concern. If so, confounding by these factors could have been avoided by making sure that all subjects were males between the ages of 50 and 60. This will ensure that the age distributions are similar in the groups being compared, so that confounding will be minimized.

This approach to controlling confounding is simple and effective, but it has several limitations:

- It reduces the number of subjects who are eligible (may cause sample size problem).

- Residual confounding can occur if you do not restrict narrowly enough. For example, in the study on exercise and heart disease, the investigators might have restricted the study to men aged 40 to 65. However, the age-related risk of heart disease still varies widely within this range as do levels of physical activity.

- One cannot evaluate the effects of factors that have been restricted for. For example, if the study is limited to men aged 50 to 60, one cannot use this study to examine the effects of gender or age (because these factors do not vary within your sample).

- Restriction limits generalizability. For example, if we restrict the study to men, we may not be able to generalize the findings to women.

### 10.2.3.2 Matching

Instead of restriction, one could also ensure that the study groups do not differ with respect to possible confounders such as age and gender by matching the two comparison groups (Stuart, 2010). For example, for every active male between the ages of 50 and 60, we could find and enroll an inactive male between the ages of 50 and 60. In this way, the groups we are comparing can artificially be made similar with respect to these factors, so they cannot confound the relationship. This method actually requires the investigators to control confounding in both the design and analysis stages of the study, because the analysis of matched study groups differs from that of unmatched studies. Like restriction, this approach is straightforward, and it can be effective. However, it has the following disadvantages:

- It can be time-consuming and expensive.
- It limits sample size.
- We cannot evaluate the effect of the factors that we matched for.

Nevertheless, matching is useful in the following circumstances:

- When one needs to control for complex, multifaceted variables (e.g., heredity, environmental factors).
- When doing a case-control study in which there are many possible controls, but a smaller number of cases matching can be useful.

## 10.3 Propensity Score Methods

Propensity score matching, stratification, and weighting have several advantages over conditioning on covariates. First, they separate the process of reduction of selection bias with propensity score methods at the design phase of a study. This design stage consists of the determination of matched observations, strata, or weights that achieve balance of covariate distributions between treated and untreated groups, and should be performed independently and without any knowledge of the outcomes. Second, matching, stratification, and weighting allow for smaller outcome models where fewer parameters are estimated, because covariates are not included in the model unless they are of theoretical interest. Third, because the process of balancing covariates between treated and untreated groups is done independently of the outcome, no assumptions are made about the functional form of the relationship between covariates and the outcome.

### 10.3.1 Propensity Scores

The most recent development and complete account of this topic is covered in the book by Leite (2017). The propensity score (PS) is defined as a conditional probability of treatment assignment, given observed covariates (Rosenbaum and Rubin, 1983): $e(X) = P(Z = 1 \mid X)$. The propensity score reduces all the information in the predictors to one number, which greatly simplifies analysis. For example, matching based on multiple covariates to reduce selection bias can be simplified to matching based on the propensity score. Rosenbaum and Rubin (1983, 1985) showed that if treatment selection is strongly ignorable given a set of observed covariates $X$, then it is also strongly ignorable given the propensity score $e(X)$ that is a function of these covariates. More specifically, the same authors proved that if potential outcomes $Y^0$ and $Y^1$ are independent of the treatment assignment given observed covariates $X$, they are also independent of the treatment assignment given the propensity score $e(X)$, and the treatment assignment is independent of covariates given the propensity score. The assumption of strongly ignorable treatment assignment is satisfied when alternative explanations have been accounted for and there is no hidden bias in treatment effect. This assumption requires the assignment to condition be independent of the outcome and not associated with other factors (Holland, 1986).

### 10.3.2 Propensity Score Estimation and Covariate Balance

Once the primary outcome of the study and covariates and possible confounders are made, available data are prepared for analysis, and estimation of propensity scores can be performed with a variety of methods, such as logistic regression. The selection of covariates for the propensity score model is critical, because the nonidentifiability of treatment assignment propensity score methods requires that there be no omitted confounders (Leite, 2017). Therefore, researchers should attempt to identify all true confounders, which are covariates that affect the treatment assignment and the outcome. Besides true confounders, the propensity score model can also include predictors of the outcome that are unrelated to treatment assignment, because these covariates will increase power to test the treatment effect. However, the propensity score model should not include covariates that are related to treatment assignment but not the outcome, because doing so would decrease power (Brookhart et al., 2006).

The question is how can we be sure that covariate balance has been achieved. Evaluation of covariate balance is the main measure of success of the propensity score method and entails comparing characteristics of the distribution of treated and untreated after the propensity score method of the choice has been applied. Evaluation of covariate balance has been performed by graphical, descriptive, and inferential measures. A graphical balance diagnostic can be performed with empirical QQ-plots for continuous covariates and with bar plots for categorical covariates. Inferential measures used for covariate balance evaluation include $t$ tests comparing group means, Hotelling's $T$ (a multivariate $t$ test), and Kolmogorov-Smirnov tests. With inferential measures, obtaining no statistical significance indicates adequate covariate balance. However, inferential measures are not recommended for evaluation of covariate balance, first because covariate balance is a property of the sample, and hypothesis tests refer to the population (Ho et al., 2013). Second, inferential measures depend on sample size, and underpowered tests may fail to indicate substantial covariate unbalance with small samples, and high levels of power may make it hard to achieve balance with very large samples, even if covariate differences between groups are very small.

## 10.4 Methods for Propensity Score Estimation

The two broad classes of methods used for propensity score estimation are parametric models and data mining methods. Parametric models include logistic regression, which will be our focus in this chapter mainly because it is the most popular method to estimate propensity scores. We detailed the data fitting using the logistic model in previous chapters.

Recall from Chapter 5 that under the logit transformation we have

$$\log \mathrm{it}(Z_i = 1|X) = \beta_0 + \beta_1 X_{1i} + \ldots \beta_k X_{ki} \tag{10.1}$$

As we know, the logit is the log odds of the probability of receiving treatment:

$$\log it(Z_i = 1|X) = \log\left(\frac{P(Z_i = 1)}{1 - P(Z_i = 1)}\right) \tag{10.2}$$

The covariate vector includes both risk factors and confounders. The logistic regression model is estimated with maximum likelihood estimation. The propensity scores are estimated probabilities of treatment assignment given covariates (that is, $e_i(X) = P(Z_i = 1 | X)$), which can be obtained from the estimated logits:

$$e_i(X) = \frac{\exp\left(\log it(Z_i = 1|X)\right)}{1 + \exp\left(\log it(Z_i = 1|X)\right)} \tag{10.3}$$

Propensity score stratification or subclassification consists of using the propensity score distribution to divide the sample of treatment participants and nonparticipants into groups that are similar with respect to the distribution of covariates. In fact, propensity score stratification is a method of coarse matching of treated and untreated participants based on propensity scores. The most popular method of propensity score stratification is to divide the sample based on quintiles of the propensity score distribution (Thoemmes and Kim, 2011). This practice originated from seminal works on covariate stratification by Cochran (1968) and propensity score stratification by Rosenbaum and Rubin (1983, 1984), and Rubin (1986) who proposed that the propensity score can be used as a balancing score instead of covariates, thus solving the dimensionality problem of stratifying by multiple covariates. With propensity score stratification, increasing the number of strata also reduces bias. In a review of applications of propensity score stratification, Thoemmes and Kim (2011) found studies that used between 5 and 20 strata are most appropriate for bias reduction.

There are two approaches to using propensity score stratification to estimate treatment effects: (1) pool strata-specific treatment effects and (2) obtain a marginal treatment effect across strata. The first approach was used in the original application of propensity score stratification presented by Rosenbaum and Rubin (1984). The second approach was promoted by Hong (2010), who refers to it as marginal mean weighting through stratification.

Once propensity scores are estimated, the next step of analysis is to divide the sample into strata based on propensity scores. Strata can be created by converting the propensity score variable into a categorical variable using thresholds defined by the quantiles of the propensity score. The most common choice for number of strata is five (Thoemmes and Kim, 2011).

If a researcher is interested in estimating the average treatment effect on the treated (ATT), it may be preferable to stratify based on quintiles of the treated, so that each group has approximately the same number of treated cases, because better covariate balance and power to test the treatment effect may be obtained than when stratification is based on the propensity scores of both groups. The next step in obtaining a treatment effect is to calculate the stratum weights $w_k$. For estimating the average treatment effect (ATE), the stratum weight is $w_k = n_k/n$, which is the stratum size divided by the total sample size. For estimating the ATT, the stratum weight is $w_k = n_{1k}/n_1$. This last weight is the treated sample size within the stratum divided by the total treated sample size.

Once stratum weights are obtained, it is possible to estimate the treatment effect $\Delta_k$ for each stratum $k$. The simplest estimator of the stratum-specific treatment effect is $\Delta_k = \bar{Y}_{1k} - \bar{Y}_{0k}$, which is the difference between the means of treated and untreated observations within each stratum. The treatment effect $\Delta$ for the entire sample can be obtained with a weighted sum of within-stratum treatment effects:

$$\Delta = \sum_{k=1}^{K} W_k \Delta_k \qquad (10.4)$$

The standard error of the treatment effect estimate can be obtained as

$$SE(\Delta) = \sqrt{\sum_{k=1}^{K} w_k^2 \mathrm{var}(\Delta_k)} \qquad (10.5)$$

## 10.5 Propensity Score Estimation When Units of Analysis Are Clusters

In research designs where the treatment assignment is at the cluster level, only confounders at the highest level of hierarchy are important. For example, if families are selected to participate in a genetic epidemiology investigation, then we may select families from communities within a specific time frame and the confounders may exist at the community level. These confounders could include individual-level covariates aggregated to the cluster level. However, in research designs where the treatment assignment is at the individual level, the propensity score model should account for both individual- and cluster-level confounders.

Two general strategies can be used for propensity score analysis of individual-level treatment effects with multilevel data: (1) estimate propensity scores within clusters, estimate a treatment effect for each cluster, and pool the treatment effect estimates; and (2) estimate propensity scores and treatment effect marginally across clusters. Estimating within-cluster effects is advantageous when cluster sizes are large (Kim and Seltzer, 2007) because it removes the confounding effects of both observed and unobserved cluster-level confounders. The cluster sizes should be large enough to allow adequate common support of propensity score distributions within clusters. However, if cluster sizes are not large, marginal estimation of the treatment effect across clusters should be used. This strategy can be implemented accounting for cluster effects in the propensity score model, in the outcome model, or both. There are many technical issues associated with propensity score estimation under cluster sampling and we shall omit the details.

With multilevel data, propensity scores can be estimated with a multilevel logistic model that includes individual-level and cluster-level covariates (Kim and Seltzer, 2007). As an example of a two-level logistic model (see Chapter 5):

$$\log \text{it} \left( Z_{ij} = 1 | X, W \right) = s_{0j} + \beta_0 + \sum_{m=1}^{M} \beta_m X_{mij} + \sum_{n=1}^{N} \pi_n W_{nj} + \sum_{m=1}^{M} s_{mj} X_{mij} \quad (10.6)$$

where $Z_{ij}$ is the treatment indicator; $\beta_0$ is the intercept; $\beta_m$ are the effects of the individual-level covariates; $X_m, \pi_n$ are the fixed effects of cluster-level covariates; $W_n, s_{0j}$ is a normally distributed random intercept; and $s_{mj}$ are normally distributed random slopes of the individual-level covariates with a mean of zero and covariance matrix $\Sigma$. This multilevel model assumes that all individual-level and cluster-level true confounders of the treatment effects are included, with appropriate specifications of the functional form of the relationship between covariates and the logit of treatment assignment. In addition, it assumes that the random intercepts and random slopes are uncorrelated with cluster-level predictors. For details see Leite (2017).

---

## 10.6 The Controversy Surrounding Propensity Score

Recently controversy surrounding the PS methodology has developed, most likely due to the increased popularity of the method and the strong endorsement it received from prominent statisticians (Rubin, 2007) and health scientists (Austin, 2009, 2011). The popularity of the method has in fact grown to the point where some federal agencies now expect program evaluators to use this approach as a substitute for experimental designs (Peikes et al., 2009). This tendency was further reinforced by empirical studies (Heckman et al., 1996; Dehejia and Wahba, 1999) in which agreement was found between propensity score analysis and randomized trials, and in which the agreement was attributed to the ability of the former to "balance" treatment and control groups on important characteristics. Rubin (2007) has encouraged such interpretations by stating: "This application uses propensity score methods to create subgroups of treated units and control units … as if they had been randomized. The collection of these subgroups then 'approximate' a randomized block experiment with respect to the observed covariates." Subsequent empirical studies, however, have taken a more critical view of propensity score, noting with disappointment that a substantial bias is sometimes measured when careful comparisons are made to results of clinical studies (Peikes et al., 2009). It should be known that no causal claim can be established by a purely statistical method, be it propensity scores, regression, stratification, or any other distribution-based design.

The difficulty that most investigators experience in comprehending what "ignorability" means, and what judgment it summons them to exercise, has tempted them to assume that it is automatically satisfied, or at least is likely to be satisfied, if one includes in the analysis as many covariates as possible. The prevailing attitude is that adding more covariates can cause no harm (Rosenbaum, 1995a) and can absolve one from thinking about the causal relationships among those covariates, the treatment, the outcome, and, most important, the confounders left unmeasured (Rubin, 2008). Another factor inflaming the controversy has been the general belief that the bias-reducing potential of propensity score methods can be assessed experimentally by running case studies and comparing effect estimates obtained by propensity scores to those obtained by controlled randomized experiments (Shadish et al., 2002; Shadish and Steiner, 2010).

Formally, Rosenbaum and Rubin were very clear in warning practitioners that propensity scores work only under "strong ignorability" conditions.

---

## 10.7  Examples

### Example 10.1: Organ Transplant (File Name: Example1_transplant_PS)

The main objective in this example is to compare the survival experience of two groups of liver transplanted patients. The first group has 2300 patients who received liver transplant by technique (0), whereas the other group of 2700 patients received transplants by another technique (1). The data were obtained from the organ transplant registry. We shall use the PS methodology in order to objectively compare the survival probabilities of patients in the two groups. The available baseline covariates are presence or absence of kidney injury, weight, gender, diabetic status, and hypertension status.

**Estimation**

It is quite often that the logistic regression is used to estimate the propensity scores (Austin, 2011). The following SAS code shows how to estimate the PS:

```
ODS RTF;
ODS GRAPHICS ON;
proc logistic descending data=transplant;
title 'Propensity Score Estimation';
model technique=kidney_injury weight gender diabetic hypertension/
lackfit outroc = ps_r;
output out = ps_p XBETA=ps_sb STDXBETA = ps_sdxb PREDICTED=ps_pred;
run;
```

**Stratification**

The majority of PS studies divide the individuals in the study into five groups based on their PS values. This grouping is based on a recommendation by Cochran (1968) and Rosenbaum and Rubin (1984). The following SAS code achieves this objective:

```
Proc rank data=ps_p out=ps_strataranks groups=5;
var ps_pred;
ranks ps_pred_rank;
run;
```

In this example we shall stratify by the PS strata and use regression adjustment by the PS score, using it as a covariate in two models, first as a categorical variable and then as a continuous variable.

```
/* survival analysis stratified by the ps_pred_rank */
proc sort data=ps_strataranks;
by ps_pred_rank;
proc ttest;
by ps_pred_rank;
class technique;
var time;
ods output statistics=strata_Out;
data weights;
 set strata_out;
run;
proc freq data= ps_strataranks ;   /* correct*/
tables ps_pred_rank;
run;
proc means data= ps_strataranks;
class ps_pred_rank;
var weight;
run;
proc freq data=ps_strataranks;
tables ps_pred_rank* (gender hypertension diabetic kidney_injury)/
chisq ;
run;
/* survival analysis stratified by the ps_pred_rank */
proc lifetest data= ps_strataranks plots=(s) graphics;
time time*status(0);
strata ps_pred_rank;
symbol1 v=none color=black line=1;
symbol2 v=none color=red line=2;
symbol3 v=none color=blue line=3;
symbol4 v=none color=green line=4;
symbol5 v=none color=yellow line=5;
run;
/* Cox regression using PS as categorical*/
proc phreg data= ps_strataranks;
class ps_pred_rank;
model time*status(0) = technique ps_pred_rank;
run;
/* Cox regression using PS as continuous*/
proc phreg data= ps_strataranks;
model time*status(0) = technique ps_pred;
run;
ODS GRAPHICS OFF;
ODS RTF CLOSE;
```

### SAS Output

Step 1—The first step to obtain PS is to fit a logistic regression model with a technique as the dependent variable. The results of fitting the logistic regression model using SAS are given in Table 10.1.

**TABLE 10.1**

Maximum Likelihood Fitting of the Logistic Regression Model

| Analysis of Maximum Likelihood Estimates | | | | | |
|---|---|---|---|---|---|
| Parameter | DF | Estimate | Standard Error | Wald Chi-Square | Pr > ChiSq |
| Intercept | 1 | 0.9307 | 0.3434 | 7.3465 | 0.0067 |
| kidney_injury | 1 | 0.3354 | 0.0608 | 30.4534 | <.0001 |
| weight | 1 | −0.0161 | 0.00407 | 15.5936 | <.0001 |
| gender | 1 | −0.2383 | 0.0841 | 8.0284 | 0.0046 |
| diabetic | 1 | −0.2860 | 0.0704 | 16.5064 | <.0001 |
| hypertension | 1 | 0.8927 | 0.0957 | 86.9377 | <.0001 |

The ROC curve is produced to test the predictive ability of the proposed logistic regression model. The ROC curve is given in Figure 10.1.

Step 2—Check the balance of the weigh distribution among the PS groups. This is done for the weight as a continuous variable. It is shown in Table 10.2 that the variation in mean weights is quite variable among the PS groups.

Table 10.3 is a cross-classification of PS with gender. The chi square test has *p*-value <0.0001 showing that the distribution of gender varies among

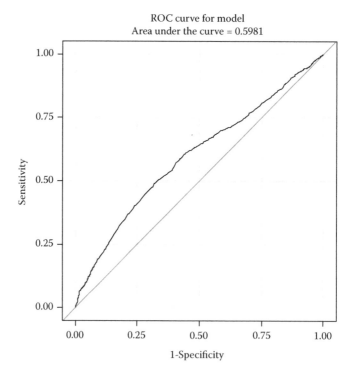

**FIGURE 10.1**

The ROC curve showing the predictive ability of the logistic regression model.

**TABLE 10.2**

Summary Statistics of Weight for the PS Groups

| Rank for Variable ps_pred | N Obs | N | Mean | Std Dev | Minimum | Maximum |
|---|---|---|---|---|---|---|
| | | | **Analysis Variable : weight** | | | |
| 0 | 991 | 991 | 82.6104945 | 9.7335020 | 55.0000000 | 103.0000000 |
| 1 | 1000 | 1000 | 79.2540000 | 8.6584324 | 61.0000000 | 105.0000000 |
| 2 | 1015 | 1015 | 79.6482759 | 10.5548924 | 51.0000000 | 95.0000000 |
| 3 | 1012 | 1012 | 76.2075099 | 7.3911771 | 50.0000000 | 97.0000000 |
| 4 | 982 | 982 | 75.7067210 | 10.7265870 | 40.0000000 | 102.0000000 |

**TABLE 10.3**

Cross-Classification of PS with Gender

| ps_pred_rank(Rank for Variable ps_pred) Frequency Percent Row Pct Col Pct | Gender 0 | Gender 1 | Total |
|---|---|---|---|
| | **Table of ps_pred_rank by gender** | | |
| 0 | 586 | 405 | 991 19.82 |
| 1 | 718 | 282 | 1000 20.00 |
| 2 | 629 | 386 | 1015 20.30 |
| 3 | 660 | 352 | 1012 20.24 |
| 4 | 709 | 273 | 982 19.64 |
| Total | 3302 66.04 | 1698 33.96 | 5000 100.00 |

the PS groups. The same results are obtained for hypertension, diabetes, and kidney injury, but the output is not shown.

Step 3—Stratified survival analysis has a significant difference among the survival curves of the PS. In Figure 10.2 we show the Kaplan-Meier survival curves for each of the five strata formed by the PS deciles.

Note that from Tables 10.4 and 10.5 that the difference in the techniques on the hazard of death is significant whether PS is entered into the Cox model as categorical or continuous covariate.

### Example 10.2: Glucose Levels (File Name: Example2_Glucose_PS)

This example illustrates the use of the PS matching using SAS.

The main objective of the study is to compare the effectiveness of two medications on lowering blood glucose levels. Medical records of patients

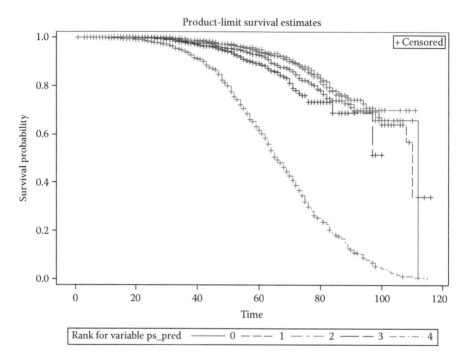

**FIGURE 10.2**
Kaplan-Meier survival curves for the five strata.

**TABLE 10.4**

Cox Regression Analysis, When PS Was Modeled As Categorical Covariate

| | | | | | | | |
|---|---|---|---|---|---|---|---|
| **Analysis of Maximum Likelihood Estimates** | | | | | | | |
| **Parameter** | **DF** | **Parameter Estimate** | **Standard Error** | **Chi-Square** | **Pr > ChiSq** | **Hazard Ratio** | **Label** |
| technique | 1 | 0.16137 | 0.06880 | 5.5016 | 0.0190 | 1.175 | |
| ps_pred_rank  0 | 1 | −1.98758 | 0.10703 | 344.8369 | <.0001 | 0.137 | Rank for Variable ps_pred 0 |
| ps_pred_rank  1 | 1 | −1.73099 | 0.10889 | 252.7023 | <.0001 | 0.177 | Rank for Variable ps_pred 1 |
| ps_pred_rank  2 | 1 | −2.02036 | 0.11569 | 304.9621 | <.0001 | 0.133 | Rank for Variable ps_pred 2 |
| ps_pred_rank  3 | 1 | −1.37687 | 0.10937 | 158.4781 | <.0001 | 0.252 | Rank for Variable ps_pred 3 |

**TABLE 10.5**

Cox Regression Analysis, When PS Was Modeled As Continuous Variable

| | | | Analysis of Maximum Likelihood Estimates | | | | |
|---|---|---|---|---|---|---|---|
| Parameter | DF | Parameter Estimate | Standard Error | Chi-Square | Pr > ChiSq | Hazard Ratio | Label |
| technique | 1 | 0.13521 | 0.06894 | 3.8470 | 0.0498 | 1.145 | |
| ps_pred | 1 | 7.96360 | 0.29440 | 731.7068 | <.0001 | 2874.411 | Estimated Probability |

involved in the study include information on cholesterol levels, glucose HDL, ratio of cholesterol to HDL, hemoglobin, age, gender, height, weight, body frame, systolic blood pressure (SBP), diastolic blood pressure (DBP), waist circumference, hip, and treatment group. The outcome of interest is the glucose levels after two weeks of treatment.

The following SAS code shows how to estimate the PS for the treatment variable:

```
/* ESTIMATION*/
proc logistic descending data=glucose;
title 'Propensity Score Estimation';
class sex frame;
model treatment = chol hdl ratio hb age height weight sbp dbp waist
frame hip sex /lackfit outroc=ps_r;
output      out=ps_p      XBETA=ps_xb      STDXBETA=      ps_sdxb
PREDICTED=ps_pred;
run;
/* TREATMENT EFFECT */
proc rank data=ps_p out=ps_strataranks group=5;
var ps_pred;
ranks ps_pred_rank;
run;
```

The maximum likelihood estimates of the odds ratios from the logistic regression are shown in Table 10.6. Again, the predictive ability of the logistic regression model is illustrated by the ROC curve as shown in Figure 10.3.

Table 10.7 summarizes the distribution of subjects by the PS strata.

Now to compute the differences between the treatment means and their standard errors we apply the following SAS code. When we used the t-test to test for treatment differences within each PS groups, the variances of the glucose variable within PS = 0, PS = 1, and PS = 4 were significantly different, but the variances of the glucose variable were not significantly different with PS = 2 and PS = 3. Therefore to compare between the treatment means using ANCOVA we need to account for the variance heterogeneity by weighting each mean by the reciprocal of the variance. The following SAS code executes this step. The ANOCOVA results are shown in Table 10.8.

**TABLE 10.6**

Maximum Likelihood Estimates of the Logistic Regression Model

| | Odds Ratio Estimates | | |
| --- | --- | --- | --- |
| Effect | Point Estimate | 95% Wald Confidence Limits | |
| chol | 0.989 | 0.979 | 0.999 |
| hdl | 1.027 | 0.995 | 1.059 |
| ratio | 1.284 | 0.898 | 1.835 |
| hb | 1.048 | 0.939 | 1.171 |
| age | 0.999 | 0.982 | 1.017 |
| height | 1.154 | 1.048 | 1.270 |
| weight | 1.002 | 0.986 | 1.018 |
| sbp | 0.999 | 0.986 | 1.013 |
| dbp | 0.997 | 0.978 | 1.017 |
| waist | 1.104 | 1.009 | 1.209 |
| frame large vs small | 3.105 | 1.450 | 6.648 |
| frame medium vs small | 1.410 | 0.797 | 2.496 |
| hip | 0.858 | 0.772 | 0.953 |
| sex 0 vs 1 | 0.242 | 0.113 | 0.516 |

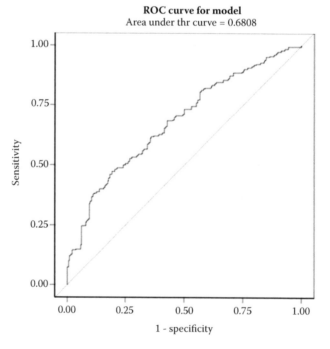

**ROC curve for model**
Area under thr curve = 0.6808

**FIGURE 10.3**
The ROC curve showing the predictive ability of the logistic regression model.

**TABLE 10.7**

Distribution of Subjects by PS Groups

| | | Rank for Variable ps_pred | | |
|---|---|---|---|---|
| ps_pred_rank | Frequency | Percent | Cumulative Frequency | Cumulative Percent |
| 0 | 74 | 19.89 | 74 | 19.89 |
| 1 | 75 | 20.16 | 149 | 40.05 |
| 2 | 74 | 19.89 | 223 | 59.95 |
| 3 | 75 | 20.16 | 298 | 80.11 |
| 4 | 74 | 19.89 | 372 | 100.00 |

**TABLE 10.8**

ANCOVA results using the inverse variances as weights

| Source | DF | Type III SS | Mean Square | F Value | Pr > F |
|---|---|---|---|---|---|
| treatment | 1 | 83396.10269 | 83396.10269 | 28.32 | <.0001 |
| ps_pred | 1 | 34438.08719 | 34438.08719 | 11.69 | 0.0007 |

```
proc sort data=ps_strataranks;
by ps_pred_rank;

proc ttest;
by ps_pred_rank;
class treatment;
var glucose;
ods output statistics=strata_out;
data weights; set strata_out;
if class ='Diff (1-2)';
wt_i = 1/(StdErr**2);
wt_diff=wt_i*mean;
proc means data=weights;
var wt_i wt_diff;
output out=total sum =sum_wt sum_diff;
data total2;
set total;
Mean_diff=sum_diff/sum_wt;
SE_Diff = SQRT(1/sum_wt);
proc print data=total2;
run;

/* ANCOVA using the propensity score as covariate*/
proc glm data=ps_p;
class treatment;
model glucose = treatment ps_pred/ss3;
lsmeans treatment;
run;
```

ODS RTF CLOSE;
**Propensity Score Matching**
    Here we shall describe the PS matching in SAS (Rheta et al., 2012), but we
shall use R for this purpose. The main objective of matching is to obtain
similar groups in the two treatments by matching individual observations on
their propensity score, which is obtained from fitting the logistic regression.
One of the matching methods commonly used in PS analysis is 1:1
matching (Thoemmes and Kim, 2011), which forms pairs of subjects from
each treatment group. Rheta et al. (2012) provided a simple 1:1 matching and
provided an SAS code. In summary, after the matched pairs have been
identified, the difference between the outcome means in the two groups can
be tested using PROC TTEST. It is important to specify a correlated-means
t-test (rather than an independent-means TTEST) because the matched pairs
impose a correlated structure to these data. Use of the PAIRED statement
with PROC TTEST will request the appropriate version of this inferential
procedure. Inverse probability of treatment weights (IPTW) individuals are
weighted by the inverse probability of receiving the treatment that they
actually received. Treated individuals receive an IPTW equal to $1/p_i$ and
control individuals receive a weight equal to $1/(1 - p_i)$ (Harder et al., 2010).
The weights are then used in a weighted least squares (WLS) regression
model along with other predictor covariates. The IPTW method is inclusive
of all subjects in a study, therefore no loss of sample occurs as in other
conditioning methods (i.e., matching, stratification). A drawback of the
IPTW method is the possibility of extreme propensity scores that can result in
very large weights that can bias the treatment effect estimates (Austin, 2011;
Shadish and Steiner, 2010). Bias from extreme weights can be adjusted using
a stabilization technique (Robins et al., 2000) that multiplies the treatment
and comparison weights by a constant or by using a trimming technique that
trims the stabilized weights within a specified range.

---

## 10.8 Propensity Score Matching in R

Using R to perform propensity score matching requires the installation of the
"**MatchIt**" package. After the installation of the package we shall export the
Excel data file "glucose". The next step is to perform the matching and
evaluate the results.

```
R Code
glucose=read.csv(file.choose())
library(MatchIt)
attach(glucose)
names(glucose)
glucose[1:5,]
```

The preceding statement will read the first 5 lines of the data file. The next statement will perform the matching:

```
m.out=matchit(treatment~chol+hdl+hb+age+sex+height+weight
+frame+dbp+sbp+
hip+waist, data=glucose,method="nearest",ratio=1)
summary(m.out)
```

The "**matchit** function, the method="nearest" means that the nearest neighbor method of matching is used. There are many other matching methods, such as exact matching, subclassification matching, and genetic matching, to choose from (see Ho et al., 2013; Randolph et al., 2014). The results of the matching are in an object named "m.out". The "summary (m.out)" object outlines the results of the matching. We omitted the output, but it can be checked that the matching has reduced the differences between the two treatment groups. Table 10.9 shows the median, mean, and maximum quartile differences between the two treatment groups. Smaller QQ values

**TABLE 10.9**

Summary Results Showing the Effectiveness of the Matching

| | Percent Balance Improvement | | | |
| --- | --- | --- | --- | --- |
| Parameter | Mean Diff. | eQQ Med | eQQ Mean | eQQ Max |
| distance | 16.6626 | 13.0078 | 18.0779 | 7.1226 |
| chol | 25.9590 | 28.5714 | 14.9733 | 0.0000 |
| hdl | 5.8824 | 0.0000 | 5.5046 | 0.0000 |
| hb | −13.1001 | −7.8947 | −1.6073 | 0.0000 |
| age | −7.6433 | 0.0000 | −1.8349 | 0.0000 |
| sex | 0.0000 | 0.0000 | 0.0000 | 0.0000 |
| height | 28.7374 | 0.0000 | 26.4706 | 0.0000 |
| weight | −9.8692 | 0.0000 | 9.0047 | 11.4286 |
| framelarge | 11.3703 | 0.0000 | 13.6364 | 0.0000 |
| framemedium | 14.5631 | 0.0000 | 8.3333 | 0.0000 |
| framesmall | 6.5693 | 0.0000 | 0.0000 | 0.0000 |
| dbp | 14.5352 | 0.0000 | 7.8431 | 0.0000 |
| sbp | 23.2877 | 0.0000 | 5.8394 | 0.0000 |
| hip | 50.2646 | 0.0000 | 10.7784 | 0.0000 |
| waist | −4.8458 | 0.0000 | 2.0833 | 20.0000 |
| Sample sizes: | | | | |
| | Control | Treated | | |
| All | 192 | 180 | | |
| Matched | 180 | 180 | | |
| **Unmatched** | 12 | 0 | | |
| Discarded | 0 | 0 | | |

indicate better matching. Since part of the output is not presented here, it is left as an exercise to verify that the QQ values are all substantially smaller after matching than before matching. The statement.

```
plot(m.out,type="jitter")
```

creates a jitter plot (Figure 10.4) and histograms (Figure 10.5) to visualize the quality of the matching. In Figure 10.4 the circle represents a case's PS. Note that the absence of cases in the uppermost stratification indicates that there were no unmatched treatment units. The middle stratification shows almost complete matching between matched treatment 1 units and matched treatment 2. The final stratification shows the unmatched control units that will not be used in the analysis (only 12 units), as shown in Table 10.9.

```
plot(m.out,type="hist")
```

The above statement produces the histograms of the PS scores before and after the matching as shown in Figure 10.5. As can be seen the histograms after the matching are very similar.

The matching will be completed using the statement

```
glucose.data=match.data(m.out)
```

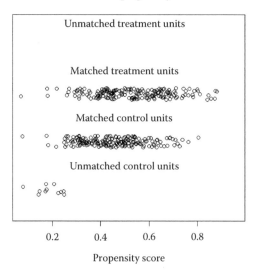

**FIGURE 10.4**
(Jitter plot.) Distribution of propensity scores.

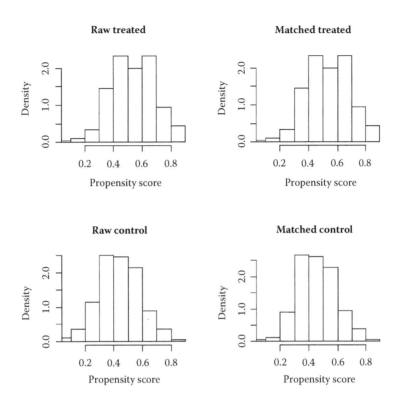

**FIGURE 10.5**
Distribution of propensity scores before and after matching.

Thereafter we export the matched data file as a csv file and use it to analyze the data:

```
glucose.data=match.data(m.out)
write.csv(glucose.data,file="D:nearest100.csv")
new.matched=read.csv(file.choose())
attach(new.matched)
names(new.matched)
fit.matched=lm(glucose~treatment+chol+hdl+hb+age+sex+height
+weight+
frame+dbp+sbp+hip+waist, data=new.matched)
summary(fit.matched)
```

**R Output**

In Table 10.10 we provide a summary of the R output for the PS matching using R.

Residual standard error: 44.65 on 345 degrees of freedom, multiple R-squared: 0.3881, adjusted R-squared: 0.3632, F-statistic: 15.63 on 14 and 345 DF, $p$-value: < 2.2e-16.

The preceding example demonstrates that the propensity score matching is an effective tool for reducing bias resulting from observational or retrospective studies.

**After the Matching**

We save the matched dataset named (nearest100) in a csv file:

```
write.csv(glucose.data,file="D:nearest100.csv").
```

Recall that a matched case-control study can be analyzed using proportion hazard regression model. We can do that using the following R commands:

```
library(survival)
matched_ps=read.csv(file.choose())
names(matched_ps)
"X"     "chol"   "glucose" "hdl"     "ratio"   "hb"
"age"    "gender"  "height"  "weight"  "frame"   "sbp"
"dbp"    "waist"   "hip"     "sex"     "treatment" "distance"
```

**TABLE 10.10**

Summary Results of PS Matching Fitting the Glucose Data

| Residuals: | | | | |
|---|---|---|---|---|
| Min | 1Q | Median | 3Q | Max |
| −160.311 | −25.004 | −8.396 | 17.637 | 257.583 |

| Coefficients: | | | | |
|---|---|---|---|---|
|  | Estimate | Std. Error | t value | Pr(>\|t\|) |
| (Intercept) | 104.839926 | 73.597577 | 1.425 | 0.1552 |
| Treatment | −31.511748 | 4.889982 | −6.444 | 3.91e-10 *** |
| chol | −0.062806 | 0.060267 | −1.042 | 0.2981 |
| hdl | −0.161332 | 0.148534 | −1.086 | 0.2782 |
| hb | 13.414945 | 1.211671 | 11.071 | < 2e-16 *** |
| age | 0.301368 | 0.188560 | 1.598 | 0.1109 |
| sex | 10.303803 | 8.142702 | 1.265 | 0.2066 |
| height | 0.154212 | 0.985992 | 0.156 | 0.8758 |
| weight | 0.425624 | 0.165838 | 2.567 | 0.0107 * |
| framemedium | −2.086686 | 6.513022 | −0.320 | 0.7489 |
| framesmall | −10.544636 | 8.338554 | −1.265 | 0.2069 |
| dbp | −0.008602 | 0.211386 | −0.041 | 0.9676 |
| sbp | −0.034036 | 0.149720 | −0.227 | 0.8203 |
| hip | −2.038516 | 1.126867 | −1.809 | 0.0713 . |
| waist | −0.919367 | 0.972460 | −0.945 | 0.3451 |

```
"weights"
attach(matched_ps)
# Fitting the Cox regression
model=coxph(Surv(glucose,weights)~treatment,data=matched_ps)
 summary(model)
```

### R Output: For the Analysis of Propensity Score Matched Data

```
Call:
coxph(formula  =  Surv(glucose,  weights)  ~  treatment,  data  =
matched_ps)
 n= 360, number of events= 360
      coef  exp(coef)  e(coef)    z   Pr(>|z|)
treatment 0.4385   1.5503  0.1081  4.057   4.98e-05 ***
      exp(coef) exp(-coef)    lower .95  upper .95
treatment   1.55    0.645    1.254      1.916
Score (logrank) test = 16.7 on 1 df,  p=4.37e-05
```

## 10.9 Propensity Score Stratification in R

```
R-Code
glucose=read.csv(file.choose())
names(glucose)
attach(glucose)
fit1=lm(glucose~treatment, data=glucose)
summary(fit1)
lm(formula = glucose ~ treatment, data = glucose)
```

**Coefficients:**
```
(Intercept)   treatment
     130.17       -25.08
```
**#1-Obtaining the Propensity Scores by fitting the logistic regression**
```
glm1=glm(treatment~chol+hdl+hb+age+sex+height+weight+frame+dbp
+sbp+hip+waist,family=binomial,data=glucose)
X=glm1$fitted
tapply(X,treatment,FUN=fivenum)
```
**#plotting the histogram of the PS for each treatment group**
```
hist(X[treatment==0],clim=c(0.2,1))
hist(X[treatment==1],clim=c(0.2,1))
```

**#Dividing the PS into five intervals**
```
breakvals=fivenum(X)
strat=cut(X,breaks=breakvals,
    labels=c('bot quart', '2nd quart', '3rd quart', 'top quart'),
    include.lowest=TRUE)
```

```
stratdf=data.frame(glucose,treatment,strat)
#Finally fitting the linear regression with glucose as dependent
variable
#while treatment and PS strat are covariates

fitted=lm(glucose~treatment+strat,data=stratdf)
summary(fitted)
lm(formula = glucose ~ treatment + strat, data = stratdf)

Residuals:
  Min    1Q Median    3Q    Max
-95.49 -37.05 -14.79 16.92 245.95

Coefficients:
          Estimate Std. Error t value Pr(>|t|)
(Intercept)    125.054    5.871 21.300  < 2e-16 ***
treatment      -30.760    5.966 -5.156 4.13e-07 ***
strat2nd quart   1.841    8.089  0.228  0.82012
strat3rd quart   5.158    8.070  0.639  0.52314
strattop quart  24.440    8.429  2.899  0.00396 **
--
```

Residual standard error: 54.46 on 367 degrees of freedom , multiple R-squared: 0.077, adjusted R-squared: 0.06694 , F-statistic: 7.654 on 4 and 367 DF, *p*-value: 6.231e-06.

# Moreover we can use student's t-test within each stratum
#to detect a significance difference between the two treatment arms.

out.file=by(stratdf,strat,function(mydataframe)
{with(mydataframe,t.test(glucose[treatment==0],
glucose[treatment==1]))})
out.file

```
strat: bot quart

    Welch Two Sample t-test

data: glucose[treatment == 0] and glucose[treatment == 1]
t = 3.2836, df = 89.117, p-value = 0.001465
alternative hypothesis: true difference in means is not equal to 0
95 percent confidence interval:
 11.46317 46.59330
sample estimates:
mean of x mean of y
 124.5882  95.5600
```

```
------------------------------
strat: 2nd quart
     Welch Two Sample t-test
data: glucose[treatment == 0] and glucose[treatment == 1]
t = 5.6101, df = 51.774, p-value = 7.982e-07
alternative hypothesis: true difference in means is not equal to 0
95 percent confidence interval:
 34.91243 73.80146
sample estimates:
mean of x mean of y
138.31250  83.95556
------------------------------
strat: 3rd quart
     Welch Two Sample t-test
data: glucose[treatment == 0] and glucose[treatment == 1]
t = -0.7712, df = 81.472, p-value = 0.4428
alternative hypothesis: true difference in means is not equal to 0
95 percent confidence interval:
 -31.35046 13.83512
sample estimates:
mean of x mean of y
 111.9400  120.6977
------------------------------
strat: top quart
     Welch Two Sample t-test
data: glucose[treatment == 0] and glucose[treatment == 1]
t = 3.0622, df = 31.124, p-value = 0.004504
alternative hypothesis: true difference in means is not equal to 0
95 percent confidence interval:
 17.35975 86.56678
sample estimates:
mean of x mean of y
 164.7692  112.8060
```

## Exercises

10.1. The data in this example represent the survival experience of a group of subjects who underwent cardiac surgery (see *Analysis of Observational Health Care Data Using SAS* edited by D.E. Faris, A.C. Leon, J.M. Haro, and R.L. Obenchain, 2010). The data is saved in the Excel sheet Exercie_10.1. The sample has patients who underwent a PCI with placement of either a drug-eluting stent (DES) or a bare-metal stent

(BMS). The outcome of interest is death. Use the PS stratification methods in SAS similar to Example 10.1 to properly analyze the data. Compare the analysis results to the analysis using Cox regression by directly accounting for the possible confounders.

10.2. This is again the survival liver transplant data saved in the Excel sheet Exercise_10.2. The analysis of this data is quite challenging. Again the main objective is to test if there is a significant difference in the survival probabilities of two groups of patients: the first group received a liver transplant using a certain technique and the other group received a liver transplant using another technique. There are two sets of covariates, the first being measured at the baseline before intervention and the other set of covariates or risk factors are measured after interventions. Fit the data using the Cox regression model. Use the R function "Matchit" to conduct propensity score matching analysis. Compare the results of the propensity score matched analysis to the direct Cox regression unmatched analysis. Report your findings.

# 11

## *Introductory Meta-Analysis*

### 11.1 Introduction

The volume of medical literature that health care professionals, researchers, and decision makers need to keep up with is expanding at a very high rate. This makes it difficult for any clinical researcher to be up to date with the recent advances in their field of knowledge. For this reason, reviews of medical literature have become essential tools for keeping abreast of the new evidence that is accumulating regarding new treatments and other interventions. Moreover, reviews also identify areas where the available evidence is insufficient and further studies are required (Higgins and Green, 2006). Over three decades ago, psychologists and social scientists drew attention to the systematic steps to be followed to minimize uncertainties when reviewing the results of scientific reports (see Light and Smith, 1971; Rosenthal, 1978; Glass et al., 1981).

It is important to distinguish a systematic review (SR) from a meta-analysis (MA) (see Oxman and Guyatt 1988). This is because it is always appropriate and desirable to systematically review a body of data, but it may sometimes be inappropriate, or even misleading, to statistically pool results from separate studies.

To examine the relative effect of different interventions, ideally systematic reviews and meta-analyses should be based on randomized controlled trials. However, systematic reviews and meta-analyses have been developed for other study types such as prospective cohorts or diagnostic studies. Recognition of the need for systematic reviews (with or without meta-analyses) has grown rapidly, and this is reflected by the number of systematic reviews and empirical studies of the review methodology that have been published, for example, in the medical literature, Cochrane Database of Systematic Reviews, and the Cochrane Methodology Register (Higgins and Green, 2006).

Systematic reviews use explicit, objective, and prospectively defined methods to collect, critically evaluate, and synthesize studies, making them less biased and more reproducible than traditional reviews. A systematic review may, or may not, include a meta-analysis. A meta-analysis is a group of statistical techniques that are concerned with the pooling of the results from individual studies to obtain a single overall estimate of treatment effect

(Chalmers and Altman, 1995) and provide summary uncertainty. A meta-analysis can provide more precise estimates of the effects of an intervention than those derived from the individual studies included in a review (Cooper, 1998). This allows decisions that are based on the pooled evidence with summary measures of uncertainty to be more rational.

Typically, a meta-analysis has two fundamental goals: (1) testing the homogeneity of the studies from which the results are obtained, and (2) obtaining a combined summary measure of the effect size of the studied relation, together with measures of uncertainty, a confidence interval, and its statistical significance. Basically, there can be two sources of variability that explain the heterogeneity in a set of studies in a meta-analysis. One of them is the variability due to sampling error, also named within-study variability. The sampling error variability is always present in a meta-analysis, because every single study uses a different sampling strategy. The other source of heterogeneity is the between-studies variability, which can appear in a meta-analysis when there is true heterogeneity among the population effect sizes estimated by the individual studies (National Research Council, 1992; Erez et al., 1996; Hunter and Schmidt, 2000; Brockwell and Gordon, 2001; Field, 2003). A good tutorial on meta-analysis was given by Normand (1999).

## 11.2  Definition and Goals of Meta-Analysis

The term *meta-analysis* was coined by Glass (1976) who defined it as "the statistical analysis of a large collection of analyzed results from individual studies for the purpose of integrating the findings."

The great majority of meta-analyses one can find in perusing medical literature concern results of controlled clinical trials. In fact, it was emphasized by Bulpitt (1988) that only the latter should be meta-analyzed. However, there are many meta-analyses of epidemiological data arising from case-control and cohort studies.

## 11.3  How Is a Meta-Analysis Done?

There are generally several steps to performing a meta-analysis:

1. Decide on the topic and provide an adequate background.
2. Decide on the hypothesis being tested.
3. Review the literature for all studies that test that hypothesis including careful study of the references in articles, examination of papers,

abstracts, and presentations not published, and other sources of unpublished work (including government agencies and rejected submissions). This review needs to be carefully done to minimize bias in reporting the effect of interventions.

4. Evaluate each study carefully to decide whether it is of sufficient quality to be worthy of inclusion, and whether it includes sufficient information to be included. Studies should have the same end point.

5. Aggregate data to be extracted and create a database containing the information necessary for the analyses.

6. Perform the meta-analysis.

7. Interpret the results.

We will discuss each of these steps separately.

### 11.3.1 Decide on a Research Topic and the Hypothesis to be Tested

Generally, a topic that might be appropriate is one for which the question is clearly focused, for which literature is available, and is of some importance. The research question may be broad, for example, the effect of chemotherapy on patients with breast cancer. This may entail examining all types of chemotherapy in patients with breast cancer. We may be more specific and focus on a specific question (e.g., the effect of cisplatin-based chemotherapy on patients with advanced non-small-cell breast cancer). The hypothesis to be tested should be formulated while the topic is being selected.

### 11.3.2 Inclusion Criteria

The inclusion criteria should explicitly describe the type of study design to be included (see Chapter 1), the types of intervention, the type of patient, and possibly the time period in order to address the question of interest. For a question concerning the effect of chemotherapy on patients with advanced non-small-cell breast cancer inclusion, criteria might be

- Randomized controlled trials since 1970
- Comparison of chemotherapy with no chemotherapy
- Patients with advanced non-small-cell breast cancer

### 11.3.3 Searching Strategy and Data Extraction

A good meta-analysis should begin with a thorough review of the literature. We should make an effort to obtain studies from all sources, such as the published literature, unpublished literature, presentations, and nonpublic sources.

In order to be "comprehensive" and not overly "selective," searches should ideally be based on a number of study sources (e.g., MEDLINE, EMBASE, Cochrane Central Register of Controlled Trials, the scientific literature index portal such PubMed, Internet engine search such as Google). Optimal strategies for identifying trials in MEDLINE and EMBASE are available.

Most systematic reviews and meta-analyses rely on extracting aggregate trial data from trial reports in the form of peer-reviewed articles or presentations. Data on the participants and outcomes of interest should be extracted and placed in a standard form. This should ideally be done independently by two individuals, who together should resolve any discrepancies.

### 11.3.4 Study Evaluation

The quality of a meta-analysis depends ultimately on the choice of the studies that comprise it. Each study should pass some minimal preset requirements, and a careful assessment of the quality of the study should be made. Emphasis should be placed on assessment of the heterogeneity of the studies and on identifying potential confounders. It should be noted that many studies will use similar but not identical endpoints, and we have to translate them into some common measure of effect size, along with a measure of precision, and sample size justifications.

For continuous endpoints, the effect size is often taken as the standardized mean difference. For binary endpoints that usually means odds ratios or relative risks. In survival studies, the effect size is measured by the hazard ratios. In the medical meta-analysis literature, a measure sometimes used is the "number needed to treat," or NNT, which is the number of patients needed to be treated to prevent an additional adverse event or for one additional patient to benefit. For the binary outcome the NNT is the reciprocal of the absolute risk difference. In survival analytic studies, there is no one unique NNT, rather there is an NNT for each specified time point. For example if $S_c(3)$ and $S_t(3)$ are, respectively, the three years survival probabilities for the cases and the controls, then the NNT to prolong the life of one patient by three years is

$$NNT = 1/(S_t(3) - S_c(3)) \tag{11.1}$$

One of the major objectives of study evaluation is to reduce bias, which could result from publication bias (the probability of publication may be higher for a positive study than for a negative study). Studies that have been published multiple times may be more likely to be discovered. The problem is exacerbated if it is not apparent that several different articles may in fact all come from the same study. Several "studies" may share a control group, and that may not be obvious. Bias may also result from the manner in which subjects were recruited or ascertained. Trials that are observational may be more subject to bias than experimental trials. There are diagnostic tools, such as funnel plots, to evaluate publication bias.

Quality may be variable among studies. A study with a large sample size is, all other things being equal, of higher quality, because measures of effect sizes have higher precision. An evaluator who knows the area, knows the literature, and knows the methodology used in the research may be able to make judgments about quality. One possibility to recognize the study quality is to assign weight to each study. Sample size is most often used as a weight, but there are many other things besides sample size that affect quality.

### 11.3.5 Establish Database

Many meta-analysis tools include a form that can be handy to use to record information about the studies. Some of these are incorporated directly into the software. To utilize either SAS or R it is recommended to have the data stored in an Excel sheet or TXT file format.

### 11.3.6 Performing the Analysis

We shall present several examples on an MA using SAS, a commercial software, and R, a free ware, for two types of end points. The first is measured on the continuous scale where the effect size is the mean difference, or hazard ratios, for time-to-event studies, and the other is a binary outcome where the odds ratio will be the effect size of choice.

---

## 11.4 Issues in Meta-Analysis

### 11.4.1 Design Issues

We should not expect all studies that address the same question to use similar study designs. It is therefore necessary to examine the designs and determine whether they are sufficiently similar that it is reasonable to assume they are testing "the same" hypothesis. One would like to have a variety of different types of studies, all of which test similar hypotheses, so that the results of the meta-analysis are generalizable.

### 11.4.2 Positive Studies Are More Likely to be Published (Publication Bias)

This publication bias can lead to a selection bias in the studies included in the meta-analysis. Careful searching of sources other than simply publications may reduce this selection bias, and tools such as funnel plots may aid in learning whether such bias may exist. Publication bias is a difficult issue to deal with (Macaskill et al., 2001).

Meta-analyses are subject to publication bias because studies with negative results are less likely to be published and, therefore, results from meta-analyses may overstate a treatment effect. One strategy to minimize publication bias is to contact well-known investigators in the field of interest to discover whether they have conducted a negative study that remains unpublished.

A funnel plot is a graphic representation in which the size of the study on the $y$-axis is plotted against the measure of effect on the $x$-axis. Sampling error decreases as sample size increases and, therefore, larger studies should provide more precise estimates of the true treatment effect.

### 11.4.3 Funnel Plot

A funnel plot shows the estimated treatment effect versus its measure of uncertainty, usually measured by its estimated standard error (Sterne et al., 2001). For example, if we take the odds ratio (or its natural logarithm) as the effect size and its standard error (standard error of log-odds ratio) as its measure of uncertainty, then we place the effect size on the $x$-axis and the measure of uncertainty on the $y$-axis. Therefore, in the absence of bias, results from small studies will scatter widely at the bottom of the graph, with the spread narrowing among larger studies.

Effect estimates from small studies will therefore scatter more widely at the bottom of the graph, with the spread narrowing among larger studies. In the absence of bias, the plot should approximately resemble a symmetrical (inverted) funnel. This is illustrated in Figure 11.1, in which the effect estimates in the larger studies are close to the true intervention odds ratio of 0.5.

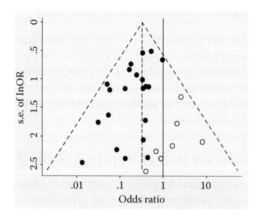

**FIGURE 11.1**
Funnel plot.

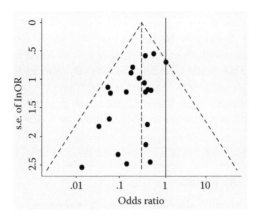

**FIGURE 11.2**
Funnel plot.

If there is bias, for example, because smaller studies without statistically significant effects (shown as open circles) remain unpublished, this will lead to an asymmetrical appearance of the funnel plot with a gap in a bottom corner of the graph, as shown in Figure 11.2. In this situation the effect calculated in a meta-analysis will tend to overestimate the intervention effect (Egger and Smith, 1997; Egger et al., 1997, 2001). The more the asymmetry, the more likely it is that the amount of bias will be substantial.

### 11.4.4 Studies May Be Heterogeneous

Studies may differ widely in quality. The most obvious case of this is when studies differ grossly with respect to the number of experimental units or subjects. All other considerations being equal, a study with more experimental units will be of higher quality, because estimates of effect size will be more precise, and hypothesis tests will be more powerful. It is possible to weight studies by sample size in a meta-analysis. However, there are many other ways in which studies can differ with respect to quality, and studies with the same sample size can range in quality from low to high. It is possible to assign quality scores to studies, and weight the meta-analysis using such quality scores, but such use is not without controversy. Restriction of the studies in a meta-analysis to only "high-quality" studies may introduce bias.

### 11.4.5 Confounding

Important confounders could be used for stratification purposes. However, it is unlikely that all studies will use such confounders, and it is also unlikely that sufficient information will be reported to use them in the meta-analysis. Restriction to only the studies using these confounders may introduce bias.

### 11.4.6 Modeling

One must make a decision whether to treat the likely variations among the studies as a fixed or random effect. Treating a study as a fixed effect takes the position that these particular studies are the only studies to which we wish to generalize and that other studies are not of interest. Treating a study as a random effect assumes that these studies are a sample from all possible similar studies, and we wish to be able to generalize the results to the larger population.

### 11.4.7 Evaluating the Results

Data abstraction results should be clearly presented. The meta-analysis should provide a table outlining the features of the studies, such as subjects' demographics, study design, sample size, and intervention, including the dose and durations of any drugs. Substantial differences in the study design or patient populations will exacerbate heterogeneity.

The typical graphic displaying meta-analysis data is a *forest plot*, in which the point estimate for the selected effect size is represented by a square or circle and the confidence interval for each study is represented by a horizontal line. The size of the circle or square corresponds to the weight of the study in the meta-analysis, with larger shapes given to studies with larger sample sizes or data of better quality or both. The 95% confidence interval is represented by a horizontal line except for the summary statistic, which can be shown by a diamond, the length of which represents the confidence interval. In all the examples presented in the chapter a forest plot will be displayed.

## 11.5  Assessing Heterogeneity in Meta-Analysis

### 11.5.1  Sources of Heterogeneity

Differences among studies may be categorized broadly into those related to the phenomenon being studied and those unrelated. Those differences related to the phenomenon being studied are mostly beyond the original investigator's control and constitute clinical incomparability. For example, the treatment may work differently for specific populations, the treatment may have a different effect on mortality measures as compared to morbidity, or the treatment effect may depend on exposure level (Thompson and Pocock, 1991; Petitti, 2001). The original investigator may focus on a particular patient subgroup to reduce such incomparability. The investigator may control the design dimension of incomparability. For example, he/she may control whether the study is prospective or retrospective, how long to follow the patients, what outcome to measure given measurement error issues, whether to analyze an odds ratio or a risk difference, and how to analyze that statistical outcome. Choice of study design, or how certain problems such as attrition are dealt with analytically, may induce differential biases in the results as well.

### 11.5.2 Measuring Heterogeneity

We consider the randomized controlled trial setting, restricting attention to dichotomous study outcomes and choosing as the summary statistic the odds ratio. For example, the outcome might be mortality within a specified follow-up time, and the summary statistic would be the odds of death in the treatment group as compared to the control group. The usual first step is to assess whether heterogeneity exists using a chi-squared test (a Q-statistic). This test is known to have low statistical power, which means that the probability is small that the null hypothesis of homogeneity of study treatment effects is rejected given that the alternative hypothesis of heterogeneity is true. Thus nonrejection of the null hypothesis does not necessarily mean that heterogeneity does not exist, and the meta-analyst is well-served to consider that heterogeneity exists regardless and attempt to estimate it.

### 11.5.3 Measures of Heterogeneity

Let $M$ denote the number of studies in a meta-analysis. Further, let $\hat{\theta}_k$ be the treatment effect estimate (e.g., a log-odds ratio), $\hat{\sigma}_k^2$ its estimated variance, and $w_k = 1/\hat{\sigma}_k^2$ the corresponding weight from study $k$, $k = 1, \ldots, M$. Several measures of statistical heterogeneity are widely used:

1. Cochran's $Q$ statistic, which under the null hypothesis of no heterogeneity follows a $\chi^2$-distribution with $M-1$ degrees of freedom (Cochran, 1954). $Q$ is given by

$$Q = \sum_{k=1}^{M} w_k \left( \theta_k - \frac{\sum_{k=1}^{M} w_k \hat{\theta}_k}{\sum_{k=1}^{M} w_k} \right)^2 \tag{11.2}$$

2. Higgins and Thompson's $I^2$ (2002), derived from Cochran's $Q$ is defined as

$$I^2 = \max\left\{ 0, \frac{Q - (M-1)}{Q} \right\} \tag{11.3}$$

3. The between-study variance, $\tau^2$, as estimated in a random effects meta-analysis. There are several proposals of estimating $\tau^2$ in a

meta-analysis (Petitti, 2001). Nevertheless, most reviewers use the moment-based estimate of $\tau^2$ (DerSimonian and Laird, 1986), calculated as

$$\hat{\tau}^2 = \max\left\{0, \frac{Q - (M-1)}{\sum_{k=1}^{M} w_k - \frac{\sum_{k=1}^{M} w_k^2}{\sum_{k=1}^{M} w_k}}\right\} \tag{11.4}$$

4. $H^2$, derived by Higgins and Thompson (2002):

$$H^2 = \frac{Q}{M-1} \tag{11.5}$$

5. $R^2$ similar to $H^2$ and calculated from $\tau^2$ and a so-called "typical" within-study variance $\sigma^2$ (which must be estimated), and defined as

$$R^2 = \frac{\tau^2 + \sigma^2}{\sigma^2} \tag{11.6}$$

Note that $R^2 = 1$ indicates perfect homogeneity among the studies. Table 11.1 shows key properties of the various measures.

Notice that, in contrast to $\tau^2$, the measures $Q$, $I^2$, $H^2$, and $R^2$ all depend on the precision, which is proportional to study size. Thus, given an underlying model, if the study sizes are large the confidence intervals become smaller and the heterogeneity, measured for example using $I^2$, increases. This is reflected in the interpretation: $I^2$ is the percentage of variability that is due to between-study heterogeneity and $1-I^2$ is the percentage of variability that is due to sampling error. When the studies become very large, the sampling error tends to 0 and $I^2$ tends to 1. Such heterogeneity may not be clinically relevant.

**TABLE 11.1**

Measures of Heterogeneity and Their Domain of Definition

| Measure | Scale | Range |
| --- | --- | --- |
| $Q$ | Absolute | $[0, \infty)$ |
| $I^2$ | Percent | $[0, 100\%]$ |
| $\tau^2$ | Outcome | $[0, \infty)$ |
| $H^2$ | Absolute | $[1, \infty)$ |
| $R^2$ | Absolute | $[1, \infty)$ |

## 11.6 Statistical Methods

### 11.6.1 Fixed Effect Approach

A parameter $\mu$ of clinical interest for assessing the effect of two treatments, independent estimates of this parameter computed "within" each of $M$ trials, and the weighted means of these statistics "across" trials are the basic ingredients of every meta-analysis.

Let $\hat{\mu}_m$ ($m = 1, 2, ..., M$} be the estimate of $\mu$ calculated in the $m$th trial; it is assumed for the moment that $\hat{\mu}_m$ is normally distributed around $\mu$ with variance $\sigma^2_{\varepsilon m}$ based upon the internal variability of the $m$th trial and let $v_m$ be its estimate. After defining the weight $w_m = v_m^{-1}$, we consider the statistic $\sqrt{w_m}\hat{\mu}_m$, which is asymptotically normally distributed with expectation

$$E\left(\sqrt{w_m}\hat{\mu}_m\right) = \sqrt{w_m}\mu$$

and variance

$$var(\sqrt{w_m}\hat{\mu}_m) = 1 \tag{11.7}$$

By the standard least-squares analysis, the estimate of $\mu$ pooled among the trials is found to be

$$\bar{\hat{\mu}} = \frac{\sum_{m=1}^{M} w_m \hat{\mu}_m}{\sum_{m=1}^{M} w_m} \tag{11.8}$$

with variance

$$var\left(\bar{\hat{\mu}}\right) = \left(\sum_{m=1}^{M} w_m\right)^{-1} \tag{11.9}$$

Since $\bar{\mu} \sim N(\mu, (\sum_{m=1}^{M} w_m)^{-1})$, it is easy to

1. Test the null hypothesis, $H_0: \mu = 0$, by means of the standardized normal deviate:

$$z = \frac{\sum_{m=1}^{M} w_m \mu_m}{\sqrt{\sum_{m=1}^{M} w_m}} \tag{11.10}$$

2. Calculate the $100(1-\alpha)\%$ confidence interval of $\mu$ by

$$\bar{\mu} \pm \frac{Z_{1-\alpha/2}}{\sqrt{\sum_{m=1}^{M} w_m}} \tag{11.11}$$

In this approach we assumed the "fixed effects" model:

$$\hat{\mu}_m = \mu + \varepsilon_m \tag{11.12}$$

where $\varepsilon_m$ is a random component with average equal zero and variance $\sigma_{\varepsilon m}^2$. However, before merging information from the $M$ trials, the null hypothesis of homogeneity of treatment effect across all trials, $H_0 : \mu_1 = \mu_2 = \ldots = \mu_m = \ldots = \mu_M = \mu$, should be tested.

Since in a meta-analysis each trial is thought of as a stratum, the pertinent test statistic is equal to

$$Q_{\text{hom}} = \sum_{m=1}^{M} w_m \left(\hat{\mu}_m - \bar{\mu}\right)^2 \tag{11.13}$$

which is asymptotically distributed as $\chi^2$ with $(M-1)$ degrees of freedom (Hedges, 1994; Marubini and Valsecchi, 1995).

### 11.6.2 Binary Data

Two measures of treatment effect are usually adopted in clinical trials: the difference in the outcome probability and the log-odds ratio. Let results or the $m$th trial can be arranged in a $2 \times 2$ contingency table as shown in Table 11.2.

Let $\pi_{Am}$ be the true failure probability under treatment A, trial $m$; and $\pi_{Bm}$ be the true failure probability under treatment B, trial $m$ ($m = 1,2,\ldots.M$).

First, consider the probability difference: $p_m = \pi_{Am} - \pi_{Bm}$.

$$\hat{p}_m = \frac{f_{Am}}{n_{Am}} - \frac{f_{Bm}}{n_{Bm}} \tag{11.14}$$

and

$$v_m = \left(\frac{f_{Am} \cdot s_{Am}}{n_{Am}^3} + \frac{f_{Bm} \cdot s_{Bm}}{n_{Bm}^3}\right) \tag{11.15}$$

**TABLE 11.2**

Sampling Outcome for Binary End Point in the $m$th Study

| Treatment | Failure | Success | Total | |
|-----------|---------|---------|-------|---|
| A | $f_{Am}$ | $s_{Am}$ | $n_{Am}$ | $p_{Am} = f_{Am}/n_{Am}$ |
| B | $f_{Bm}$ | $s_{Bm}$ | $n_{Bm}$ | $p_{Bm} = f_{Bm}/n_{Bm}$ |
| Total | $f_{\cdot m}$ | $s_{\cdot m}$ | $n_{\cdot m}$ | |

The logarithm of the odds ratio is

$$\eta_m = \log \theta_m = \log \frac{\pi_{Am}(1 - \pi_{Bm})}{\pi_{Bm}(1 - \pi_{Am})}$$

and by the maximum likelihood method we obtain the estimates

$$\hat{\eta}_m = \log \frac{f_{Am} \cdot s_{Am}}{s_{Am} \cdot f_{Bm}} \tag{11.16}$$

and

$$v_m = \left[ f_{Am}^{-1} + s_{Am}^{-1} + f_{Bm}^{-1} + s_{Bm}^{-1} \right] \tag{11.17}$$

$$v_m = \frac{f_{\cdot m} S_{\cdot m} n_{Am} n_{Bm}}{n_{\cdot m}^2 (n_{\cdot m} - 1)} \tag{11.18}$$

## 11.7 Random Effect Model

Suppose that the analyses accomplished as mentioned in previous sections convinced the investigator that the separate estimates $\hat{\mu}_m$ differ by more than they should under the hypothesis of treatment effect homogeneity. The reasons for presence of heterogeneity could then be investigated.

A model alternative to Equation 11.12 was proposed by DerSimonian and Laird (1986), Raudenbush (1994), Hedge and Vevea (1998), Hunter and Schmidt (2000), and Overton (1998) to deal with these situations. Basically, in the absence of a specific explanation of the variation in $\xi$, this is taken as random and the model becomes

$$\left\{ \begin{array}{l} \hat{\mu}_m = \mu_m + \varepsilon_m \\ \mu_m = \mu + \phi_m \end{array} \right\}$$

where $\phi_m$ are assumed to be independent normally distributed random variables with mean zero and variance $\sigma_\phi^2$. This implies assuming that the $M$ $\hat{\mu}_m$ estimates are independently normally distributed with mean $\mu$ and variance $\sigma_{\varepsilon m}^2 + \sigma_\phi^2$ and, consequently, the weights to be used in the random effect model are

$$w_m' = \left( \sigma_{\varepsilon m}^2 + \sigma_\phi^2 \right)^{-1} \tag{11.19}$$

As in the previous section, $\sigma_{\varepsilon m}^2$ is estimated by $v_m$. DerSimonian and Laird (1986) suggested estimating $\sigma_\phi^2$ the method of moments using $Q_{\text{hom}}$ from Equation 11.13:

$$S_\phi^2 = \max\left\{0, [Q_{\text{hom}} - (M-1)]/\left[\sum_{m=1}^M w_m - \left(\sum_{m=1}^M w_m^2\right)\left(\sum_{m=1}^M w_m\right)^{-1}\right]\right\} \quad (11.20)$$

The estimator of $\mu$ is now

$$\bar{\mu} = \sum_{m=1}^M \frac{\hat{\mu}_m}{v_m + S_\phi^2}\left[\sum_{m=1}^M (v_m + S_\phi^2)^{-1}\right]^{-1} \quad (11.21)$$

with variance

$$var(\bar{\mu}) = \left[\sum_{m=1}^M (v_m + S_\phi^2)^{-1}\right]^{-1} \quad (11.22)$$

Therefore, in the presence of real heterogeneity and aiming at summarizing the difference of treatment effects in a clinically relevant measure, the researcher can assume that the additional variability, "between trials," in a given sense be random and incorporate the heterogeneity of effects in the estimation of the overall treatment effect by Equations 11.21 and 11.22. However, it appears sensible to interpret with caution the results of meta-analyses carried out according to the random effect model.

The applications of meta-analytic methods are widely illustrated in the medical research literature, for example, Sacks et al. (1987) and L'Abbe et al. (1987).

In the next section we illustrate meta-analysis methods using actual data, for continuous, binary, and time-to-event problems.

---

## 11.8 Examples

### Example 11.1: Effects and Side Effects of Low-Dosage Tricyclic Antidepressants in Depression (Furukawa et al., 2002)

The main objective of this meta-analysis is to compare the effect and the side effects of low-dosage tricyclic antidepressants (<100 mg/day) with placebo and with standard dosage tricyclics in the acute phase treatment of depression. Effectiveness of intervention was measured by basing the analyses on continuous measurements. The data are from 17 studies and the scores were at four weeks after intervention. The results are given in Figure 11.3.

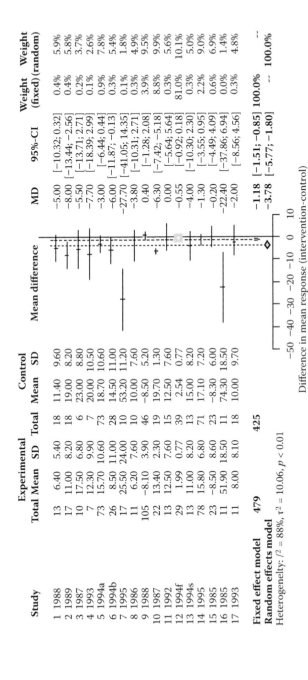

| Study | Experimental Total | Mean | SD | Control Total | Mean | SD | Mean difference | MD | 95%-CI | Weight (fixed) | Weight (random) |
|---|---|---|---|---|---|---|---|---|---|---|---|
| 1 1988 | 13 | 6.40 | 5.40 | 18 | 11.40 | 9.60 | | −5.00 | [−10.32; 0.32] | 0.4% | 5.9% |
| 2 1989 | 17 | 11.00 | 8.20 | 18 | 19.00 | 8.20 | | −8.00 | [−13.44; −2.56] | 0.4% | 5.8% |
| 3 1987 | 10 | 17.50 | 6.80 | 6 | 23.00 | 8.80 | | −5.50 | [−13.71; 2.71] | 0.2% | 3.7% |
| 4 1993 | 7 | 12.30 | 9.90 | 7 | 20.00 | 10.50 | | −7.70 | [−18.39; 2.99] | 0.1% | 2.6% |
| 5 1994a | 73 | 15.70 | 10.60 | 73 | 18.70 | 10.60 | | −3.00 | [−6.44; 0.44] | 0.9% | 7.8% |
| 6 1994b | 26 | 8.50 | 11.00 | 28 | 14.50 | 11.00 | | −6.00 | [−11.87; −0.13] | 0.3% | 5.4% |
| 7 1995 | 17 | 25.50 | 24.00 | 10 | 53.20 | 11.20 | | −27.70 | [−41.05; 14.35] | 0.1% | 1.8% |
| 8 1986 | 11 | 6.20 | 7.60 | 10 | 10.00 | 7.60 | | −3.80 | [−10.31; 2.71] | 0.3% | 4.9% |
| 9 1988 | 105 | −8.10 | 3.90 | 46 | −8.50 | 5.20 | | 0.40 | [−1.28; 2.08] | 3.9% | 9.5% |
| 10 1987 | 22 | 13.40 | 2.30 | 19 | 19.70 | 1.30 | | −6.30 | [−7.42; −5.18] | 8.8% | 9.9% |
| 11 1992 | 13 | 12.50 | 7.60 | 15 | 12.50 | 7.60 | | 0.00 | [−5.64; 5.64] | 0.3% | 5.6% |
| 12 1994f | 29 | 1.99 | 0.77 | 39 | 2.54 | 0.77 | | −0.55 | [−0.92; 0.18] | 81.0% | 10.1% |
| 13 1994s | 13 | 11.00 | 8.20 | 13 | 15.00 | 8.20 | | −4.00 | [−10.30; 2.30] | 0.3% | 5.0% |
| 14 1995 | 78 | 15.80 | 6.80 | 71 | 17.10 | 7.20 | | −1.30 | [−3.55; 0.95] | 2.2% | 9.0% |
| 15 1985 | 23 | −8.50 | 8.60 | 23 | −8.30 | 6.00 | | −0.20 | [−4.49; 4.09] | 0.6% | 6.9% |
| 16 1985 | 11 | 51.90 | 18.50 | 11 | 74.30 | 18.50 | | −22.40 | [−37.86; 6.94] | 0.0% | 1.4% |
| 17 1993 | 11 | 8.00 | 8.10 | 18 | 10.00 | 9.70 | | −2.00 | [−8.56; 4.56] | 0.3% | 4.8% |
| **Fixed effect model** | **479** | | | **425** | | | | **−1.18** | **[−1.51; −0.85]** | **100.0%** | -- |
| **Random effects model** | | | | | | | | **−3.78** | **[−5.77; −1.80]** | -- | **100.0%** |

Heterogeneity: $I^2 = 88\%$, $\tau^2 = 10.06$, $p < 0.01$

−50 −40 −30 −20 −10 0 10

Difference in mean response (intervention-control)

**FIGURE 11.3**

Forest plot for the data of Example 11.1.

The effect size in this meta-analysis is the standardized mean difference. The test of heterogeneity shows that there is significant variation among the studies ($p$-value < 0.01). Moreover, whether the model adopted is fixed effect or random effect, there is significant difference effect of intervention on the outcome as compared to the placebo.

### R-Code for Example 11.1

```
furukawa=read.csv(file.choose())
library(meta)
#Fitting Fixed effect model
model1=metacont(Ne,Me,Se,Nc,Mc,Sc,data=furukawa,studlab=paste
(study,year))
figure1=forest(model1,comb.random=FALSE,xlab="difference in mean
response (intervention-control)",
xlim=c(-50,10),xlab.pos=-20,smlab.pos=-20)
#Fitting Random effects model
model2=metacont(Ne,Me,Se,Nc,Mc,Sc,data=furukawa,studlab=paste
(study,year))
figure2=forest(model2,comb.random=TRUE,xlab="difference in mean
response (intervention-control)",
xlim=c(-50,10),xlab.pos=-20,smlab.pos=-20)
```

### Example 11.2: Ovarian Cancer and Oral Contraceptive Use

Oral contraceptives were introduced almost 60 years ago and over 100 million women currently use them. It is believed that oral contraceptives can reduce ovarian cancer, but the eventual public-health effects of this reduction will depend on how long the protection lasts. The data in this example (Table 11.3) are extracted from a much larger data set from 45 epidemiological case-control studies (Collaborative Group on Epidemiological Studies of Ovarian Cancer, 2008).

**TABLE 11.3**

Data for the Studies Constituting the Meta-Analysis of Ovarian Cancer and Oral Contraceptive Use

**Study Name: BCCD (USA)**

| Exposure | Status | |
|---|---|---|
| | Cases | Controls |
| Ever used | 72 | 323 |
| Never used | 293 | 1076 |

*(Continued)*

**TABLE 11.3 (CONTINUED)**
Data for the Studies Constituting the Meta-Analysis
of Ovarian Cancer and Oral Contraceptive Use

**Study Name: Nurse-Health (USA)**

| Exposure | Status | |
|---|---|---|
| | Cases | Controls |
| Ever used | 267 | 1152 |
| Never used | 413 | 1568 |

**Study Name: RCGP (UK)**

| Exposure | Status | |
|---|---|---|
| | Cases | Controls |
| Ever used | 86 | 422 |
| Never used | 90 | 282 |

**Study Name: Netherland Health Study**

| Exposure | Status | |
|---|---|---|
| | Cases | Controls |
| Ever used | 35 | 434 |
| Never used | 218 | 1343 |

**Study Name: CPSII-Mortality (USA)**

| Exposure | Status | |
|---|---|---|
| | Cases | Controls |
| Ever used | 437 | 2330 |
| Never used | 2160 | 8648 |

**Study Name: CPSII-Nutrition (USA)**

| Exposure | Status | |
|---|---|---|
| | Cases | Controls |
| Ever used | 105 | 508 |
| Never used | 245 | 897 |

**Study Name: RCGP: UK**

| Exposure | Status | |
|---|---|---|
| | Cases | Controls |
| Ever used | 1350 | 6245 |
| Never used | 90 | 5013 |

(*Continued*)

**TABLE 11.3 (CONTINUED)**
Data for the Studies Constituting the Meta-Analysis
of Ovarian Cancer and Oral Contraceptive Use

**Study Name: Other**

| Exposure | Status | |
|---|---|---|
| | Cases | Controls |
| Ever used | 241 | 1147 |
| Never used | 249 | 815 |

**R-Code for Example 11.2**

```
ovarian=read.csv(file.choose())
library(meta)
summary(ovarian$Ee/ovarian$Ne)
summary(ovarian$Ec/ovarian$Nc)
ov=metabin(Ee,Ne,Ec, Nc, sm="OR",method="I", data=ovarian,
studlab=study)
print(summary(ov),digits=2)
forest(ov,comb.random=TRUE, hestate=FALSE)
#to produce funnel plot, we use the function funnel in the R "meta"
package.
funnel(ov)
```

The odds ratio estimates and their 95% confidence limits together with the Forest plot are given in Figure 11.4.

**R Output**
Number of studies combined: k = 8

| | OR | 95%-CI | z | *p*-value |
|---|---|---|---|---|
| Fixed effect model | 0.75 | [0.71; 0.79] | −10.36 | <0.0001 |
| Random effects model | 0.74 | [0.68; 0.81] | −6.99 | <0.0001 |

Quantifying heterogeneity:

$tau^2 = 0.0049$; $H = 1.28$ [1.00; 1.93]; $I^2 = 39.2\%$ [0.0%; 73.2%]

Test of heterogeneity:

| Q | d.f. | *p*-value |
|---|---|---|
| 11.51 | 7 | 0.1178 |

Note that based on the Q-test, the data does not support the hypothesis of heterogeneity among the studies. The conclusion is women who use oral contraceptives are at a significant lower risk of ovarian cancer as compared to the controls. The funnel plot is given in Figure 11.5.

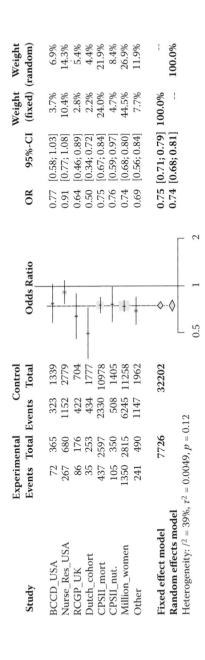

| Study | Experimental Events | Total | Control Events | Total | Odds Ratio | OR | 95%-CI | Weight (fixed) | Weight (random) |
|---|---|---|---|---|---|---|---|---|---|
| BCCD_USA | 72 | 365 | 323 | 1339 | | 0.77 | [0.58; 1.03] | 3.7% | 6.9% |
| Nurse_Res_USA | 267 | 680 | 1152 | 2779 | | 0.91 | [0.77; 1.08] | 10.4% | 14.3% |
| RCGP_UK | 86 | 176 | 422 | 704 | | 0.64 | [0.46; 0.89] | 2.8% | 5.4% |
| Dutch_cohort | 35 | 253 | 434 | 1777 | | 0.50 | [0.34; 0.72] | 2.2% | 4.4% |
| CPSII_mort | 437 | 2597 | 2330 | 10978 | | 0.75 | [0.67; 0.84] | 24.0% | 21.9% |
| CPSII_nut. | 105 | 350 | 508 | 1405 | | 0.76 | [0.59; 0.97] | 4.7% | 8.4% |
| Million_women | 1350 | 2815 | 6245 | 11258 | | 0.74 | [0.68; 0.80] | 44.5% | 26.9% |
| Other | 241 | 490 | 1147 | 1962 | | 0.69 | [0.56; 0.84] | 7.7% | 11.9% |
| **Fixed effect model** | | 7726 | | 32202 | | **0.75** | **[0.71; 0.79]** | **100.0%** | -- |
| **Random effects model** | | | | | | **0.74** | **[0.68; 0.81]** | -- | **100.0%** |

Heterogeneity: $I^2 = 39\%$, $\tau^2 = 0.0049$, $p = 0.12$

**FIGURE 11.4**
Forest plot for the oral contraceptive and ovarian cancer data.

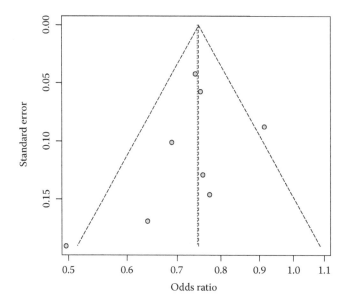

**FIGURE 11.5**
Funnel plot of the meta-analysis of oral contraceptive use and ovarian cancer.

## 11.9 Meta-Analysis of Diagnostic Accuracy

To evaluate the diagnostic accuracy of a new test we assume that $n_1 + n_2$ are enrolled in a study that compares the performance of a new test relative to a gold standard. The data are summarized in Table 11.4.

The two most common measures to evaluate the performance of a diagnostic (Table 11.3) or a screening test are the sensitivity and specificity.

**TABLE 11.4**

Evaluating the Diagnostic Accuracy of a Test

| Test | Gold | |
|------|------|------|
|      | +    | −    |
| +    | TP   | FN   |
| −    | FP   | TN   |
| Total | $n_1$ | $n_2$ |

The sensitivity is defined as *Sens.* = $TP/n_1$, and the specificity is defined by *Spec.* = $TN/n_2$.

There are two error rates: False negative rate = 1 – *Sens*, and False positive rate = 1 – *Spec*.

We now define the diagnostic odds ratio (ODR) as an omnibus measure of diagnostic accuracy. The odds that the patient tests positive versus negative is simply the quotient among true positives (TP) and false negatives (FN): TP/FN. Moreover, the odds that a healthy person tests positive versus negative is the quotient among false positives (FP) and true negatives (TN): FP/TN. With these definitions, we define the ratio of the two odds:

$$\text{DOR} = \frac{TP}{FN} \bigg/ \frac{FP}{TN} = \frac{Sens}{1 - sens} \bigg/ \frac{1 - spec}{spec}$$

$$= \frac{PPV}{1 - PPV} \bigg/ \frac{1 - NPV}{NPV} = \frac{Likelihood\ ratio\ positive}{Likelihood\ ratio\ negative}$$

As any odds ratio, the possible values of DOR range from zero to infinity. The null value is one, which means that the test has no discriminatory capacity among healthy and sick. A value greater than one indicates discriminatory ability, which will be greater the higher the value is. Finally, values between zero and one will indicate that the test does not only not discriminate well among healthy and sick, but that it incorrectly classifies them and yield more negative values among sick than among healthy people. DOR is a global measure that is easy to interpret and does not depend on the prevalence of the disease, although it must be said that it can vary among groups of patients with different severity of their disease.

$$SE\big(\log(DOR)\big) = \left( \frac{1}{TP} + \frac{1}{TN} + \frac{1}{FP} + \frac{1}{FN} \right)^{1/2}$$

We can now construct a 95% confidence interval on the DOR by first constructing the interval on the log-DOR, and then take the anti-log:

$$\ln(DOR) \pm SE\big(\log(DOR)\big)$$

**Example 11.3: Meta-Analysis for Sensitivity, Specificity, and Diagnostic Odds Ratio (Lee et al., 2015)**

The data for this example are given in Table 11.5.

**TABLE 11.5**

Results of 11 Studies to Evaluate the Diagnostic
Accuracy of Radiologic Diagnostic Tests

| Test | Gold | |
|------|:----:|----:|
| | + | − |
| + | 34 | 1 |
| − | 31 | 34 |
| Total | 65 | 35 |

| Test | Gold | |
|------|:----:|----:|
| | + | − |
| + | 17 | 4 |
| − | 15 | 43 |
| Total | 32 | 47 |

| Test | Gold | |
|------|:----:|----:|
| | + | − |
| + | 17 | 1 |
| − | 9 | 13 |
| Total | 26 | 14 |

| Test | Gold | |
|------|:----:|----:|
| | + | − |
| + | 227 | 38 |
| − | 49 | 188 |
| Total | 276 | 226 |

| Test | Gold | |
|------|:----:|----:|
| | + | − |
| + | 33 | 5 |
| − | 7 | 14 |
| Total | 40 | 19 |

| Test | Gold | |
|------|:----:|----:|
| | + | − |
| + | 65 | 5 |
| − | 7 | 67 |
| Total | 72 | 72 |

(*Continued*)

**TABLE 11.5 (CONTINUED)**
Results of 11 Studies to Evaluate the Diagnostic
Accuracy of Radiologic Diagnostic Tests

| Test | Gold | |
|---|---|---|
| | + | − |
| + | 38 | 4 |
| − | 4 | 13 |
| Total | 42 | 17 |

| Test | Gold | |
|---|---|---|
| | + | − |
| + | 26 | 5 |
| − | 2 | 9 |
| Total | 28 | 14 |

| Test | Gold | |
|---|---|---|
| | + | − |
| + | 55 | 17 |
| − | 3 | 113 |
| Total | 58 | 130 |

| Test | Gold | |
|---|---|---|
| | + | − |
| + | 46 | 1 |
| − | 1 | 13 |
| Total | 47 | 14 |

| Test | Gold | |
|---|---|---|
| | + | − |
| + | 27 | 40 |
| − | 0 | 62 |
| Total | 27 | 102 |

## R Code

```
#R-Code for example #3 diagnostic accuracy
data=read.csv(file.choose())
attach(data)
n1=tp+fn
```

```
n2=fp+tn
data=data.frame(data,n1,n2)
#The package that will be used for the MA
library(mada)
#calculate sensitivity
sens=round(data$tp/data$n1,4)
#calculate specificity
spec=round(data$tn/data$n2,4)
#Here we use the from the R package "meta" to calculate confidence
intervals on sensitivity
library(meta)
#Here we use from the R package "meta" to calculate confidence
intervals specificity
metaprop(tp,n1,data=data,comb.fixed=FALSE,comb.random=FALSE,
studlad=paste(study))
forest(metaprop(tp,n1,data=data,comb.fixed=FALSE,
comb.random=FALSE,studlad=paste(study)))
metaprop(tn,n2,data=data,comb.fixed=FALSE,comb.random=FALSE,
studlad=paste(study))
forest(metaprop(tn,n2,data=data,comb.fixed=FALSE,
comb.random=FALSE,studlad=paste(study)))
# Now we use the "mada" package to obtain meta-analysis on diag-
nostic accuracy measured by DOR
data$TP=data$tp
data$FN=data$n1 -data$tp
data$FP=data$n2 - data$tn
data$TN=data$tn
attach(data)
print(data)
data.short=data.frame(TP,FN,FP,TN)
print(data.short)
madad(data.short,correction.control="none")
diagnostic=madad(data.short)
print(diagnostic,digit=2)
dor=diagnostic$DOR
#IMPORTANT REMARK: Note that the DOR can also be calculated using
the "metabin" function
fit=metabin(tp,n1,n2-tn,n2,data=data,sm="OR",comb.fixed=FALSE,
comb.random=FALSE,correction.control="none")
forest(fit)
```

The results from the R code are listed in Table 11.6.

The estimated DOR and its 95% confidence limits together with the LR positive and LR ratio negative are give in Table 11.7.

The 95% confidence limits and the Forest plots for sensitivities and specificities are given respectively in Figure 11.6a and b.

## TABLE 11.6

Estimated Sensitivities and Specificities and Their 95% Confidence Levels

|        | Sens. | 2.5%  | 97.5% | spec  | 2.5%  | 97.5% |
|--------|-------|-------|-------|-------|-------|-------|
| [1,]   | 0.523 | 0.404 | 0.640 | 0.971 | 0.855 | 0.995 |
| [2,]   | 0.531 | 0.364 | 0.691 | 0.915 | 0.801 | 0.966 |
| [3,]   | 0.654 | 0.462 | 0.806 | 0.929 | 0.685 | 0.987 |
| [4,]   | 0.822 | 0.773 | 0.863 | 0.832 | 0.778 | 0.875 |
| [5,]   | 0.825 | 0.681 | 0.913 | 0.737 | 0.512 | 0.882 |
| [6,]   | 0.903 | 0.813 | 0.952 | 0.931 | 0.848 | 0.970 |
| [7,]   | 0.905 | 0.779 | 0.962 | 0.765 | 0.527 | 0.904 |
| [8,]   | 0.929 | 0.774 | 0.980 | 0.643 | 0.388 | 0.837 |
| [9,]   | 0.948 | 0.859 | 0.982 | 0.869 | 0.801 | 0.917 |
| [10,]  | 0.979 | 0.889 | 0.996 | 0.929 | 0.685 | 0.987 |
| [11,]  | 1.000 | 0.875 | 1.000 | 0.608 | 0.511 | 0.697 |

**Test for equality of sensitivities:**

$\chi$-squared = 90.0059, df = 10, p-value = 5.34e-15

**Test for equality of specificities:**

$\chi$-squared = 55.1942, df = 10, p-value = 2.9e-08

## TABLE 11.7

Diagnostic OR and Likelihood Ratios

|        | DOR     | 2.5%   | 97.5%     | pos.LR | 2.5%  | 97.5%   | neg.LR | 2.5%  | 97.5% |
|--------|---------|--------|-----------|--------|-------|---------|--------|-------|-------|
| [1,]   | 37.290  | 4.814  | 288.880   | 18.308 | 2.616 | 128.122 | 0.491  | 0.378 | 0.637 |
| [2,]   | 12.183  | 3.534  | 42.002    | 6.242  | 2.314 | 16.837  | 0.512  | 0.351 | 0.748 |
| [3,]   | 24.556  | 2.752  | 219.091   | 9.154  | 1.357 | 61.771  | 0.373  | 0.216 | 0.645 |
| [4,]   | 22.919  | 14.387 | 36.512    | 4.891  | 3.641 | 6.571   | 0.213  | 0.164 | 0.277 |
| [5,]   | 13.200  | 3.573  | 48.768    | 3.135  | 1.458 | 6.743   | 0.238  | 0.115 | 0.490 |
| [6,]   | 124.429 | 37.579 | 412.000   | 13.000 | 5.562 | 30.383  | 0.104  | 0.052 | 0.212 |
| [7,]   | 30.875  | 6.737  | 141.489   | 3.845  | 1.623 | 9.110   | 0.125  | 0.047 | 0.328 |
| [8,]   | 23.400  | 3.843  | 142.491   | 2.600  | 1.278 | 5.290   | 0.111  | 0.028 | 0.447 |
| [9,]   | 121.863 | 34.258 | 433.490   | 7.252  | 4.637 | 11.341  | 0.060  | 0.020 | 0.179 |
| [10,]  | 598.000 | 34.961 | 10228.606 | 13.702 | 2.072 | 90.621  | 0.023  | 0.003 | 0.160 |
| [11,]  | Inf     | NaN    | Inf       | 2.550  | 2.003 | 3.247   | 0.000  | 0.000 | NaN   |

Correlation of sensitivities and false positive rates:

rho 2.5 % 97.5 %

0.557 - 0.064 0.867

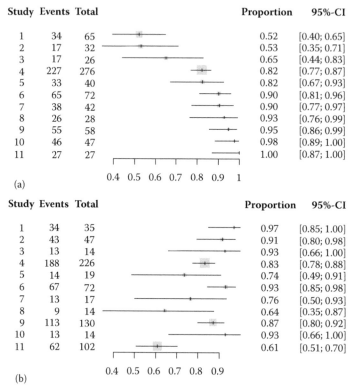

**FIGURE 11.6**
Forest plot for (a) sensitivity and (b) for specificity.

Important Remark: Note also that DOR can be calculated using the "metabin" function.

It is straightforward to get the "Forest plot" for DOR using "metbin" (Table 11.8).The results of using the "metabin" function are shown in Table 11.8, and its Forest plot is given in Figure 11.7.

### Example 11.4: Meta-Analysis of Survival Studies (Smedslund and Ringdal, 2004)

This example includes eight studies for the meta-analysis. The main objective is to provide a quantitative summary of effects of psychological interventions on cancer survival. When the outcome involves the comparison of two survival curves, the appropriate statistic to use is the log-hazard ratio. For each study, log-hazard and the standard errors are provided to complete the meta-analyses. (Also see Figure 11.9.)

### R Code

```
hazard=read.csv(file.choose())
library(meta)
```

**TABLE 11.8**

Similar Values of DOR When Estimated by "metbin" to Those Obtained from "madad" As Shown in Table 11.7

| | DOR | 95%-CI |
|---|---|---|
| 1 | 37.2903 | [ 4.8137; 288.8800] |
| 2 | 12.1833 | [ 3.5340; 42.0019] |
| 3 | 24.5556 | [ 2.7522; 219.0913] |
| 4 | 22.9194 | [14.3869; 36.5124] |
| 5 | 13.2000 | [ 3.5728; 48.7679] |
| 6 | 124.4286 | [37.5788; 411.9999] |
| 7 | 30.8750 | [ 6.7374; 141.4888] |
| 8 | 23.4000 | [ 3.8428; 142.4911] |
| 9 | 121.8627 | [34.2581; 433.4902] |
| 10 | 598.0000 | [34.9612; 10228.6059] |
| 11 | 84.8765 | [ 5.0354; 1430.6672] |

Quantifying heterogeneity:

tau^2 = 0.4446; H = 1.42 [1.00; 2.00] ; I^2 = 50.2% [0.6%; 75.0%]

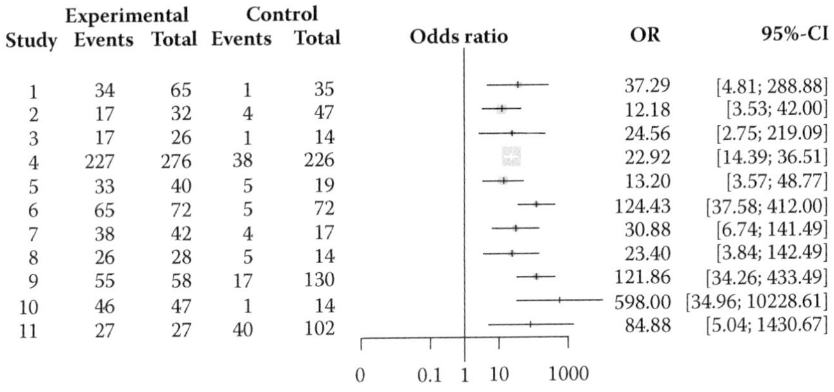

| Study | Experimental Events | Total | Control Events | Total | Odds ratio | OR | 95%-CI |
|---|---|---|---|---|---|---|---|
| 1 | 34 | 65 | 1 | 35 | | 37.29 | [4.81; 288.88] |
| 2 | 17 | 32 | 4 | 47 | | 12.18 | [3.53; 42.00] |
| 3 | 17 | 26 | 1 | 14 | | 24.56 | [2.75; 219.09] |
| 4 | 227 | 276 | 38 | 226 | | 22.92 | [14.39; 36.51] |
| 5 | 33 | 40 | 5 | 19 | | 13.20 | [3.57; 48.77] |
| 6 | 65 | 72 | 5 | 72 | | 124.43 | [37.58; 412.00] |
| 7 | 38 | 42 | 4 | 17 | | 30.88 | [6.74; 141.49] |
| 8 | 26 | 28 | 5 | 14 | | 23.40 | [3.84; 142.49] |
| 9 | 55 | 58 | 17 | 130 | | 121.86 | [34.26; 433.49] |
| 10 | 46 | 47 | 1 | 14 | | 598.00 | [34.96; 10228.61] |
| 11 | 27 | 27 | 40 | 102 | | 84.88 | [5.04; 1430.67] |

0  0.1  1  10  1000

**FIGURE 11.7**

Forest plot for DOR.

```
hazard1=metagen(logHR,selogHR,studlab=paste(study,year),
data=hazard,sm="HR")
print(hazard1,digits=2)
forest(hazard1)
HR                         95%-CI        %W(fixed)    %W(random)
fawzy 2003      0.3465 [0.1216; 0.9867]    2.1          6.1
McCorkle 2000   0.4902 [0.3197; 0.7515]   12.8         13.5
Spiegel 1989    0.5056 [0.3128; 0.8173]   10.1         12.7
Kuchler 1999    0.6126 [0.4460; 0.8416]   23.2         15.1
Cunninm 1998    0.7634 [0.4316; 1.3503]    7.2         11.3
```

```
Ilnyckyi 1994  1.1829 [0.7656; 1.8278]   12.3        13.4
goodwin 2001   1.2300 [0.8797; 1.7197]   20.8        14.8
Edelman 1999   1.3231 [0.8413; 2.0808]   11.4        13.1
Number of studies combined: k = 8
                        HR          95%-CI        z      p-value
Fixed effect model    0.8023 [0.6885; 0.9348] -2.82    0.0047
Random effects model  0.7693 [0.5582; 1.0601] -1.60    0.1089
Quantifying heterogeneity:
 tau^2 = 0.1511; H = 2.00 [1.41; 2.84]; I^2 = 74.9% [49.5%; 87.6%]
Test of heterogeneity:
   Q d.f.  p-value
 27.93  7  0.0002
Details on meta-analytical method:
- Inverse variance method
- Der-Simonian-Laird estimator for tau^2. The Forest plot is given
in Figure 11.8.
```

As can be seen there is significant heterogeneity among the studies. The overall hazard ratio is significant under the fixed effect model but it is not under the random effects model. As we mentioned before, the choice of the model depends on the objectives of the study.

### Example 11.5

The data for the meta-analysis in this example is part of meta-analytic study by Furukawa et al. (2002).

The main outcome measures odds of response in depression, defined as 50% or greater reduction in severity of depression. Relative risks of overall dropouts and dropouts due to side effects is the outcome measure. Fifteen studies from figure 2 in the cited paper, page 6, were selected. The SAS code for this example is provided to produce the forest plot. This is shown in (Figure 11.9).

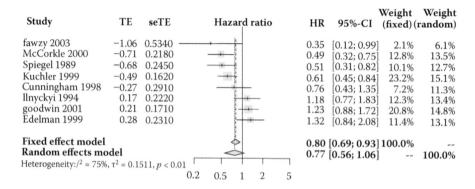

| Study | TE | seTE | Hazard ratio | HR | 95%-CI | Weight (fixed) | Weight (random) |
|---|---|---|---|---|---|---|---|
| fawzy 2003 | −1.06 | 0.5340 | | 0.35 | [0.12; 0.99] | 2.1% | 6.1% |
| McCorkle 2000 | −0.71 | 0.2180 | | 0.49 | [0.32; 0.75] | 12.8% | 13.5% |
| Spiegel 1989 | −0.68 | 0.2450 | | 0.51 | [0.31; 0.82] | 10.1% | 12.7% |
| Kuchler 1999 | −0.49 | 0.1620 | | 0.61 | [0.45; 0.84] | 23.2% | 15.1% |
| Cunningham 1998 | −0.27 | 0.2910 | | 0.76 | [0.43; 1.35] | 7.2% | 11.3% |
| Ilnyckyi 1994 | 0.17 | 0.2220 | | 1.18 | [0.77; 1.83] | 12.3% | 13.4% |
| goodwin 2001 | 0.21 | 0.1710 | | 1.23 | [0.88; 1.72] | 20.8% | 14.8% |
| Edelman 1999 | 0.28 | 0.2310 | | 1.32 | [0.84; 2.08] | 11.4% | 13.1% |
| **Fixed effect model** | | | | 0.80 | [0.69; 0.93] | 100.0% | -- |
| **Random effects model** | | | | 0.77 | [0.56; 1.06] | -- | 100.0% |

Heterogeneity: $I^2$ = 75%, $\tau^2$ = 0.1511, $p < 0.01$

0.2  0.5  1  2  5

### FIGURE 11.8
Forest plot for the meta-analysis of the survival studies of Example 11.4.

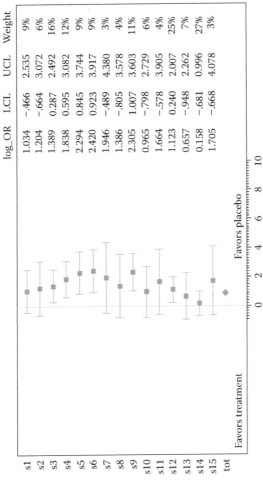

**FIGURE 11.9**

Forest plot for the 15 studies analyzed in Example 11.5.

## Exercises

11.1 Explain what is meant by "publication" bias? Why can publication bias be a problem in meta-analysis?

11.2 How can a publication bias be detected graphically?

11.3 If in meta-analysis we detect heterogeneity among the studies, can we still produce an overall summary measure of effect size? How do we interpret the effect size and its confidence interval in the presence of heterogeneity?

11.4 Suppose that we have M studies studying the possible association between maternal smoking (number of cigarettes smoked per day) and birth weight. This association is measured by Pearson's correlation coefficient $\rho$. The sample correlation coefficient in the $i$th study is $r_i$. If $n_i$ is the sample size of the $i$th study, then it is well known, when the sample size is sufficiently large, that $r_i$ has mean $\rho$ and standard deviation $(1 - \rho^2)/\sqrt{n_i}$. Moreover, under Fisher's variance stabilizing transformation, and as the sample size gets large:

$$z(r) = 1/2\log\left(\frac{(1 + r)}{(1 - r)}\right)$$

has normal distribution with mean $z(\rho)$ and variance $\dfrac{1}{(n_i - 2)}$.

Now, to perform meta-analysis using the M studies with the correlation as an effect size, comment on the merits of using Fisher's variance stabilizing transformation as compared to the analysis that does not use the transformation.

11.5 Write R code to produce the forest plot in Figure 11.9 for the data given in Example 11.5.

# 12

## Missing Data

### 12.1 Introduction

As indicated in Chapter 1, randomized controlled trials (RCTs) are considered the gold standard for establishing causality between exposure and intervention. To achieve its objectives, that it is to report an unbiased treatment effect, complete data should be available during and at the conclusion of the study. Missing or incomplete data are a serious problem in many fields of research and is ubiquitous in medical research, even RCTs. What we mean by missing data points is unavailable data, whether for the response or one or more covariate values (item missing) or an entire questionnaire. The issue of missing data is in particular very frequent in longitudinal or repeated measures studies. Often we use the term *missing* to include both attrition and noncompliance (Schafer, 1997).

There are several statistical and epidemiological problems caused by missing data. The first is the loss of statistical power to detect clinically or biologically meaningful treatment effect due to the reduction in the planned sample size. The second problem is the possible bias in the estimated treatment effect. For example, patients who suffer from severe side effects attributed to treatment are more likely to miss their subsequent clinical visits. Likewise, patients whose health improve and do not experience side effects may also miss the clinical visits or may drop out and discontinue their enrollment.

Given that in many studies the problem of missing data is unavoidable, researchers are forced to accept a certain amount of missing data points. An important question is how much missing data is acceptable. There are situations when 10% to 20% missing data may be acceptable and will have no or little effect on the study objectives. The seriousness of the problem with missing data depends on the reason it's missing. If we ignore the reasons for missing data and base the analysis only on the available data, the treatment effect will be overestimated and will not accurately reflect the entire population of patients in the study. The problem is further complicated due to the absence of a unified strategy to deal with the problem.

**TABLE 12.1**

Patterns of Missing Data

| Time Pattern | 1 | 2 | 3 | 4 | n |
|---|---|---|---|---|---|
| $t_1$ | y | y | y | . | y |
| $t_2$ | y | . | y | . | . |
| $t_3$ | . | . | y | . | y |
| \| | y | y | y | y | y |
| \| | y | y | y | y | . |
| $t_m$ | y | . | . | y | y |
| | y | . | . | y | . |
| $t_k$ | y | . | . | y | . |

Note:   Patient 1, single missing; patient 2, both patterns; patient 3, dropped-out after time point $t_{m-1}$; patient 4, late entry into the study.

## 12.2 Patterns of Missing Data

Missing data follow specific patterns, and some of the commonly occurring patterns are

- Intermittent (nonmonotone missing)—An observation on a subject is missing, but we do observe subsequent observations.
- Dropout (monotone missing)—No observation on a patient is available after a certain point in time.

Table 12.1 shows different patterns for missing data.

## 12.3 Mechanisms of Missing Data

The most appropriate way to handle missing will depend upon how data points became missing. Little and Rubin (1987) define three unique mechanisms of missing data: data missing completely at random, data missing at random, and data missing not at random.

### 12.3.1 Data Missing Completely at Random (MCAR)

Cases with complete data are indistinguishable from cases with incomplete data. This is a strong assumption about the missing data mechanism. The basic assumption is that missing data is completely unrelated to the outcome of interest. Heitjan (1997) provides an example of MCAR data: Imagine a research associate shuffling raw data sheets and arbitrarily discarding some of the sheets. Another example of MCAR data arises when investigators randomly assign research participants to complete two-thirds of a survey instrument. Other examples of MCAR are

- Accidental death of patient
- Patient changing his/her address
- Nurse forgets to administer questionnaire
- Some patients may not be able to participate because of a language barrier

The following notations will be used to formalize the definition of MCAR:

Denote the complete data vector with Y (all responses that one would have observed if all participants have complete responses).

Denote the observed data with $Y^o$ (for all observed responses). Denote the nonobserved responses with the vector $Y^m$.

A vector is defined as $D_i$, a vector of indicators of the missing data for subject $i$, so that $D_{ij} = 1$ if $Y_{ij}$ is missing, and $D_{ij} = 0$ if $Y_{ij}$ is observed. Therefore, we express the MCAR mechanism by

$$P_r[D_i = 1|Y_i^0, Y_i^m, X_i] = P_r[D_i = 1] \tag{12.1}$$

We also define a covariate dependent dropout as

$$P_r[D_i = 1|Y_i^0, Y_i^m, X_i] = P_r[D_i = 1|X_i] \tag{12.2}$$

### 12.3.1.1 Remarks on MCAR

If the missing mechanism is MCAR, or even covariate-dependent dropout, then

- ANOVA based on the observed $(Y_i^0)$ data only for the repeated measures will give UNBIASED estimates of treatment means. Suppose that we use the $(Y_i^0)$ data, and we want to model the subject-to-subject variability, the time effect, and treatment by time interaction. If we use PROC-MIXED in SAS, we will get UNBIASED estimates of the above effects provided that there is only one component of variance (subject is random). We also have to assume constant intrasubject correlation that does not change overtime.
- With regard to the aforementioned, if the attrition is heavy, statistical power will be lost unless we have enough $(Y_i^0)$.

**TABLE 12.2**

Effect of MCAR on OR and SE

| Calcium Scoring | | Complete Data | | | | MCAR | |
| | | | | | | 25% | Dropped Out |
| --- | --- | --- | --- | --- | --- | --- | --- |
| Low: | | Y | N | | | Y | N |
| Log (OR) = 1.1 | a | 60 | 20 | Log (OR) = 1.1 | | 45 | 15 |
| se = .342 | b | 40 | 40 | se = .394 | | 30 | 30 |
| | | | | | | | |
| High: | | Y | N | | | Y | N |
| Log (OR) = 1.1 | a | 120 | 40 | Log (OR) = 1.1 | | 90 | 30 |
| se = .365 | b | 20 | 20 | se = .422 | | 15 | 15 |
| | | | | | | | |
| Combined: | | Y | N | | | Y | N |
| Log (OR) = 1.1 | a | 180 | 60 | Log (OR) = 1.1 | | 135 | 45 |
| se = .236 | b | 60 | 60 | se = .309 | | 45 | 45 |

### Example 12.1

The data presented is hypothetical to illustrate the MCAR mechanism.

Patients were randomized to receive one of two treatments (a or b). The outcome of interest is achievement of 20% reduction in systolic blood pressures within two weeks. In Table 12.2, the participants were stratified according to their calcium scoring levels (high versus low) as determined by the CT scan. The results are shown in the left pane of Table 12.2. The log-odds ratio estimates and their standard errors are shown as well. Also assume that 25% of the results were discarded due to reasons unrelated neither to the treatment nor to the outcome of interest. The results are in the right side of the table. There were no changes in the estimated log-odds ratios. However, because of the reduction in the sample size the standard errors became higher. This shows that the treatment effect is unbiased under the MCAR mechanism but is less precise.

## 12.3.2 Missing at Random (MAR)

When the missing data mechanism is related to the observed response variable, but not to the unobserved, the missing data mechanism is said to be "missing at random," or MAR. Formally, the missing data indicator has the probabilistic structure given in the following equation:

$$P_r\left[D_i = 1 | Y_i^0, Y_i^m, X_i\right] = P_r\left[D_i = 1 | Y_i^0, X_i\right] \tag{12.3}$$

To illustrate the MAR, we shall consider another hypothetical situation similar to that given in Example 12.1.

### Example 12.2

We shall assume that the probability of missing data depends on both the treatment and the calcium score levels. For example, at the end of the

**TABLE 12.3**

Effect of MAR

| Calcium Scoring | | Complete Data | | | MAR | | |
|---|---|---|---|---|---|---|---|
| Low: | | Y | N | | Y | N | |
| Log (OR) = 1.1 | a | 300 | 100 | Log (OR) = 1.1 | 270 | 90 | 10% drop in both |
| se = .153 | b | 200 | 200 | se = .162 | 180 | 180 | treatments |
| High: | | Y | N | | Y | N | |
| Log (OR) = 1.1 | a | 120 | 280 | Log (OR) = 1.1 | 96 | 224 | 20% drop from a |
| se = .186 | b | 50 | 50 | se = .246 | 25 | 25 | 50% drop from b |
| Combined: | | Y | N | | Y | N | |
| Log (OR) = 0.1 | a | 420 | 380 | Log (OR) = .702 | 366 | 314 | |
| se = .114 | b | 250 | 250 | se = .117 | 205 | 355 | |

study, 10% of the low calcium scores stratum dropped from treatment a, while 10% dropped from treatment b. In the high calcium score stratum, 20% dropped from the treatment a group, and 50% dropped from the treatment b group. As can be seen in Table 12.3, the percentage of missing data depends on the observed data. Despite the differential missing, stratum specific log-odds ratio remains the same (log-odds = 1.1). However, the direct pooling of the two strata will produce an effect size (log-odds = 0.702) that is smaller.

### 12.3.3 Nonignorable, or Missing Not at Random (MNAR)

The pattern of data missingness is nonrandom and it is not predictable from other variables in the database. If a participant in a hypertension study does not attend a clinical visit due to concerns about his or her high systolic blood pressure, then the data are missing due to nonignorable factors. In contrast to the MAR situation outlined earlier where data missingness is explainable by other measured variables in a study, nonignorable missing data arise due to the data missingness pattern being explainable by the very variable(s) on which the data are missing.

$$P_r\left[D_i = 1 | Y_i^0, Y_i^m, X_i\right] = P_r\left[D_i = 1 | Y_i^0, Y_i^m, X_i\right] \tag{12.4}$$

**Example 12.3: Gay Men Cohort (GMC) Study**

Liu et al. (1999) reported the attrition rates in a study named the gay men cohort (GMC) study. The study involved a cohort of 178 gay men recruited in New York City in late 1987 and early 1988. Among them, 55 were HIV negative and 123 HIV positive but did not meet the 1986 Center for Disease Control criteria for acquired immune deficiency syndrome at baseline. The outcome of interest was neurological impairment (NI). Subjects were followed for five years and assessed every six months, that is, 10 assessments per participant.

**TABLE 12.4**

Rates of NI by HIV Status at Each Visit

|  | 1 | 2 | 3 | 4 | 5 | 6 | 7 | 8 | 9 | 10 | Total |
|---|---|---|---|---|---|---|---|---|---|---|---|
| HIV– | 1/55 | 1/50 | 0/52 | 0/46 | 0/51 | 2/45 | 0/47 | 0/42 | 1/42 | 0/37 | |
| HIV+ | 3/123 | 0/110 | 4/102 | 2/104 | 6/94 | 5/88 | 8/81 | 4/67 | 3/67 | 4/57 | |
| Total assessments | 178 | 160 | 154 | 150 | 145 | 133 | 128 | 109 | 105 | 94 | |
| # Dropping out | 0 | 18 | 6 | 4 | 5 | 12 | 5 | 19 | 4 | 11 | 84 |

**TABLE 12.5**

Summary Information for the Cohort Study

|  | Dropout Rate | Missing Rate |
|---|---|---|
| HIV– | 18/55 (33%) | 83/550 (15%) |
| HIV+ | 66/123 (54%) | 343/1230 (28%) |
| Total | 84/178 (47%) | |

The main aim of this study was to investigate the effect of HIV status on "Neurological Impairment" (NI status). This NI, the outcome of interest, is a dichotomous outcome that varies over time: Impairment may be present at one assessment and absent at a later assessment of the same patient. Table 12.4 shows the rates of NI by HIV status at each visit.

From Table 12.5 we can see that nearly 50% (84/178) dropped out during the study, with the majority of dropouts in the HIV+ group (66/123). Although not shown in the previous tables, some of the HIV+ subjects were medically ill on the visit before drop out and about half of them (30/66) died after withdrawal. This suggests that NI at the time of missed assessments may be related to the subjects' withdrawal from the study. Therefore, the withdrawal is "INFORMATIVE" and therefore should not be ignored.

The lesson to be learned is that we must understand the reasons for dropping out, how many patients will go into the numerator and how many in the denominator in order to accurately estimate the attrition rate.

In practice it is usually difficult to meet the MCAR assumption. MAR is an assumption that is more often but not always tenable. The more relevant and related predictors one can include in statistical models, the more likely it is that the MAR assumption will be met.

## 12.4 Methods of Handling Missing Data

The following sections describe some of the commonly used methods for handling missing data. This list is not exhaustive, but it covers some of the more widely recognized approaches to handling databases with incomplete cases.

### 12.4.1 Listwise or Casewise Data Deletion

If a record has missing data for any one variable used in a particular analysis, omit that entire record from the analysis.

### 12.4.2 Pairwise Data Deletion

For bivariate correlations or covariances, compute statistics based upon the available pairwise data. Pairwise data deletion is available in a number of SAS and R statistical procedures.

### 12.4.3 Mean Substitution

Substitute a variable's mean value computed from available cases to fill in missing data values on the remaining cases. This option appears in several SPSS procedures.

### 12.4.4 Regression Methods

Develop a regression equation based on complete case data for a given variable, treating it as the outcome and using all other relevant variables as predictors. Then, for cases where Y is missing, plug the available data into the regression equation as predictors and substitute the equation's predicted Y value into the database for use in other analyses. An improvement to this method involves adding uncertainty to the imputation of Y so that the mean response value is not always imputed.

### 12.4.5 Maximum Likelihood Methods

Use all available data to generate maximum likelihood-based sufficient statistics. Usually these consist of a covariance matrix of the variables and a vector of means. This technique is also known as full information maximum likelihood (FIML).

### 12.4.6 Multiple Imputation (MI)

Similar to the maximum likelihood method, except that multiple imputation generates actual raw data values suitable for filling in gaps in an existing database. Typically, five to ten databases are created in this fashion. The investigator than analyzes these data matrices using an appropriate statistical analysis method, treating these databases as if they were based on complete case data. The results from these analyses are then combined into a single summary finding. A good review of the multiple imputation methods is given by Allison (2005b).

Among all the techniques designed to handle missing data, the multiple imputation is the most attractive. Rubin (1976, 1996) introduced the MI technique and its implementation is summarized in the following steps:

1. Assume a model that incorporates random variation and impute (substitute) the missing values.
2. Repeat the above step M times (between three and five times) to produce M complete data sets.
3. The required data analyses would then be performed on the complete data sets.
4. A simple estimate of the target parameter is obtained by averaging its estimates from the M samples.
5. Square the standard errors of the M estimates and then add the between samples variance estimates to get the overall variance estimator:

$$1/M \sum_j^M s_j^2 + (1 + 1/M)(1/(M-1)) \sum_j^M \left(b_j - \bar{b}\right)^2 \qquad (12.5)$$

Allison (2002) outlined several advantages for the MI procedure:

- Introduces appropriate element of uncertainty in the imputation process.
- Repeated imputations allow us to estimate the standard error of the estimates from each of the M samples.

Rubin (1994) described situations when the MI procedure should have desirable properties, the most important of which are that the missing mechanism should be MAR. Moreover the model used to fit the data should markedly deviate from the model used for the imputation process.

In summary, MI has several advantages: It is fairly well understood and robust to violations of nonnormality of the variables used in the analysis. Like hot deck imputation, it outputs complete raw data matrices. It is clearly superior to listwise, pairwise, and mean substitution methods of handling missing data in most cases. Disadvantages include the time intensiveness in imputing five to ten databases, testing models for each database separately, and recombining the model results into one summary. Furthermore, summary methods have been worked out for linear and logistic regression models.

### 12.4.7 Expectation Maximization (EM)

The expectation maximization (EM) approach to missing data handling was documented extensively by Little and Rubin (1987). The EM approach is an iterative procedure that proceeds in two discrete steps. First, in the expectation (E) step the procedure computes the expected value of the complete data log likelihood based upon the complete data cases and the algorithm's best guess as to what the sufficient statistical functions are for the missing data based upon the model specified and the existing data points; actual imputed values for the missing data points need not be generated. In the maximization (M) step it substitutes the expected values (typically means and covariances) for the missing data obtained from the E step and then maximizes the likelihood function as if no data were missing to obtain new parameter estimates. The new parameter estimates are substituted back into the E step and a new M step is performed. The procedure iterates through these two steps until convergence is obtained. Convergence occurs when the change of the parameter estimates from iteration to iteration becomes negligible.

The strength of the EM approach is that it has well-known statistical properties and it generally outperforms popular ad hoc methods of incomplete data handling such as listwise and pairwise data deletion and mean substitution because it assumes incomplete cases have data missing at random rather than missing completely at random. The primary disadvantage of the EM approach is that it adds no uncertainty component to the estimated data. Practically speaking, this means that while parameter estimates based upon the EM approach are reliable, standard errors and associated test statistics (e.g., $t$-tests) are not. This shortcoming led statisticians to develop two newer likelihood based methods for handling missing data: the raw maximum likelihood approach and multiple imputations.

The maximum likelihood (Graham et al., 1996), also known as full information maximum likelihood (FIML), methods use all available data points in a database to construct the best possible first- and second-order moment estimates under the MAR assumption. Put less technically, if the MAR assumption can be met, maximum likelihood-based methods can generate a vector of means and a covariance matrix among the variables in a database that is superior to the vector of means and covariance matrix produced by commonly used missing data handling methods such as listwise deletion, pairwise deletion, and mean substitution.

Software packages that use the maximum likelihood approach to handle incomplete data include the MIXED procedure in SAS, which can fit ANOVA, ANCOVA, and repeated measures models with time-constant and time-varying covariates. We should consider using PROC MIXED instead of SAS PROC GLM whenever we have repeated measures data with missing data points. PROC MIXED can also fit hierarchical linear models (HLMs), also known as multilevel or random coefficient models.

Raw maximum likelihood has the advantage of convenience/ease of use and well-known statistical properties. Unlike EM, it also allows for the direct computation of appropriate standard errors and test statistics. Disadvantages include an assumption of joint multivariate normality of the variables used in the analysis and the lack of a raw data matrix produced by the analysis. Recall that the raw maximum likelihood method only produces a covariance matrix and a vector of means for the variables; the statistical software then uses these as imputes for further analyses.

Raw maximum likelihood methods are also *model based.* That is, they are implemented as part of a fitted statistical model. The investigator may want to include relevant variables (e.g., reading comprehension) that will improve the accuracy of parameter estimates, but not include these variables in the statistical model as predictors or outcomes. Although it is possible to do this, it is not always easy or convenient, particularly in large or complex models.

Finally, Roth (1994) reviews these methods and concludes, as did Little and Rubin (1987) and Wothke (1998), that listwise, pairwise, and mean substitution missing data handling methods are inferior when compared with maximum likelihood–based methods such as raw maximum likelihood or multiple imputation. Regression methods are somewhat better, but not as good as hot deck imputation or maximum likelihood approaches. The EM method falls somewhere in between: It is generally superior to listwise, pairwise, and mean substitution approaches, but it lacks the uncertainty component contained in the raw maximum likelihood and multiple imputation methods.

It is important to understand that these missing data handling methods and the discussion that follows deal with incomplete data primarily from the perspective of *estimation* of parameters and computation of test statistics rather than prediction of values for specific cases.

## 12.5 Pattern-Mixture Models for Nonignorable Missing Data

All the methods of missing data handling considered earlier require that the data meet the Little and Rubin (1987) MAR assumption. There are circumstances, however, when this assumption cannot be met to a satisfactory degree; cases are considered missing due to nonignorable causes (Heitjan, 1997). In such instances the investigator may want to consider the use of a *pattern-mixture model*, a term used by Hedeker and Gibbons (1997). Earlier works dealing with pattern-mixture models include Little and Schenker (1995), Little (1993), and Glynn et al. (1986).

Pattern-mixture models categorize the different patterns of missing values in a data set into a predictor variable, and this predictor variable is incorporated into the statistical model of interest. The investigator can then determine if the missing data pattern has an predictive power in the model, either by itself (a main effect) or in conjunction with another predictor (an interaction effect). The chief advantage of the pattern-mixture model is that it does not assume the incomplete data are MAR or MCAR. The primary disadvantage of the pattern-mixture model approach is that it requires some custom programming on the part of the data analyst to obtain one part of the pattern-mixture analysis: the pattern-mixture averaged results.

## 12.6 Strategies to Cope with Incomplete Data

The best advice is to minimize the possibility for missing values. One should plan an appropriate data collection procedure and design interviews and questionnaires so that subjects have little reason to refuse an answer. Adequate planning can also help to avoid differential missingness with respect to disease status or exposure status. The same data collection procedures should be used for cases and controls in case-control studies, and exposed and unexposed subjects should be followed using similar procedures and effort in cohort studies. Usually one knows in advance which variables are most likely to have missing values. Then a fruitful strategy can be to collect data on a surrogate variable that is available on most subjects and to collect the variable of interest with additional effort only in a randomly selected subsample, ensuring that the MAR assumption holds. Then it is possible to use statistical methods very similar to the sophisticated methods discussed earlier, except that the surrogate variable is not included in the regression model. A general idea is to collect additional data to predict missingness. By incorporating such variables in the analysis, the MAR assumption may become more reliable. Finally, one can try to recontact a representative sample of the nonresponders, and then try to collect the missing data. If this succeeds, a valid analysis becomes possible in principle.

If these approaches are infeasible or unsuccessful, then one should at least discuss the possible impact of the missing values on the analysis. The first step is to report the missing rates for all variables, stratified by disease status and exposure levels, and to summarize major associations of missingness with other variables. The second step is to justify the analytical approach. If a complete case analysis is applied in a case-control study, then one should give arguments to exclude an important difference in the missing value mechanism between cases and controls. If one uses methods relying on the MAR assumption, then the latter must be justified or a sensitivity analysis should be conducted.

In summary, missing values are a common problem in the analysis of epidemiologic studies. We need to address the possibility of missing data occurring, and this problem should be addressed in the planning stage of the study so that we can minimize the its potential negative effect on the outcome of the study. Keep in mind that no matter how sophisticated the analytic statistical methods we intend to use, we will not be able to salvage a poorly planned and executed study.

## 12.7 Missing Data in SAS

In release 9.4 the procedures PROC MI and PROC MIANALYZE were made available in SAS, which implemented multiple imputation. The imputation step is carried out by PROC MI, which allows use of either mono-tone (predictive mean matching, denoted by REGRESSION, or propensity, denoted by PROPENSITY) or nonmonotone (using MCMC) missingness methods. The MCMC methods can also be used in a hybrid model where the dataset is divided into monotone and non-monotone parts, and a regression method is used for the monotone component. Extensive control and graphical diagnostics of the MCMC methods are provided. Once PROC MI has been run, use of complete data methods is straightforward; the only addition is the specification of a BY statement to repeat these methods (i.e., PROC GLM, PROC PHREG, or PROC LOGISTIC) for each value of the variable.

_Imputation_. This approach is attractive, since it allows the full range of regression models available within SAS to be used in imputation.

The results are combined using PROC MIANALYZE, which provides a clear summary of the results. SAS provides an option (EDF) to use the adjusted degrees of freedom suggested by Barnard and Rubin (1999), and it displays estimates of the fraction of missing information for each parameter.

---

## 12.8  Missing Data in R: MICE

Multiple imputation by chained equations (MICE) is a package for R. MICE provides a variety of imputation models, including forms of predictive mean matching and regression methods logistic and polytomous regression, and discriminant analysis.

---

## 12.9  Examples

### Example 12.4: Medical Insurance

This example has data on medical insurance. The dependent is the total amount of coverage, which depends on the household income and the number of individuals covered. The missing data are only in the dependent variable "amount". The dependent variable is measured on the continuous scale, and a linear multiple regression model is appropriate for the analysis of the pooled data. For this example, we shall use the R package MICE in R. The SAS and the R codes are provided.

The output from R is quite lengthy but we present only the relevant segments.

1. The pattern of missing data (Figure 12.1)
2. The results of each imputation
3. The pooled imputations to obtain the complete data ready for the appropriate analysis

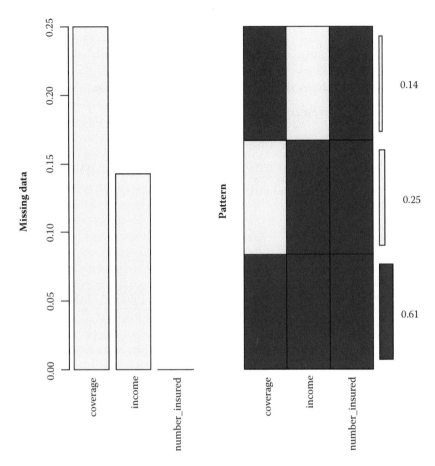

**FIGURE 12.1**
Pattern of missing data produced by the package "MICE."

### R Code

```
data=read.csv(file.choose())
summary(data)
sapply(data,function(x) sum(is.na(x)))
sapply(data, function(x) length(unique(x)))
require(mice)
library(VIM)
mice_plot=aggr(data,col=c ('navyblue','yellow'),numbers=TRUE,
sortVars=TRUE,
labels=names(data),cex.axis=.7,gap=3,ylab=c("Missing
data","Pattern"))

#Let us impute the missing values
imputed_data=mice(data,m=5,maxit=50,method='pmm',seed=600)
summary(imputed_data)
#Building a predictive model
fit=with(data=imputed_data,exp=lm(as.numeric(covrerage) ~income
+number_insured))
```

```
summary(fit)
#combine results of all 5 models
combine=pool(fit)

summary(pool(fit))
```

Figure 12.1 shows the pattern of missingness for this data.

**Summary of Imputation 1**

```
Call:
lm(formula = as.numeric(coverage) ~ income + number-insured)
```

The "MICE" package produces the results of the five imputations. These results are summarized in Table 12.6.

Combining the five imputations produces a complete data set ready for analysis.

```
combine=pool(fit)
summary(pool(fit))
```

The combined imputations are produced in Table 12.7.

**Remarks**

Data exploration is important before using the "Multiple Imputations" procedure to obtain a complete data set. In the exploratory analysis we find out that the relationship between the amount of coverage and the number insured is positive, therefore the results produced by "MICE" make sense. Unfortunately, this is not the case when using PROC MI in SAS. Each of the five imputations in SAS produced a negative value for the regression coefficient estimate of the same covariate (number insured) and that did not make sense. For this reason, we did not include the SAS analysis for this example. The message is that one has to be very careful when using any of the available "MI" procedures.

**Example 12.5: Epilepsy Count Data**

The data of this example was analyzed in Chapter 7. The response variable is a count and the covariates are treatment, age, and other factors. The missing mechanism affects the response variable only. We shall use SAS to analyze this data in three steps:

1. Detect the pattern of missing (Table 12.8)
2. Use the "MI" procedure to impute the missing data (Tables 12.9 to 12.13)
3. Use the "MIANALYZE" procedure to analyze the complete data set (Table 12.14)

SAS depicts the pattern of missingness in a table (see Table 12.8). It is identical to the graphical pattern of missingness produced by "MICE".

In Table 12.9 through 12.13 we list the results of the imputations produced by SAS "MI" procedure. Table 12.14 lists the results of combining the five imputations using the "MIANALYZE" procedure.

Note that in all five imputations, the significant factors associated with the response variable are baseline count, age, and treatment.

We now use "MIANALYZE" to analyze the complete data pooled from the five imputations (Table 12.14).

**TABLE 12.6**

Results of Five Imputations by "MICE"

Coefficients:

|  | Estimate | Std. Error | t Value | Pr(>\|t\|) |
|---|---|---|---|---|
| (Intercept) | -57653.772 | 18747.288 | -3.075 | 0.00504 ** |
| income | 40.831 | 8.117 | 5.030 | 3.45e-05 *** |
| number-insured | 10727.087 | 4352.174 | 2.465 | 0.02093 * |

Multiple R-squared: 0.7286, Adjusted R-squared: 0.7068
F-statistic: 33.55 on 2 and 25 DF, p-value: 8.334e-08

### Summary of Imputation 2

Coefficients:

|  | Estimate | Std. Error | t Value | Pr(>\|t\|) |
|---|---|---|---|---|
| (Intercept) | -60537.835 | 18493.577 | -3.273 | 0.0031 ** |
| income | 39.602 | 8.155 | 4.856 | 5.41e-05 *** |
| number-insured | 11915.850 | 4306.270 | 2.767 | 0.0105 * |

Multiple R-squared: 0.7361, Adjusted R-squared: 0.7149
F-statistic: 34.86 on 2 and 25 DF, p-value: 5.875e-08

### Summary of Imputation 3

Coefficients:

|  | Estimate | Std. Error | t Value | Pr(>\|t\|) |
|---|---|---|---|---|
| (Intercept) | -78635.03 | 24914.65 | -3.156 | 0.00414 ** |
| income | 61.58 | 10.96 | 5.621 | 7.54e-06 *** |
| number-insured | 5100.00 | 5795.99 | 0.880 | 0.38728 |

Multiple R-squared: 0.694, Adjusted R-squared: 0.6696
F-statistic: 28.35 on 2 and 25 DF, p-value: 3.723e-07

### Summary of Imputation 4

|  | Estimate | Std. Error | t Value | Pr(>\|t\|) |
|---|---|---|---|---|
| (Intercept) | -67406.32 | 17550.10 | -3.841 | 0.000745 *** |
| income | 41.16 | 7.53 | 5.466 | 1.12e-05 *** |
| number-insured | 12873.68 | 4061.07 | 3.170 | 0.003999 ** |

Multiple R-squared: 0.7767, Adjusted R-squared: 0.7589
F-statistic: 43.48 on 2 and 25 DF, p-value: 7.256e-09

### Summary of Imputation 5

Coefficients:

|  | Estimate | Std. Error | t Value | Pr(>\|t\|) |
|---|---|---|---|---|
| (Intercept) | -64808.88 | 27641.73 | -2.345 | 0.027292 * |
| income | 45.70 | 11.96 | 3.820 | 0.000786 *** |
| number-insured | 9721.49 | 6414.36 | 1.516 | 0.142168 |

Multiple R-squared: 0.5798, Adjusted R-squared: 0.5462
F-statistic: 17.25 on 2 and 25 DF, p-value: 1.965e-05

**TABLE 12.7**

Results of the Combined Models from the Five Imputations

|                  | Estimate   | se       | Pr(>\|t\|) |
|------------------|------------|----------|-----------|
| (Intercept)      | -65808.367 | 23578.85 | 0.012     |
| income           | 45.77335   | 13.804   | 0.015     |
| number-insured   | 10067.621  | 6058.018 | 0.122     |

**TABLE 12.8**

Pattern of Missing in the Epilepsy Data

| | | | | Missing Data Patterns | | | | | |
|---|---|---|---|---|---|---|---|---|---|
| Group | treat | baseline | age | count1 | count2 | count3 | count4 | Freq | Percent |
| 1 | X | X | X | X | X | X | X | 41 | 64.06 |
| 2 | X | X | X | X | X | X | . | 4 | 6.25 |
| 3 | X | X | X | X | X | . | X | 2 | 3.13 |
| 4 | X | X | X | X | . | X | X | 5 | 7.81 |
| 5 | X | X | X | X | . | X | . | 1 | 1.56 |
| 6 | X | X | X | X | . | . | X | 1 | 1.56 |
| 7 | X | X | X | . | X | X | X | 6 | 9.38 |
| 8 | X | X | X | . | . | X | X | 2 | 3.13 |
| 9 | X | X | X | . | . | . | . | 2 | 3.13 |

**TABLE 12.9**

Summary Estimates of Imputation 1

| | | | Analysis of GEE Parameter Estimates | | | | |
|---|---|---|---|---|---|---|---|
| | | | Empirical Standard Error Estimates | | | | |
| Parameter | | Estimate | Standard Error | 95% Confidence Limits | | Z | Pr > \|Z\| |
| Intercept |   | 0.8673  | 0.2923 | 0.2943  | 1.4402  | 2.97  | 0.0030 |
| time | 1 | 0.0195  | 0.1365 | -0.2480 | 0.2870  | 0.14  | 0.8866 |
| time | 2 | 0.0329  | 0.0567 | -0.0782 | 0.1439  | 0.58  | 0.5622 |
| time | 3 | 0.0955  | 0.1109 | -0.1218 | 0.3128  | 0.86  | 0.3890 |
| time | 4 | 0.0000  | 0.0000 | 0.0000  | 0.0000  | .     | .      |
| age  |   | 0.0220  | 0.0090 | 0.0044  | 0.0395  | 2.45  | 0.0142 |
| treat |   | -0.2908 | 0.1353 | -0.5560 | -0.0256 | -2.15 | 0.0316 |
| baseline |   | 0.0794 | 0.0046 | 0.0703  | 0.0885  | 17.12 | <.0001 |

**TABLE 12.10**

Summary Estimates of Imputation 2

| | | | Analysis of GEE Parameter Estimates | | | | |
|---|---|---|---|---|---|---|---|
| | | | Empirical Standard Error Estimates | | | | |
| Parameter | | Estimate | Standard Error | 95% Confidence Limits | | Z | Pr > \|Z\| |
| Intercept | | 0.9156 | 0.2740 | 0.3785 | 1.4526 | 3.34 | 0.0008 |
| time | 1 | 0.0708 | 0.1421 | -0.2077 | 0.3494 | 0.50 | 0.6181 |
| time | 2 | 0.0559 | 0.0626 | -0.0668 | 0.1786 | 0.89 | 0.3715 |
| time | 3 | 0.0782 | 0.1143 | -0.1457 | 0.3021 | 0.68 | 0.4936 |
| time | 4 | 0.0000 | 0.0000 | 0.0000 | 0.0000 | . | . |
| age | | 0.0207 | 0.0085 | 0.0040 | 0.0373 | 2.43 | 0.0151 |
| treat | | -0.3439 | 0.1527 | -0.6432 | -0.0446 | -2.25 | 0.0243 |
| baseline | | 0.0798 | 0.0052 | 0.0696 | 0.0900 | 15.32 | <.0001 |

**TABLE 12.11**

Summary Estimates of Imputation 3

| | | | Analysis of GEE Parameter Estimates | | | | |
|---|---|---|---|---|---|---|---|
| | | | Empirical Standard Error Estimates | | | | |
| Parameter | | Estimate | Standard Error | 95% Confidence Limits | | Z | Pr > \|Z\| |
| Intercept | | 0.9479 | 0.2827 | 0.3939 | 1.5019 | 3.35 | 0.0008 |
| time | 1 | 0.0364 | 0.1249 | -0.2085 | 0.2812 | 0.29 | 0.7710 |
| time | 2 | 0.0605 | 0.0605 | -0.0580 | 0.1790 | 1.00 | 0.3168 |
| time | 3 | 0.0769 | 0.1118 | -0.1423 | 0.2961 | 0.69 | 0.4917 |
| time | 4 | 0.0000 | 0.0000 | 0.0000 | 0.0000 | . | . |
| age | | 0.0199 | 0.0086 | 0.0030 | 0.0368 | 2.31 | 0.0210 |
| treat | | -0.3207 | 0.1388 | -0.5927 | -0.0486 | -2.31 | 0.0209 |
| baseline | | 0.0794 | 0.0048 | 0.0700 | 0.0889 | 16.41 | <.0001 |

It is interesting to see from Table 12.14 that there are negligible variations among the parameter estimates from the five imputations.

**Remarks**

For the correlated counts response variables, we use SAS to fit the GEE model. For each imputation, we fitted the appropriate model using PROC GENMOD, with Exchangeable correlation structure. In this situation, the SAS program does have the capability to produce a pooled data from multiple imputations. Unfortunately, our attempt to produce "pooled" estimates from the five imputations using "MICE" failed. An error message was given by R indicating that the "GEE" model does not belong to the "MIRA" class. There are ways to overcome this problem as shown in Example 12.6.

**TABLE 12.12**

Summary Estimates of Imputation 4

| Parameter | | Estimate | Standard Error | 95% Confidence Limits | | Z | Pr > \|Z\| |
|---|---|---|---|---|---|---|---|
| **Intercept** | | 1.0080 | 0.2900 | 0.4396 | 1.5764 | 3.48 | 0.0005 |
| **time** | 1 | 0.0188 | 0.1446 | −0.2646 | 0.3022 | 0.13 | 0.8964 |
| **time** | 2 | 0.0038 | 0.0702 | −0.1338 | 0.1414 | 0.05 | 0.9569 |
| **time** | 3 | 0.0838 | 0.1117 | −0.1351 | 0.3028 | 0.75 | 0.4530 |
| **time** | 4 | 0.0000 | 0.0000 | 0.0000 | 0.0000 | . | . |
| **age** | | 0.0187 | 0.0091 | 0.0009 | 0.0365 | 2.06 | 0.0399 |
| **treat** | | −0.2495 | 0.1329 | −0.5099 | 0.0110 | −1.88 | 0.0605 |
| **baseline** | | 0.0766 | 0.0049 | 0.0671 | 0.0862 | 15.76 | <.0001 |

*Analysis of GEE Parameter Estimates — Empirical Standard Error Estimates*

**TABLE 12.13**

Summary Estimates Imputation 5

| Parameter | | Estimate | Standard Error | 95% Confidence Limits | | Z | Pr > \|Z\| |
|---|---|---|---|---|---|---|---|
| **Intercept** | | 0.9227 | 0.2845 | 0.3651 | 1.4802 | 3.24 | 0.0012 |
| **time** | 1 | 0.0720 | 0.1022 | −0.1283 | 0.2723 | 0.70 | 0.4810 |
| **time** | 2 | 0.0404 | 0.0541 | −0.0656 | 0.1465 | 0.75 | 0.4549 |
| **time** | 3 | 0.1026 | 0.1092 | −0.1114 | 0.3166 | 0.94 | 0.3474 |
| **time** | 4 | 0.0000 | 0.0000 | 0.0000 | 0.0000 | . | . |
| **age** | | 0.0189 | 0.0090 | 0.0014 | 0.0365 | 2.11 | 0.0346 |
| **treat** | | −0.2850 | 0.1339 | −0.5473 | −0.0226 | −2.13 | 0.0333 |
| **baseline** | | 0.0817 | 0.0046 | 0.0728 | 0.0907 | 17.95 | <.0001 |

*Analysis of GEE Parameter Estimates — Empirical Standard Error Estimates*

**Example 12.6: Correlated Binary Data**

This example uses the Mial-Oldham family data when the sibs-response is binary. There are two covariates: age (measured at the sibling level) and father's status (1/0, hypertensive/normotensive) and this is measured at the cluster (family) level. Note that we did not include the mother status as a predictor to avoid possible multicollinearity, because this covariate is correlated with the father-status. SAS pattern of missingness is depicted in Table 12.15. The results of each of the five imputations are given in Tables 12.16 through 12.20.
**SAS Output**

**TABLE 12.14**

Analysis of Pooled Data from the Five Imputations

| | | | | Parameter Estimates | | | | |
|---|---|---|---|---|---|---|---|---|
| | | | | 95% Confidence | | | | |
| Parameter | Time | Estimate | Std. Error | Limits | | DF | Minimum | Maximum |
| Intercept | | 0.932286 | 0.290285 | 0.36309 | 1.50148 | 2820.8 | 0.867266 | 1.008000 |
| time | 1 | 0.043500 | 0.134145 | −0.21959 | 0.30659 | 1839.1 | 0.018833 | 0.072001 |
| time | 2 | 0.038706 | 0.065854 | −0.09114 | 0.16855 | 203.12 | 0.003795 | 0.060510 |
| time | 3 | 0.087406 | 0.112260 | −0.13263 | 0.30744 | 27756 | 0.076889 | 0.102606 |
| time | 4 | 0 | 0 | . | . | . | 0 | 0 |
| age | | 0.020041 | 0.008954 | 0.00249 | 0.03759 | 5513.2 | 0.018690 | 0.021976 |
| treat | | −0.297951 | 0.144415 | −0.58148 | −0.01442 | 715.48 | −0.343883 | −0.249459 |
| baseline | | 0.079407 | 0.005220 | 0.06911 | 0.08970 | 189.76 | 0.076641 | 0.081730 |

**TABLE 12.15**

Pattern of Missing Observation

| | | | | | | Missing Data Patterns | | |
|---|---|---|---|---|---|---|---|---|
| | | | | | | | Group Means | |
| Group | Father-Status | Age | Sib-Status | Freq. | Percent | Father-Status | Age | Sib-Status |
| 1 | X | X | X | 528 | 84.35 | 0.446970 | 25.479167 | 0.458333 |
| 2 | X | X | . | 56 | 8.95 | 0.392857 | 12.839286 | . |
| 3 | X | . | X | 3 | 0.48 | 0 | . | 0.333333 |
| 4 | X | . | . | 14 | 2.24 | 0.500000 | . | . |
| 5 | . | X | X | 24 | 3.83 | . | 35.625000 | 0.458333 |
| 6 | . | X | . | 1 | 0.16 | . | 55.000000 | . |

**TABLE 12.16**

Summary Estimates of Imputation 1

| | | Analysis of GEE Parameter Estimates | | | | | |
|---|---|---|---|---|---|---|---|
| | | Empirical Standard Error Estimates | | | | | |
| Parameter | Estimate | Standard Error | 95% Confidence Limits | | Z | Pr > \|Z\| |
| Intercept | −2.1864 | 0.2548 | −2.6859 | −1.6870 | −8.58 | <.0001 |
| Father-status | 0.3448 | 0.1935 | −0.0344 | 0.7240 | 1.78 | 0.0747 |
| Age | 0.0685 | 0.0084 | 0.0520 | 0.0850 | 8.13 | <.0001 |

**TABLE 12.17**

Summary Estimates Imputation 2

| | | Analysis of GEE Parameter Estimates | | | | |
|---|---|---|---|---|---|---|
| | | Empirical Standard Error Estimates | | | | |
| Parameter | Estimate | Standard Error | 95% Confidence Limits | | Z | Pr > \|Z\| |
| Intercept | −1.9450 | 0.2416 | −2.4186 | −1.4714 | −8.05 | <.0001 |
| Father-status | 0.2381 | 0.1867 | −0.1278 | 0.6040 | 1.28 | 0.2023 |
| Age | 0.0627 | 0.0083 | 0.0465 | 0.0789 | 7.57 | <.0001 |

**TABLE 12.18**

Summary Estimates of Imputation 3

| | | Analysis of GEE Parameter Estimates | | | | |
|---|---|---|---|---|---|---|
| | | Empirical Standard Error Estimates | | | | |
| Parameter | Estimate | Standard Error | 95% Confidence Limits | | Z | Pr > \|Z\| |
| Intercept | −1.8365 | 0.2312 | −2.2897 | −1.3832 | −7.94 | <.0001 |
| Father-status | 0.2805 | 0.1824 | −0.0770 | 0.6379 | 1.54 | 0.1241 |
| Age | 0.0588 | 0.0078 | 0.0434 | 0.0742 | 7.49 | <.0001 |

**TABLE 12.19**

Summary Estimates of Imputation 4

| | | Analysis of GEE Parameter Estimates | | | | |
|---|---|---|---|---|---|---|
| | | Empirical Standard Error Estimates | | | | |
| Parameter | Estimate | Standard Error | 95% Confidence Limits | | Z | Pr > \|Z\| |
| Intercept | −1.9059 | 0.2477 | −2.3913 | −1.4205 | −7.70 | <.0001 |
| Father-status | 0.2151 | 0.1858 | −0.1491 | 0.5794 | 1.16 | 0.2470 |
| Age | 0.0627 | 0.0082 | 0.0466 | 0.0788 | 7.62 | <.0001 |

**TABLE 12.20**

Summary Estimates of Imputation 5

| | | Analysis of GEE Parameter Estimates | | | | |
|---|---|---|---|---|---|---|
| | | Empirical Standard Error Estimates | | | | |
| Parameter | Estimate | Standard Error | 95% Confidence Limits | | Z | Pr > \|Z\| |
| Intercept | −2.0495 | 0.2459 | −2.5315 | −1.5676 | −8.34 | <.0001 |
| Father-status | 0.2689 | 0.1868 | −0.0972 | 0.6349 | 1.44 | 0.1500 |
| Age | 0.0652 | 0.0084 | 0.0487 | 0.0816 | 7.75 | <.0001 |

**Remarks**

Note that for each of the five imputations the GEE model for the analysis of correlated binary outcomes was used to estimate the model parameters. This model is an appropriate model to fit this type of data. However, we could not use "MIANALYZE" to pool the estimates from the five imputations. The approach that should be followed to produce the pooled model after obtaining the complete data will be outlined. Also note that when the covariate "father-status" was entered as a categorical variable, "MI" produced results that did not make sense, because this covariate has some missing data points. However, when this covariate was entered as a quantitative variable $(0/1)$, the model fitting was problem free.

Let $\theta_i(i = i, 2,...M)$ denote the parameter estimate from the $i$th imputation $S_i^2 = (SE)^2$ be the square of its standard error

$$\bar{\theta} = \frac{1}{M} \sum_{i=1}^{M} \theta_i.$$

With $M = 5$ imputations, the variance of $\theta_i$ is taking into account the within-imputations variance $\frac{1}{M-1} \sum_{j=1}^{M} (\theta_i - \bar{\theta})^2$,

and we can construct a 95% confidence interval on the "pooled" estimate avoiding the difficulties faced by using either "pool" function in R or the "MIANALYE" in SAS. This is illustrated in Example 12.6 (see SAS and R Codes).

A summary of the five imputations from SAS is shown in Table 12.21. The variance of the pooled estimate is

$$\text{Var}\left(\theta_{\text{pooled}}\right) = \frac{1}{M} \sum_{i=1}^{M} S_i^2 + \left(1 + \frac{1}{M}\right)\left(\frac{1}{M-1} \sum_{i=1}^{M} (\theta_i - \bar{\theta})^2\right) \quad (12.6)$$

Therefore

Mean (intercept) = −1.98, SD (intercept) = .244
Mean (father-status) = .271, SD (father-status) = .187
Mean (age) = .0625,  SD (age) = .008

**TABLE 12.21**

Summary Results of Five Imputations for Correlated Binary Outcomes Using GEMOD

| | IMPUTATION ($i$) | | | | | | | | | |
|---|---|---|---|---|---|---|---|---|---|---|
| | 1 | | 2 | | 3 | | 4 | | 5 | |
| | $\theta$ | Se | $\theta$ | Se | $\theta$ | Se | $\theta$ | Se | $\theta$ | Se |
| Intercept ($\theta_1$) | −2.18 | .254 | −1.94 | .242 | −1.84 | .231 | −1.91 | .247 | −2.05 | .25 |
| Father_status ($\theta_2$) | .345 | .194 | .24 | .187 | .28 | .182 | .215 | .186 | .269 | .187 |
| Age ($\theta_3$) | .068 | .008 | .063 | .008 | .059 | .008 | .063 | .008 | .065 | .008 |

Note that for all estimates of the model parameters, the estimates across imputations were quite close to each other, therefore the between-imputations variance component $\left(\dfrac{1}{M-1}\displaystyle\sum_{i=1}^{M}(\theta_i - \bar{\theta})^2\right)$ is almost zero. Hence, for this data, the standard error of the estimates is basically the square root of $\dfrac{1}{M}\displaystyle\sum_{i=1}^{M}s_i^2$. One can then construct a 95% confidence interval on the population parameters using the mean and the standard errors shown earlier.

Finally, from the pooled information we have the required model:

$$\text{Logit(sib-status)}$$

$$= -1.985 + .0625\,(\text{age}) + .271\,(\text{father-status}) \qquad (12.7)$$

The second remark on this example pertains to the attempt to use "MICE". Again, neither using the GEE function to fit a model for correlated binary outcomes nor the pooling was successful. Alternatively, we fitted a "GLM" model for binary data, ignoring the correlation structure. The next step is to produce a "pooled" model using the aforementioned approach. Obviously, the "GLM" does not account for the correlation. To obtain correct estimates of the standard errors of the estimates, we inflate each standard error by the standard error of the pooled estimate by the square root of the DEFF (the design effect), which depend on both the intracluster correlation and the average cluster size.

**R-Code**

```
data=read.csv(file.choose())
summary(data)
names(data)
sapply(data,function(x) sum(is.na(x)))
sapply(data, function(x) length(unique(x)))
require(mice)
library(VIM)
mice_plot=aggr(data,col=c('navyblue','yellow'),
numbers=TRUE,sortVars=TRUE,
labels=names(data),cex.axis=.7,gap=3,ylab=c("Missing
data","Pattern"))

#Let us impute the missing values
imputed_data=mice(data,m=5,maxit=50,method='pmm',
seed=600)
summary(imputed_data)
#Building a predictive model
fit2=with(data=imputed_data,exp=glm
(sib_status~as.numeric(age)+father_status,
family=binomial("logit")))
summary(fit2)
#combine results of all 5 models
```

```
combine=pool(fit2)
summary(pool(fit2))
```

### R-Output

Figure 12.2 shows the pattern of missing produced by "MICE". This graphical pattern is identical to the pattern of missing in Table 12.15 produced by SAS.

In Table 12.22 through 12.26 we show the summary of the five imputations.

### Summary of Imputation 1

```
Call:
glm(formula   =   sib_status   ~   as.numeric(age)   +
father_status, family = binomial("logit"))
```

```
combine=pool(fit2)
summary(pool(fit2))
```

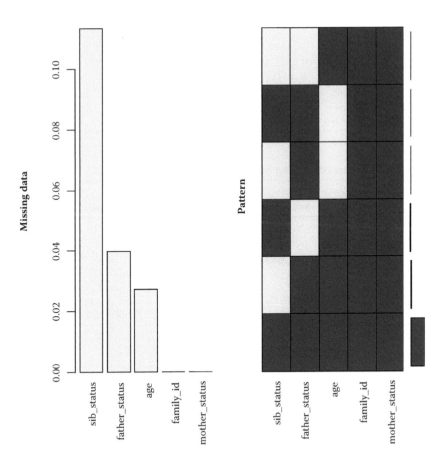

**FIGURE 12.2**
Pattern of missing observation from "MICE."

**TABLE 12.22**

Summary Estimates of Imputation 1

Coefficients:

|  | Estimate | Std. Error | z Value | Pr(>\|z\|) |
|---|---|---|---|---|
| (Intercept) | -1.941875 | 0.232516 | -8.352 | <2e-16 *** |
| as.numeric(age) | 0.061183 | 0.007636 | 8.012 | 1.13e-15 *** |
| father-status | 0.324845 | 0.173684 | 1.870 | 0.0614 . |

Null deviance: 857.05 on 625 degrees of freedom
Residual deviance: 781.89 on 623 degrees of freedom
AIC: 787.89

**TABLE 12.23**

Summary Estimates of Imputation 2

Coefficients:

|  | Estimate | Std. Error | z Value | Pr(>\|z\|) |
|---|---|---|---|---|
| (Intercept) | -2.250205 | 0.243099 | -9.256 | <2e-16 *** |
| as.numeric(age) | 0.069626 | 0.007935 | 8.775 | <2e-16 *** |
| father-status | 0.401946 | 0.177357 | 2.266 | 0.0234 * |

Null deviance: 852.42 on 625 degrees of freedom
Residual deviance: 758.63 on 623 degrees of freedom
AIC: 764.63

**TABLE 12.24**

Summary Estimates of Imputation 3

Coefficients:

|  | Estimate | Std. Error | z Value | Pr(>\|z\|) |
|---|---|---|---|---|
| (Intercept) | -2.286117 | 0.244144 | -9.364 | <2e-16 *** |
| as.numeric(age) | 0.071685 | 0.008034 | 8.923 | <2e-16 *** |
| father-status | 0.329697 | 0.177654 | 1.856 | 0.0635 * |

Null deviance: 850.46 on 625 degrees of freedom
Residual deviance: 753.87 on 623 degrees of freedom
AIC: 759.87

**TABLE 12.25**

Summary Estimates of Imputation 4

Coefficients:

|  | Estimate | Std. Error | z Value | Pr(>\|z\|) |
|---|---|---|---|---|
| (Intercept) | -1.944876 | 0.230809 | -8.426 | <2-16 *** |
| as.numeric(age) | 0.061957 | 0.007642 | 8.107 | 5.17e-16 *** |
| father-status | 0.271780 | 0.173559 | 1.566 | 0.117 |

Null deviance: 856.51 on 625 degrees of freedom
Residual deviance: 779.60 on 623 degrees of freedom
AIC: 785.6

**TABLE 12.26**

Summary Estimates of Imputation 5

Coefficients:

|  | Estimate | Std. Error | z Value | Pr (>|z|) |
|---|---|---|---|---|
| (Intercept) | -2.028051 | 0.235337 | -8.618 | <2e-16 *** |
| as.numeric(age) | 0.063563 | 0.007786 | 8.164 | 3.24e-16 *** |
| father-status | 0.349347 | 0.174540 | 2.002 | 0.0453 * |

Null deviance: 855.41 on 625 degrees of freedom
Residual deviance: 776.42 on 623 degrees of freedom
AIC: 782.42

### Model for Pooled Imputations

|  | est | se | t | Pr(>|t|) |
|---|---|---|---|---|
| (Intercept) | -2.090 | 0.299 | -6.983 | 1.672442e-07 |
| as.numeric(age) | 0.065 | 0.009 | 6.992 | 2.329540e-08 |
| father_status | 0.335 | 0.182 | 1.836 | 6.734940e-02 |

We now compare the f imputations produced by SAS and R for the covariates age and father-status. The results are summarized in Table 12.27.

The standard errors of the estimated parameters obtained from the "MICE" are smaller since the within-family correlation has not been accounted for, unlike the GEE procedure. One way to correct the standard errors is to multiply each standard error in the R column by the square root of the design effect $\sqrt{1 + (n_0 - 1)\rho}$ to obtain $s_i^* = s_i\sqrt{1 + (n_0 - 1)\rho}$. The variance of the estimated parameter pooled over the M imputations becomes

$$\text{Var}\left(\theta_{\text{pooled}}\right) = \frac{1}{M}\sum_{i=1}^{M}s_i^{*2} + \left(1 + \frac{1}{M}\right)\left(\frac{1}{M-1}\sum_{i=1}^{M}(\theta_i - \bar{\theta})^2\right)$$

$$(12.8)$$

It is left as an exercise to show that $\rho = 0.226$ and $n_0 = 3.14$.

**TABLE 12.27**

Comparing the Standard Errors of SAS and R for Example 12.6

|  | Imputation | | | | | | | | | |
|---|---|---|---|---|---|---|---|---|---|---|
|  | 1 | | 2 | | 3 | | 4 | | 5 | |
| Estimate | SAS | R | SAS | R | SAS | R | SAS | R | SAS | R |
| Intercept | .255 | .233 | .241 | .243 | .231 | .244 | .248 | .231 | .246 | .235 |
| Age | .008 | .008 | .008 | .008 | .008 | .008 | .008 | .008 | .008 | .008 |
| Father-status | .194 | .174 | .187 | .177 | .182 | .177 | .186 | .174 | .187 | .174 |

# References

Aalen, O.O., and Tretli, S. (1999). Analyzing incidence of testis cancer by means of a frailty model. *Cancer Causes and Control*, 10, 285–292.

Abraham, B., and Ledolter, J. (1983). *Statistical Methods for Forecasting*, John Wiley, New York.

Abrams, D., Goldman A., Launer, C., Korvick, J.A., Neaton, J.D., Crane, L.R., Grodesky, M. et al. (1994). A comparative trial of didanosine and zalcitabine in patients with human immunodeficiency virus infection who are intolerant of or have failed zidovudine. *New England Journal of Medicine*, 330, 657–662.

Agarwal, R., Bunay, Z., Bekele, D.M., and Light, R.P. (2008). Competing risk factor analysis of end-stage renal disease and mortality in chronic kidney disease. *American Journal of Nephrology*, 569–575.

Agresti, A. (1990). *Categorical Data Analysis*, John Wiley, New York.

Ahn, C., and Odom-Maryon, T. (1995). Estimation of a common odds ratio under binary cluster sampling. *Statistics in Medicine*, 14, 1567–1577.

Akaike, H. (1973). Information theory and extension of the maximum likelihood principle, in *Second International Symposium on Information Theory*, B.N. Petrov and F. Csaki, eds., 267–281, Budapest, Akademiai Kaido.

Albers, W. (1978). Testing the mean of a normal population under dependence. *The Annals of Statistics*, 6(6), 1337–1344.

Allison, D. (1996). Fixed effects partial likelihood for repeated events. *Sociological Methods & Research*, 25, 207–222.

Allison, P. (2002). *Missing Data*, Sage, Thousand Oaks, CA.

Allison, P.D. (2005a). *Fixed Effects Regression Methods for Longitudinal Data Using SAS*. SAS Institute INC., Cary, NC.

Allison, P. (2005b). Imputation of categorical variables with PROC MI. SUGI 30 Proceedings, Philadelphia, April 10–13.

Allison, P.D. (2010). *Survival Analysis Using SAS: A Practical Guide*, 2nd ed., SAS Institute, Cary, NC.

Altman, D.G., Schulz, K.F., Moher, D., Davidoff, F., Elbourne, D., Gotzsche, P.C., Lang, T., for the CONSORT Group. (2001). The revised CONSORT statement for reporting randomized trials: Explanation and elaboration. *Annals of Internal Medicine*, 134, 663–694.

Andersen, P.K., Klein, J.P., and Zhang, M.-J. (1999). Testing for center effects in multi-center survival studies: A Monte Carlo comparison of fixed and random effects tests. *Statistics in Medicine*, 18, 1489–1500.

Anderson, T.W. (1971). *The Statistical Analysis of Time Series*, John Wiley, New York.

Andrews, D., and Herzberg, A. (1985). *Data: A Collection of Problems from Many Fields for Students and Research Workers*, Springer-Verlag, New York.

Armitage, P., Berry, G., and Matthews, J.N.S. (2002). *Statistical Methods in Medical Research*. Blackwell, London.

Austin, P.C. (2009). Balance diagnostics for comparing the distribution of baseline covariates between treatment groups in propensity-score matched samples. *Statistics in Medicine*, 28, 3083–3107.

Austin, P.C. (2011). Optimal caliper widths for propensity-score matching when estimating differences in means and differences in proportions in observational studies. *Pharmaceutical Statistics*, 10(2), 150–161.

Barnard, J., and Rubin, D.B. (1999). Small-sample degrees of freedom with multiple imputation. *Biometrika*, 86(4), 948–955.

Bartlett, M. (1937). Properties of sufficiency and statistical tests, *Proceedings of the Royal Society, A*, 160, 268–282.

Bartlett, M. (1946). On the theoretical justification of sampling properties of an autocorrelated time series, *Journal of Royal Statistical Society: B*, 8, 27–41.

Bartolucci, A.A., Tendera, M., and Howard, G. (2011). Meta-analysis of multiple primary prevention trials of cardiovascular events using aspirin. *American Journal of Cardiology*, 107(12), 1796–1801.

Benjamini, Y., and Hochberg, Y. (1995). Controlling the false discovery rate: A practical and powerful approach to multiple testing. *Journal of the Royal Statistical Society, Series B (Methodological)*, 57(1), 289–300.

Benjamini, Y., and Yekutieli, D. (2001). The control of the false discovery rate in multiple testing under dependency. *Annals of Statistics*, 29(4), 1165–1188.

Berg, K. (1981). Twin research in coronary heart disease, twin research 3: Epidemiological and clinical studies. Proceedings of the Second International Congress on Twin Studies, Alan R. Liss, New York, 117–130.

Bishop, Y.M., Fienberg, S.E., and Holland, P.W. (1975). *Discrete Multivariate Analysis*, MIT Press, Cambridge, MA.

Bloomfield, P. (1976). *Fourier Analysis of Time Series: An Introduction*, Wiley Interscience, New York.

Box, G.E.P. (1954a). Some theorems on quadratic forms applied in the study of analysis of variance problems II. Effects of inequality of variances and of correlation between errors in the two-way classification. *Annals of Mathematics and Statistics*, 25, 484–498.

Box, G.E.P. (1954b). Some theorems on quadratic forms applied in the study of analysis of variance problems. *Annals of Mathematical Statistics*, 25, 290–302.

Box, G.E.P., and Jenkins, G.M. (1970). *Time Series Analysis, Forecasting, and Control*, Holden Day, San Francisco.

Box, G.E.P., and Pierce, D.A. (1970). Distribution of residual autocorrelations in autoregressive-integrated moving average time series models. *Journal of the American Association*, 70, 1509–1526.

Breslow, N. (1972). Contribution to the discussion of a paper by D. R. Cox. *Journal of the Royal Statistical Society, B*, 34, 216–217.

Breslow, N., Day, N., Halvorsen, K., Prentice, R., and Sabai, C. (1978). Estimation of multiple relative risk functions in matched case-control studies. *American Journal of Epidemiology*, 108, 299–307.

Brockwell, S.E., and Gordon, I.R. (2001). A comparison of statistical methods for meta-analysis. *Statistics in Medicine*, 20, 825–840.

Brookhart, M.A., Schneeweiss, S., Rothman, K.J., Glynn, R.J., Avorn, J., and Stürmer, T. (2006). Variable selection for propensity score models. *American Journal of Epidemiology*, 163(12), 1149–1156.

Brown, B., and Forsythe, A. (1974a). Robust tests for the equality of variances. *Journal of the American Statistical Association*, 69, 364–367.

Brown, B., and Forsythe, A. (1974b). The small sample behavior of some statistics which test the equality of several means. *Technometrics*, 16, 129–132.

Bryk, A.S., and Raudenbush, S.W. (1992). *Hierarchical linear models: Applications and data analysis methods*. Newbury Park, CA; Sage.

Bulpitt, C.J. (1988). Meta-analysis. *The Lancet*, 2(8602), 93–94.

Cameron, A.C., and Trivedi, P.K. (1998). *Regression Analysis of Count Data*. Cambridge University Press, UK.

Campbell, M.K., Thomson, S., Ramsay, C.R., MacLennan, G.S., and Grimshaw, J.M. (2004). Sample size calculator for cluster randomized trials. *Computers in Biology and Medicine*, 34, 113–125.

Chalmers, I., and Altman, D. (1995). *Systematic reviews*. BMJ, London.

Clayton, D.G. (1978). A model for association in bivariate life tables and its application in epidemiological studies of familial tendency in chronic disease incidence. *Biometrika*, 65, 141–151.

Cochran, W.G. (1937). Problems arising in the analysis of a series of similar experiments. *Journal of the Royal Statistical Society*, suppl. 4, 102–118.

Cochran, W.G. (1968). The effectiveness of adjustment by sub-classification in removing bias in observational studies. *Biometrics*, 24(2), 295–313.

Cochran, W.G. (1977). *Sampling Techniques*, 3rd ed., Wiley and Sons, New York.

Cochran, W.G. (1954). The combination of estimates from different experiments. *Biometrics*, 10, 101–129.

Cohen, J. (1965). Some statistical issues in psychological research. In *Handbook of Clinical Psychology*, B.B. Wolman, ed., 95–121, McGraw Hill, New York.

Cohen, J. (1988). *Statistical Power Analysis for the Behavioral Sciences*, 2nd ed., New York, Academic Press.

Cohen, J. (1990). Things I have learned (so far). *American Psychologists*, 45, 1304–1312.

Collaborative Group on Epidemiological Studies of Ovarian Cancer. (2008). Ovarian cancer and oral contraceptives: Collaborative reanalysis of data from 45 epidemiological studies including 23,257 women with ovarian cancer and 87303 controls. *The Lancet*, 371, 303–314.

Collett, D. (2003). *Modeling Binary Data*, 2nd ed., Chapman & Hall/CRC Press, London.

Collings, B.J., and Margolin, B.H. (1985). Testing goodness of fit for the Poisson assumption when observations are not identically distributed. *Journal of the American Statistical Association*, 74, 411–418.

Connor, R.J. (1987). Ample size for testing differences in proportions for the paired sample design. *Biometrics*, 43(1), 207–211.

Consul, P.C., and Jain, G.C. (1973). A generalization of the Poisson distribution. *Technometrics*, 15, 791–799.

Cooper, H.M. (1998). *Integrating Research: A Guide for Literature Reviews*, 3rd ed., Sage, Thousand Oaks, CA.

Cornfield, J. (1978). Randomization by group: A formal analysis. *American Journal of Epidemiology*, 108, 100–102.

Cox, D.R. (1970). *Analysis of Binary Data*, Chapman & Hall, London.

Cox, D.R. (1972). Regression Models and Life-Tables (with discussion), *Journal of the Royal Statistical Society, Series B*, 34, 187–220.

Cox, D.R., and Hinkley, D. (1974). *Theoretical Statistics*, Chapman & Hall, London.

Cox, D.R., and Snell, E. (1989). *Analysis of Binary Data*, 2nd ed., Chapman & Hall, London.

Crowder, M., and Hand, D. (1990). *Analysis of Repeated Measures*, Chapman & Hall, London.

Cryer, J. (1986). *Time Series Analysis*, Duxbury Press, Boston.

D'Agostino, R., Belanger, A., and D'Agostino, J.R. (1990). A suggestion for powerful and informative tests of normality. *The American Statistician*, 44, 316–321.

D'Agostino, R., and Pearson, E. (1973). Testing for the departures from normality. I. Fuller experimental results for the distribution of $b_2$ and $b_1$. *Biometrika*, 60, 613–622.

D'Agostino, R., and Stephens, M. (1986). *Goodness-of-Fit Techniques*, Marcel Dekker, New York.

Dean, C., Lawless, J.F., and Wilmot, G.E. (1989). A mixed Poisson-inverse Gaussian regression model. *Canadian Journal of Statistics*, 17, 171–182.

DeCosse, J.J., Miller, H.H., and Lesser, M.L. (1989). Effect of wheat fiber and vitamins C and E on rectal polyps in patients with familial adenomatous polyposis. *Journal of the National Cancer Institute*, 81, 1290–1297.

DeGruttola, V., and Tu, X.M. (1994). Modeling progression of CD-4 lymphocyte count and its relationship to survival time. *Biometrics*, 50, 1003–1014.

Dehejia, R., and Wahba, S. (1999). *An oversampling algorithm for non-experimental causal studies with incomplete matching and missing outcome variables*. Harvard University Mimeograph.

DerSimonian, R., and Laird, N. (1986). Meta-analysis in clinical trials. *Controlled Clinical Trials*, 7, 177–188.

Diehr, P., Martin, D.C., Koepsell, T., and Cheadle A. (1995). Breaking the matches in a paired t-test for community interventions when the number of pairs is small. *Statistics in Medicine*, 14(13), 1491–1504.

Diggle, P. (1988). An approach to the analysis of repeated measures. *Biometrics*, 44, 959–971.

Diggle, P. (1989). Testing for random dropouts in repeated measurement data. *Biometrics*, 45, 1255–1258.

Diggle, P. (1990). *Time Series: A Biostatistical Introduction*, Oxford Science Publications, Oxford.

Diggle, P., Liang, K.-Y., and Zeger, S. (1994). *The Analysis of Longitudinal Data*, Oxford Science Publications, Oxford.

Donald, A., and Donner, A. (1987). Adjustments to the Mantel-Haenszel chi-squared statistic and odds ratio estimator when the data are clustered. *Statistics in Medicine*, 6, 491–499.

Donner, A. (1982). An empirical study of cluster randomization. *International Journal of Epidemiology*, 11, 283–286.

Donner, A. (1986). A review of inference procedures for the intraclass correlation in the one-way random effects model. *International Statistical Review*, 54(1), 67–82.

Donner, A. (1989). Statistical methods in ophthalmology: An adjusted chi-squared approach. *Biometrics*, 45, 605–611.

Donner, A., Birkett, N., and Buck, C. (1981). Randomization by cluster: Sample size requirements and analysis. *American Journal of Epidemiology*, 114, 906–914.

Donner, A., and Klar, N. (2000). *Design and Analysis of Cluster Randomized Trials in Health Research*, Arnold, London.

Donner, A., and Klar, N. (2004). Pitfalls of and controversies in cluster randomization trials. *American Journal of Public Health*, 94(3), 416–422.

Donner, A. Eliasziw, M., and Klar, N. (1994). A comparison of methods for testing homogeneity of proportions in teratologic studies. *Statistics in Medicine*, 13, 1253–1264.

Draper, N., and Smith, H. (1981). *Applied Regression Analysis*, 2nd ed., John Wiley, New York.

Duffy, S.W., Tabar, L., Vitak, B., Yen, M.F., Warwick, J., Smith, R.A., and Chen, H.H. (2003). The Swedish two-county trial of mammographic screening: Cluster randomization and end point evaluation. *Annals of Oncology*, 14, 1196–1198.

Dupont, W.D. (1988). Power calculations for matched case-control studies. *Biometrics*, 44(4), 1157–1168.

Durbin, J., and Watson, G. (1951). Testing for serial correlation in leas squares regression II. *Biometrika*, 38, 159–178.

Egger, M., and Smith, G.D. (1997). Meta-analysis: Potentials and promise. *British Medical Journal*, 315, 1371–1374.

Egger, M., Smith, G.D., and Altman, D.G. (eds.). (2001). *Systematic Reviews in Health Care: Meta-Analysis in Context*, 2nd ed., BMJ Publishing, London.

Egger, M., Smith, G.D., Schneider, M., and Minder C. (1997). Bias in meta-analysis detected by a simple, graphical test. *BMJ*. https://doi.org/10.1136/bmj.315.7121.1533

Elashoff, R.M., Li, G., and Li, N. (2008). A joint model for longitudinal measurements and survival data in the presence of multiple failure types. *Biometrics*, 64(3), 762–771.

Elston, R.C. (1977). Response to query: Estimating "heritability" of a dichotomous trait. *Biometrics*, 33, 232–233.

Emslie, C., Gremshaw, J., and Templeton, A. (1993). Do clinical guidelines improve general-practice management and referral of infertile couples? *British Medical Journal*, 306, 1728–1731.

Erez, A., Bloom, M.C., and Wells, M.T. (1996). Using random rather than fixed effects models in meta-analysis: Implications for situational specificity and validity generalization. *Personnel Psychology*, 49, 275–306.

Faucett, C.L., and Thomas, D.C. (1996). Simultaneously modelling censored survival data and repeatedly measured covariates: A Gibbs sampling approach. *Statistics in Medicine*, 15, 1663–1685.

Field, A.P. (2003). The problems in using fixed-effects models of meta-analysis on real-world data. *Understanding Statistics*, 2, 77–96.

Fine, J.P., and Gray, R.J. (1999). A proportional hazards model for the sub-distribution of a competing risk. *Journal of the American Statistical Association*, 94, 496–509.

Fisher, R. (1932). *Statistical Methods for Research Workers*, 4th ed., Oliver and Boyd, Edinburgh.

Fitzmaurice, G.M. (1995). A caveat concerning independence estimating equations with multivariate binary data. *Biometrics*, 51, 309–317.

Fleiss, J. (1979). Confidence intervals for the odds ratio in case-control studies: The state of the art. *Journal of Chronic Diseases*, 32, 69–77.

Fleiss, J. (1981). *Statistical Methods for Rates and Proportions*, John Wiley, New York.

Frison, I., and Pocock, S.J. (1992). Repeated measures in clinical trials: Analysis using mean summary statistics and its implications for design. *Statistics in Medicine*, 11, 1685–1704.

Furukawa, T.A., McGuire, H., and Barbui, C. (2002). Meta-analysis of effects and side effects of low dosage tricyclic antidepressants in depression: Systematic review. *BMJ*, 325(7371), 991.

Gail, M. (1973). The determination of sample sizes for trials involving several independent 2 × 2 tables. *Journal of Chronic Diseases*, 26, 669–673.

Gangnon, R.E., and Kosorok, M.R. (2004). Sample size formula for clustered survival data using weighted log-rank statistics. *Biometrics*, 91(2), 263–275.

Gart, J.J., and Thomas, D. (1982). The performance of three approximate confidence limit methods for the odds ratio. *American Journal of Epidemiology*, 115, 453–470.

Geisser, S. (1963). Multivariate analysis of variance of a special covariance case. *Journal of the American Statistical Association*, 58, 660–669.

Glass, G.V. (1976). Primary, secondary, and meta-analysis of research. *Educational Researcher*, 5(10), 3–8.

Glass, G.V., McGaw, B., and Smith, M.L. (1981). *Meta-analysis in Social Research*. Sage, Beverly Hills, CA.

Glynn, R., Laird, N.M., and Rubin, D.B. (1986). Selection modelling versus mixture modelling with non-ignorable nonresponse. In H. Wainer (ed.), *Drawing Inferences from Self-Selected Samples*, 119–146. New York: Springer-Verlag.

Graham, J.W., Hofer, S.M., and MacKinnon, D.P. (1996). Maximizing the usefulness of data obtained with planned missing value patterns: An application of maximum likelihood procedures. *Multivariate Behavioral Research*, 31(2), 197–218.

Gray, R.J. (1988). A class of K-sample tests for comparing the cumulative incidence of a competing risk. *The Annals of Statistics*, 16, 1141–1154.

Greenwood, M., and Yule, G.U. (1920). An inquiry into the nature of frequency distributions representative of multiple happenings with particular reference to the occurrence of multiple attacks of disease or repeated accidents. *Journal of the Royal Statistical Society*, 83, 255–279.

Haldane, J. (1956). The estimation and significance of the logarithm of a ratio of frequencies. *Annals of Human Genetics*, 20, 309–311.

Hand, D.J., and Crowder, M. (1996). *Practical Longitudinal Data Analysis*, Chapman & Hall, London.

Harder, V.S., Stuart, E.A., and Anthony, J.C. (2010). Propensity score techniques and the assessment of measure covariate balance to test causal associations in psychological research. *Psychological Methods*, 15, 234–249.

Hauck, W. (1979). The large sample variance of the Mantel-Haenszel estimator of a common odds ratio. *Biometrics*, 35, 817–819.

Hayes R.J., and Bennett S. (1999). Simple sample size calculation for cluster randomized trials. *International Journal of Epidemiology*, 28, 319–326.

Hays, W.L. (1963). *Statistic*. Holt, Reinhart & Winston, New York.

Heckman, J.J., Ichimura, H., Smith, J., and Todd, P. (1996). Sources of selection bias in evaluating social programs: An interpretation of conventional measures and evidence on the effectiveness of matching as a program evaluation method. *Proceedings of the National Academy of Sciences*, 93(23), 13416–13420.

Hedeker, D., and Gibbons, R.D. (1997). Applications of the random effects pattern-mixture models for missing data in longitudinal studies. *Psychological Methods*, 2(1), 64–78.

Hedges, L.V. (1994). Fixed effects models. In *The Handbook of Research Synthesis*, H. Cooper and L.V. Hedges (eds.), 285–299, Russell Sage Foundation, New York.

Hedges, L.V., and Vevea, J.L. (1998). Fixed- and random-effects models in meta-analysis. *Psychological Methods*, 3, 486–504.

Heitjan, D.F. (1997). Annotation: What can be done about missing data? Approaches to imputation. *American Journal of Public Health*, 87(4), 548-550.

Hemming, K., Girling, A.J., Sitch, A.J., Marsh, J., and Lilford, R.J. (2011). Sample size calculations for cluster randomized controlled trials with a fixed number of clusters. *BMC Medical Research Methodology*, 11, 102. http://www.biomedcentral.com/1471-2288/11/102.

Henderson, R., Diggle, P., and Dobson A. (2000). Joint modeling of longitudinal measurements and event time data. *Biostatistics*, 4, 465–480.

Higgins, J.P.T., and Green, S., eds. (2006). *Cochrane Handbook for Systematic Reviews of Interventions, version 4.2.6*. The Cochrane Collaboration.

Higgins, J.P.T., and Thompson, S.G. (2002). Quantifying heterogeneity in a meta-analysis. *Statistics in Medicine*, 21, 1539–1558.

Ho, D., Kosuke, I., King, G., and Stuart, E. (2013). MatchIt: Nonparametric preprocessing for parametric causal inference [software]. Retrieved from http://gking.harvard.edu/matchit.

Hochberg, Y. (1988). A sharper Bonferroni procedure for multiple tests of significance. *Biometrika*, 75(4), 800–802.

Hochberg, Y., and Benjamini, Y. (1990). More powerful procedures for multiple significance testing. *Statistics in Medicine*, 9(7), 811–818.

Hogan, J.W., and Laird, N.M. (1997). Model-based approaches to analyzing incomplete longitudinal and failure time data. *Statistics in Medicine*, 16, 259–272.

Holland, P.W. (1986). Statistics and causal inference. *Journal of the American Statistical Association*, 81(396), 945–960.

Holm, S. (1979). A simple sequentially rejective multiple test procedure. *Scandinavian Journal of Statistics*, 6(2), 65–70.

Hommel, G. (1988). A stage-wise rejective multiple test procedure based on a modified Bonferroni test. *Biometrika*, 75(2), 383.

Hong G. (2010). Marginal mean weighting through stratification: Adjustment for selection bias in multilevel data. *Journal of Educational and Behavioral Statistics*, 35(5), 495–531.

Hong, G., and Raudenbush, S. (2006). Evaluating kindergarten retention policy: A case study of causal inference for multilevel observational data. *Journal of the American Statistical Association*, 101, 901–910.

Hosmer, D., and Lemeshow, S. (1989). *Applied Logistic Regression*, John Wiley, New York.

Hougaard, P. (1995). Frailty models for survival data. *Lifetime Data Analysis*, 1, 255–273.

Hsieh, F., Tseng, Y.K., and Wang, J.L. (2006). Joint modeling of survival and longitudinal data: Likelihood approach revisited. *Biometrics*, 62(4), 1037–1043.

Huber, P.J. (1967). The behavior of maximum likelihood estimators under nonstandard conditions. *Proceedings of the Fifth Berkeley Symposium on Mathematical Statistics and Probability*, Vol. A, 221–233, University of California Press, Berkeley.

Hunter, J.E., and Schmidt, F.L. (2000). Fixed effects vs. random effects meta-analysis models: Implications for cumulative research knowledge. *International Journal of Selection & Assessment*, 8, 275–292.

Janardan, K.G., Kerster, H.W., and Schaeffer, D.J. (1979). Biological applications of the Lagrangian poisson distribution. *BioScience*, 29(10), 599–602.

Jewell, N. (1984). Small-sample bias of point estimators of the odds ratio from matched sets. *Biometrics*, 40, 421–435.

Jewell, N. (1986). On the bias of commonly used measures of association for 2x2 tables. *Biometrics*, 42, 351–358.

Jones, R.H. (1993). *Longitudinal Data with Serial Correlation: A State-Space Approach*, Chapman & Hall, London.

Kalman, R. (1960). A new approach to linear filtering and prediction problems. *Transactions of the ASME–Journal of Basic Engineering*, 82, 35–45.

Kaplan, E.L., and Meier, P. (1958). Nonparametric estimation from incomplete observations. *Journal of the American Statistical Association*, 53, 457–481.

Katz, J., Carey, V., Zeger, S., and Sommer, L. (1993). Estimation of design effects and Diarrhea clustering within household and villages. *American Journal of Epidemiology*, 138(11), 994–1006.

Kendall, M., and Ord, K. (1990). *Time Series*, 3rd ed., Edward Arnold, London.

Kempthorne O., and Tandon, O.B. (1953). The estimation of heritability by regression offspring on parent. *Biometrics*, 9, 90–100.

Kerry, S.M., and Bland, J.M. (1998). The intracluster correlation coefficient in cluster randomization. *British Medical Journal*, 316, 1455.

Kerry, S.M., and Bland, M.J. (2006). Sample size in cluster randomized trials: Effect of coefficient of variation of cluster size and cluster analysis method. *International Journal of Epidemiology*, 35, 1292–1300.

Kim, J., and Seltzer, M. (2007). *Causal inference in multilevel settings in which selection processes vary across schools (CSE Technical Report 708)*. CRESST/University of California, Los Angeles.

Klein, J.P., and Andersen, P.K. (2005). Regression modeling of competing risks data based on pseudo values of the cumulative incidence function. *Biometrics*, 61, 223–229.

Klein, J.P., and Moeschberger, M.L. (2003). *Survival Analysis: Techniques for Censored and Truncated Data*, Springer Science & Business Media, New York.

Kleinbaum, D.G., Kapper, L., and Muller, K. (1988). *Applied Regression Analysis and Other Multivariable Methods*, PWS-Kent, Boston.

Kleinbaum, D.G., and Klein, M. (2005). "Competing risks survival analysis." In *Survival Analysis: A Self-Learning Text*, 391–461, Springer, New York.

Kline, R.B. (2004). *Beyond Significance Testing: Reforming Data Analysis Methods in Behavioral Research*, American Psychological Association, Washington DC, p. 95.

Kolmogorov, A. (1939). Sur L'interpolation et L'extrapolation des suites stationnaires. *Comptes rendus de l'Académie des Sciences*, 208, 2043–2045.

Kolmogorov, A. (1941). Interpolation and extrapolation von stationären Zufälligen Folgen. *Bulletin of the Academy of Sciences (Nauk), USSR, Series Mathematics*, 5, 3–14.

Korn, E.L., and Whittemore, A.S. (1979). Methods for analyzing panel studies of acute health effects of air pollution. *Biometrics*, 35, 795–802.

Krall, J.M., Uthoff, V.A., and Harley, J.B. (1975). A step-up procedure for selecting variables associated with survival. *Biometrics*, 31, 49–57.

L'Abbe, K.A., Detsky, A.S., and O'Rourke, K. (1987). Meta-analysis in clinical research. *Annals of Internal Medicine*, 107, 224–233.

Laird, N. (1988). Missing data in longitudinal studies. *Statistics in Medicine*, 7, 305–315.

Laird, N.M., and Ware, J.H. (1982). Random effects models for longitudinal data. *Biometrics* 38, 963–974.

Lambert, D. (1992). Zero-inflated Poisson regression with an application to defects in manufacturing. *Technometrics*, 34, 1–14.

Lanehart, R.E., de Gil, P.R., Kim, E.S., Bellara, A.P., and Kromrey, J.D. (2012). Propensity score analysis and assessment of propensity score approaches using SAS procedures. *SAS Global Forum 2012*, paper # 314–2012.

Lawless, J.F. (1982). *Statistical Models and Methods for Lifetime Data*, John Wiley & Sons, New York.

Lee, C.T., and Lindgren, B.R. (1996). Duration of Ventilating Tubes: A Test for Comparing Two Clustered Samples of Censored Data, *Biometrics*, 52, 328–334.

Lee, J., Kim, K.W., Choi, S.H., Huh, J., and Park, S.H. (2015). Systematic review and meta-analysis of studies evaluating diagnostic test accuracy: A practical review foe clinical researchers–Part II. Statistical methods of meta-analysis. *Korean Journal of Radiology*, 16(6), 1188–1196.

Leek, J. (2014). On the scalability of statistical procedures, *Simply Statistics* blog, available from, http://simplystatistics.org/2014/02/14/on-the-scalability-of-statistical-procedures-why-the-p-value-bashers-just-dont-get-it/.

Leite, W. (2017). *Practical Propensity Score Methods Using R*, Sage, Thousand Oaks, CA.

Levene, H. (1960). Robust tests of equality of variances: *In contributions to probability and statistics*, I. Olkin, Ed. pp. 278–292, Stanford University Press, Palo Alto, California.

Levin, M.L. (1953). The occurrence of lung cancer in man. *Acta Unio Int Contra Cancrum*, 9(3), 531–541.

Liang, K.Y., and Zeger, S.L. (1986). Longitudinal data analysis using generalized linear models. *Biometrika*, 73, 13–22.

Liang, K.Y., and Zeger, S.L. (1993). Regression analysis for correlated data. *Annual Review of Public Health*, 14, 43–68.

Liang, K.Y., Zeger S.L., and Qaqish, B. (1992). Multivariate regression analyses for categorical data. *Journal of the Royal Statistical Society, Series B: Methodological*, 54, 3–24.

Light, R.J., and Smith, P.V. (1971). Accumulating evidence: Procedures for resolving contradictions among different research studies. *Harvard Education Review*, 41, 429–471.

Lim, H.J., Zhang, X., and Dyck, R. (2010). Methods of competing risks analysis of end-stage renal disease and mortality among people with diabetes. *BMC Medical Research Methodology*, 10, 97.

Lin, J.T., Wang, L.W., Wang, J.T., Wang, T.H., Yang, C.S., and Chen, C.J. (1995). A nested case-control study on the association between *Helicobactor pylori* infection and gastric cancer risk in a cohort of 9775 in Taiwan. *Anticancer Research*, 15, 503–606.

Lindsay, J. (1993). *Models for Repeated Measurements*, Oxford Science Publications, Oxford.

Lipsey, M.W., and Wilson, D.B. (1993). The efficacy of psychological, educational, and behavioral treatment: Confirmation from meta-analysis. *American Psychologist*, 48, 1181–1209.

Littell, R.C., Milliken, G.A., Stroup, W.W., and Wolfinger, R.D. (1996). *SAS System for Mixed Models*, SAS Institute, Cary, N.C.

Little, R.J.A. (1993). Pattern-mixture models for multivariate incomplete data. *Journal of the American Statistical Association*, 88, 125–124.

Little, R.J.A., and Rubin, D. (1987). *Statistical Analysis with Missing Data*, John Wiley, New York.

Little, R.J.A., and Schenker, N. (1995). Missing data. In *Handbook of Statistical Modeling for the Social and Behavioral Sciences*, G. Arminger, C.C. Clogg, and M.E. Sobel (eds.), Plenum, New York.

Ljung, G.M., and Box, G.E.P. (1978). On the measure of lack of fit in time series models. *Biometrika*, 65, 297–304.

Lui, X., Waternaux, C., and Petkova, E. (1999). Influence of human immune-deficiency virus infection on neurological impairment: An analysis of longitudinal binary data with informative drop-out. *Applied Statistics*, 48(1), 103–115.

Macaskill, P., Walter, S.D., and Irwig, L. (2001). A comparison of methods to detect publication bias in meta-analysis. *Statistics in Medicine*, 20, 641–654.

Mantel, N. (1973). Synthetic retrospective studies and related topics. *Biometrics*, 29, 479–486.

Mantel, N., and Haenszel, W. (1959). Statistical aspects of the analysis of data from retrospective studies of disease. *Journal of the National Cancer Institute*, 22, 719–748.

Manton, K.G., and Stallard, E. (1981). Methods for evaluating the heterogeneity of aging processes in human populations using vital statistics data: Explaining the black/white mortality crossover by a model of mortality selection. *Human Biology*, 53, 47–67.

Martella, R., Nelson, J.R., and Marchand-Manella, N. (1999). *Research Methods: Learning to Become a Critical Research Consumer*, Allyn & Bacon, Boston.

Marubini, E., and Valsecchi, M.G. (1995). *Analyzing Survival Data from Clinical Trials and Observational Studies*, John Wiley, New York.

McCullagh, P., and Nelder, J. (1989). *Generalized Linear Models*, 2nd ed., Chapman & Hall, London.

McDonald, B. (1993). Estimating logistic regression parameters for bivariate binary data. *Journal of the Royal Statistical Society, Series B*, 55(2), 391–397.

Miall, W.E., and Oldham, P.D. (1955). A study of arterial blood pressure and its inheritance in a sample of the general population. *Clinical Science*, 14(3), 459-488.

Miettenen, O.S. (1968). The matched pairs design in the case of all-or-none responses. *Biometrics* 24, 339–352.

Miller, R. (1986). *Beyond ANOVA: Basics of Applied Statistics*, John Wiley, New York.

Mosteller, F., and Tukey, J. (1977). *Data Analysis and Regression: A Second Course in Statistics*, Addison-Wesley, Reading, MA.

Murray, D.M. (1998). *The Design and Analysis of Group-Randomized Trials*, Oxford University Press, London.

Murray, D.M., Perry, C.L., and Griffin, G. (1992). Results from a statewide approach to adolescent tobacco use prevention. *Preventive Medicine*, 21, 449–472.

National Research Council. (1992). *Combining Information: Statistical Issues and Opportunities for Research*. National Academy Press, Washington DC.

Neuhaus, J.M., Kalbfleisch, J.D., and Hauck, W.W. (1991). A comparison of cluster-specific and population-averaged approaches for analyzing correlated binary data. *International Statistical Review*, 59, 25–35.

Normand, S. (1999). Tutorial in biostatistics: Meta-analysis: Formulating, evaluating, combining, and reporting. *Statistics in Medicine*, 18, 321–359.

Oakes, M. (1986). *Statistical Inference*, Epidemiology Resources Inc., Chestnut Hills, MD.

Overton, R.C. (1998). A comparison of fixed-effects and mixed (random-effects) models for meta-analysis tests of moderator variable effects. *Psychological Methods*, 3, 354–379.

Oxman, A.D., and Guyatt, G.H. (1988). Guidelines for reading literature reviews. *Canadian Medical Association Journal*, 138, 697–703.

Paul, S.R. (1982). Analysis of proportions of affected fetuses in teratological experiments. *Biometrics*, 38, 361–370.

Peikes, D., Chen, A., Schore, J., and Brown, R. (2009). Effects of care coordination on hospitalization, quality of care, and health care expenditures among Medicare beneficiaries: 15 randomized trials. *JAMA*, 301(6), 603–618.

Peng, R. (2015). The reproducibility crisis in science: A statistical counterattack. *Significance*, 12(3), 30–32.

Petitti, D.B. (2001). Approaches to heterogeneity in meta-analysis. *Statistics in Medicine*, 20, 3625–3633.

Peto, R., and Peto, J. (1972). Asymptotically efficient rank invariant procedures. *Journal of the Royal Statistical Society, A*, 135, 185–207.

Pinol, A.P.Y., and Piaggio, G. (2000). *A-Cluster, version 2.1: Design and Analysis of Cluster of Cluster Randomization Trial*, World Health Organization, Geneva, Switzerland.

Pintilie, M. (2006). *Competing Risks: A Practical Perspective*. Wiley, Chichester UK.

Potthoff, R.F., and Roy, S.N. (1964). A generalized multivariate analysis of variance model useful especially for growth curve problems. *Biometrika*, 51, 313–326.

Prentice, R. (1976). Use of the logistic model in retrospective studies. *Biometrics*, 32, 599–606.

Prentice, R. (1988). Correlated binary regression with covariates specific to each binary observation. *Biometrics*, 44, 1033–1048.

R Development Core Team. (2006). R 2.3.1: A language and environment.

R Development Core Team. (2014). *R: A language and environment for statistical computing (3.0.3) [Computer software]*. Foundation for Statistical Computing, Vienna, Austria.

R Development Core Team. (2010). R: A language and environment for statistical computing, http://www.R-project.org.

Randolph, J.J., Falbe, K., Manuel, A.K., and Balloun, J.L. (2014). A step by step guide to propensity score matching in R. *Practical Assessment, Research and Evaluation*, http://pareonline.net/getvn.asp?v=19&n=18 (accessed July 10, 2017).

Rao, J.N.K., and Scott, A.J. (1992). A simple method for the analysis of clustered binary data. *Biometrics*, 48, 577–585.

Raudenbush, S.W. (1994). Random effect models. In *The Handbook of Research Synthesis*, H. Cooper and L.V. Hedges (eds.), 301–321, Russell Sage Foundation, New York.

Rizopoulos, D. (2010). JM: An R package for the joint modelling of longitudinal and time-to-event data. *Journal of Statistical Software*, 35(9). Retrieved from http://www.R-project.org/.

Robins, J., Breslow, N., and Greenland, S. (1986). Estimators of the Mantel-Haenszel variance consistent in both sparse data and large-strata limiting models, *Biometrics*, 42, 311–323.

Roizman, B., Hoggan, D., and Cornfield, J. (1960). Linear and parabolic estimates of the titers of Herpes Simplex from pock counts on the chorioallantoic membrane of embryonated eggs. *Virology*, 11, 572–589.

Rosenbaum, P.R. (1995a). *Observational Studies*. Springer Verlag, New York.

Rosenbaum, P.R. (1995b). Quantiles in nonrandom samples and observational studies. *Journal of the American Statistical Association*, 90(432), 1424–1431. Retrieved from http://www.jstor.org/stable/2291534.

Rosenbaum, P.R., and Rubin, D.B. (1983). Assessing sensitivity to an unobserved binary covariate in an observational study with binary outcome. *Journal of the Royal Statistical Society. Series B (Methodological)*, 45(2), 212–218. Retrieved from http://www.jstor.org/stable/2345524.

Rosenbaum, P.R., and Rubin, D.B. (1984). Reducing bias in observational studies using sub-classification on the propensity score. *Journal of the American Statistical Association*, 79, 516–524.

Rosenbaum, P.R., and Rubin, D.B. (1985). Constructing a control group using multivariate matched sampling methods that incorporate the propensity score. *American Statistician*, 39, 33–38.

Rosenthal R. (1978). Combining results of independent studies. *Psychological Bulletin*, 85, 185–193.

Rosenthal, R. (1994). Parametric measures of effect size. In *The Handbook of Research Synthesis*, H. Cooper and L.V. Hedges, 231–244, Russell Sage Foundation, New York.

Rosnow, R.L., and Rosenthal, R. (1996). Computing contrasts, effect sizes, and counter nulls on other people's published data: General procedures for research consumers. *Psychological Methods*, 1, 331–340.

Roth, P. (1994). Missing data: A conceptual review for applied psychologists. *Personnel Psychology*, 47, 537–560.

Rubin, D. (1976). Inference and missing data. *Biometrika*, 63, 581–592.

Rubin, D. (1994). Modeling the drop-out mechanism in repeated measures studies. 9th International Workshop on Statistical Modeling, Exeter, UK.

Rubin, D.B. (1974). Estimating causal effects of treatments in randomized and nonrandomized studies. *Journal of Educational Psychology*, 66, 688–701.

Rubin, D.B. (1986). Statistics and causal inference: Comment: What ifs have causal answers. *Journal of the American Statistical Association*, 81(396), 961–962.

Rubin, D.B. (1987). *Multiple Imputations for Nonresponse in Surveys*, John Wiley & Sons, New York.

Rubin, D.B. (1996). Multiple imputation after 18+ years. *Journal of the American Statistical Association*, 91(434), 473–489.

Rubin, D.B. (2007). The design versus the analysis of observational studies for causal effects: Parallels with the design of randomized trials. *Statistics in Medicine*, 26, 20–30.

Rubin, D.B. (2008). For objective causal inference, design trumps analysis. *The Annals of Applied Statistics*, 2(3), 808–840.

Russell, M.A.H., Merriman, R., Stapleton, J., and Taylor, W. (1983). Effect of nicotine chewing gum as an adjunct to general practitioners advice against smoking. *British Medical Journal*, 286, 1782–1785.

Sacks, H.S., Berrier, J., Reitman, D., Ancona-Berk, V.A., and Chalmers, T.C. (1987). Meta-analyses of randomized controlled trials. *New England Journal of Medicine*, 316, 450–455.

Sahl, J.D., Kelsh, M.A., and Greenland, S. (1993). Cohort nested case-control studies of hematopoietic cancers and brain cancer among electric utility workers. *Epidemiology*, 4, 104–114.

SAS/STAT version 9.4. (2012). SAS Institute, Raleigh, NC.

Schafer, J.L. (1997). *Analysis of Incomplete Multivariate Data*. Chapman & Hall, London.

Schall, R. (1991). Estimation in generalized linear models with random effects, *Biometrika*, 78, 719–727.

Schlesselman, J. (1982). *Case-Control Studies Design, Conduct, Analysis*, Oxford University Press, Oxford.

Schoenfeld, D.A. (1983). Sample size formula for the proportional hazards regression model. *Biometrics*, 38, 499–503.

Schwarz, G. (1978). Estimating the dimension of a model. *Annals of Statistics*, 6, 461–464.

Scrucca, L., Santucci, A., and Aversa, F. (2007). Regression modeling of competing risk using R: An easy guide for clinicians. *Bone Marrow Transplant*, 40, 381–387.

Searle, R.S., Casella, G., and McCulloch, C.E. (1992). *Variance Components*. Wiley, New York.

Shadish, W.R., Cook, T.D., and Campbell, D.T. (2002). *Experimental and Quasi-Experimental Designs for Generalized Causal Inference*. Wadsworth, Cengage Learning, Belmont, CA.

Shadish, W.R., and Steiner, P.M. (2010). A primer on propensity score analysis. *Newborn and Infant Nursing Reviews*, 10(1), 19–26.

Shoukri, M.M., Asyali, M.H., Van Dorp, R., and Kelton, D. (2004). The Poisson inverse Gaussian distribution regression model in the analysis of clustered counts data. *Journal of Data Science*, 2, 17–32.

Shoukri, M.M., Collison, K., and Al-Mohanna, F. (2015). Statistical issues in evaluating clustering of metabolic syndrome in spousal pairs. *Journal of Biometrics and Biostatistics*, http://dx.doi.org/10.472/2155-6180.1000233.

Shoukri, M.M., Donner, A., and El-Dali, A. (2013). Covariate adjusted confidence interval for the intraclass correlation coefficient. *Contemporary Clinical Trials*, 36, 244–253.

Shoukri, M.M., Varghese, B., Al-Hajoj S., and Al-Mohanna, F. (2014). Prediction of the number of tuberculosis cases and estimation of its treatment cost in Saudi Arabia using proxy information. *Open Journal of Statistics*, http://dx.doi.org/10.4236/.2014.

Shoukri, M.M., and Ward, R.H. (1989). Use of linear models to estimate genetic parameters and measures of familial resemblance in man. *Applied Statistics* 3, 467–479.

Shumway, R. (1982). Discriminant analysis for time series. In *Handbook of Statistics*, Vol. II, Classification, Pattern Recognition, and Reduction of Dimensionality, P.R. Kirshnaiah, ed., 1–43, North Holland, Amsterdam.

Siegfried, T. (2010, March 27). Odds are it is wrong. Science fails to face the shortcomings of statistics. *Science News*, 177(7), 26–29.

Singer, J.D. (1998). Using SAS PROC MIXED to fit multilevel models, hierarchical models, and individual growth models. *Journal of Educational and Behavioral Statistics*, 23, 323–355.

Smedslund, G., and Ringdal, G.I. (2004). Meta-analysis of the effects of psychological interventions on survival time in cancer patients. *Journal of Psychological Research* 57, 123–131.

Snedecor, G., and Cochran, W.G. (1980). *Statistical Methods*, 7th ed., Iowa State University Press, Ames, Iowa.

Stanish, W.M., and Taylor, N. (1983). Estimation of the intraclass correlation coefficient for the analysis of covariance. *The American Statistician*, 37(3), 221–224.

Sterne, J.A.C., Egger, M., and Smith, D.G. (2001). Investigating and dealing with publication and other biases. In *Systematic Reviews in Health Care: Meta-Analysis in Context*, M. Egger, D.G. Smith, and D.G. Altman, eds., BMJ Books, London.

Stuart, E.A. (2010). Matching methods for causal inference: A review and a look forward. *Statistical Science*, 25, 1–21.

Stukel, T.A. (1993). Comparison of methods for the analysis of longitudinal data. *Statistics in Medicine*, 12, 1339–1351.

Tabar, L., Fagerberg, C.J., Gad, A., Baldetorp, L., Holmberg, L.H., Grontoft, O., Ljungquist U. et al. (1985). Reduction in mortality from breast cancer after mass screening with mammography. *Lancet* 1(8433), 892–832.

Tarone, R. (1979). Testing the goodness of fit of the binomial distribution. *Biometrika*, 66, 3, 585–590.

Tashkin, D.P., Coulson, A.H., Clark, V.A., Simmons, M., Bourque, L.B., Duann, S., Spivey, G.H., and Gong, H. (1987). Respiratory symptoms and lung function in habitual heavy smokers of marijuana alone, smokers of marijuana and tobacco, smokers of tobacco alone, and nonsmokers. *American Review of Respiratory Disease*, 135, 209–216.

Thall, P.F., and Vail, S.C. (1990). Some covariance models for longitudinal count data with over-dispersion. *Biometrics*, 46, 657–671.

Therneau, T. (1993). Using a multiple-events Cox model. *Proceedings from the Biometrics Section of the American Statistical Association*, 1–14.

Thoemmes, F.J., and Kim, E.S. (2011). A systematic review of propensity score methods in the social sciences. *Multivariate Behavioral Research*, 46(1), 90–118.

Thomas, D., and Gart, J.J. (1977). A table of exact confidence limits for differences and ratios of two proportions and their odds ratios. *Journal of the American Statistical Association*, 72, 386–394.

Thompson, S.G., and Pocock, S.J. (1991). Can meta-analysis be trusted? *Lancet*, 338, 1127–1130.

Tsiatis, A.A., DeGruttola, V., and Wulfsohn, M.S. (1995). Modeling the relationship of survival to longitudinal data measured with error: Applications to survival and CD4 counts in patients with AIDS. *Journal of the American Statistical Association*, 90, 27–37.

Tukey, J. (1977). *Exploratory Data Analysis*, Addison-Wesley, Reading, MA.

Ury, H.K. (1976). A Comparison of four procedures for multiple comparisons among means (pairwise contrasts) for arbitrary sample sizes. *Technometrics*, 18(1), 89–97.

Valent, F., Brusaferro, S., and Barbone, F.A. (2001). Case-crossover study of sleep and childhood injury. *Pediatrics*, 107, E23.

Vaupel, J.W., Manton, K.G., and Stallard, E. (1979). The impact of heterogeneity in individual frailty on the dynamics of mortality. *Demography*, 16, 439–454.

Vonesh, E.F., and Chinchilli, V.G. (1997). *Linear and Nonlinear Models for the Analysis of Repeated Measurements*, Chapman & Hall, London.

Walter, S. (1985). Small-sample estimation of log odds ratios from logistic regression and fourfold tables. *Statistics in Medicine*, 4, 437–444.

Walter, S., and Cook, R. (1991). A comparison of several point estimators of the odds ratio in a single 2x2 contingency table. *Biometrics*, 47, 795–811.

Ware, J.H., Lipsitz, S., and Speizer, F.E. (1988). Issues in the analysis of repeated categorical outcomes. *Statistics in Medicine*, 7, 95–107.

Wasserstein, R.L., and Lazar, N.A. (2016). The ASA's statement on p-values. *The American Statistician*, doi: 10.1080/00031305.2016.1154108

Wedderburn, R. (1974). Quasi-likelihood functions, generalized linear models and the Gauss-Newton method. *Biometrika*, 61, 439–447.

Weil, C.S. (1970). Selection of the valid number of sampling units and a consideration of their combination in toxicological studies involving reproduction, teratogenesis or carcinogenesis. *Food and Cosmetics Toxicology*, 8, 177–182.

Welch, B.L. (1951). On the comparison of several mean values: An alternative approach. *Biometrika*, 38(3/4), 330–336.

Whittemore, A.S. (1981). Sample size for logistic regression with small response probability. *Journal of American Statistical Association*, 76, 27–32.

Whittle, P. (1983). *Prediction and Regulations*, 2nd ed., University of Minnesota Press, Minneapolis.

Wilk, M., and Gnanadesikan, R. (1968). Probability plotting methods for the analysis of data. *Biometrika*, 55, 1–17.

Willeberg, P. (1980). The analysis and interpretation of epidemiological data, Proceedings of the 2nd International Symposium on Veterinary Epidemiology and Economics, Canberra, Australia.

Williams, D.A. (1975). The analysis of binary responses from toxicological experiments involving reproduction and teratogenicity. *Biometrics*, 31, 949–952.

Wilmot, G.E. (1987). The Poisson-inverse Gaussian distribution as an alternative to the negative binomial. *Scandinavian Journal of Statistics*, 113–127.

Wilson, S.R., and Gordon, I. (1986). Calculating sample sizes in the presence of confounding variables. *Applied Statistics*, 35(2), 207–213.

Winer, B.J. (1971). *Statistical Principles in Experimental Designs*, 2nd ed., McGraw Hill, New York.

Wolfinger, R., and O'Connell, M. (1993). Generalized linear mixed models: A pseudo-likelihood approach. *Journal of Statistical Computation and Simulation*, 48, 233–243.

Woolf, B. (1955). On estimating the relationship between blood group and disease, *Annals of Human Genetics*, 19, 251–253.

Wothke, W. (1998). Longitudinal and multi-group modeling with missing data. In *Modeling Longitudinal and Multiple Group Data: Practical Issues, Applied Approaches and Specific Examples*, T.D. Little, K.U. Schnabel, and J. Baumert, eds., Lawrence Erlbaum Associates, Mahwah, NJ.

Wu, M., and Carroll, R. (1988). Estimation and comparison of changes in the presence of informative right censoring by modeling the censoring process. *Biometrics*, 44, 175–188.

Wu, S., Crespi, C.M., and Wong, W.K. (2012). Comparison of methods for estimating the intraclass correlation coefficient for binary responses in cancer prevention cluster randomized trials. *Contemporary Clinical Trials*, 33(5), 869–880.

Wulfsohn, M.S., and Tsiatis, A.A. (1997). A joint model for survival and longitudinal data measured with error. *Biometrics*, 53(1), 330–339.

Zeger, S.L., and Karim, M.R. (1991). Generalized linear models with random effects: A Gibbs sampling approach. *Journal of the American Statistical Association*, 86, 79–86.

Zeger, S.L., and Liang, K.-Y. (1986). Longitudinal data analysis for discrete and continuous outcomes. *Biometrics*, 42, 121–130.

Zeger, S.L., Liang, K.Y., and Albert, P.A. (1988). Models for longitudinal data: A generalized estimating equation approach. *Biometrics*, 44, 1049–1060.

Zeng, D., and Cai, J. (2005). Asymptotic results for maximum likelihood estimators in joint analysis of repeated measurements and survival time. *Annals of Statistics*, 33(5), 2132–2163.

# Index

Page numbers followed by f and t indicate figures and tables, respectively.

T - #0314 - 071024 - C10 - 234/156/23 - PB - 9780367734954 - Gloss Lamination